WITHDRAWN
WRIGHT STATE UNIVERSITY LIBRARIES

Iron Chelation Therapy

ADVANCES IN EXPERIMENTAL MEDICINE AND BIOLOGY

Editorial Board:

NATHAN BACK, *State University of New York at Buffalo*

IRUN R. COHEN, *The Weizmann Institute of Science*

DAVID KRITCHEVSKY, *Wistar Institute*

ABEL LAJTHA, *N. S. Kline Institute for Psychiatric Research*

RODOLFO PAOLETTI, *University of Milan*

Recent Volumes in this Series

Volume 501
BIOACTIVE COMPONENTS OF HUMAN MILK
Edited by David S. Newburg

Volume 502
HYPOXIA: From Genes to the Bedside
Edited by Robert C. Roach, Peter D. Wagner, and Peter H. Hackett

Volume 503
INTEGRATING POPULATION OUTCOMES, BIOLOGICAL MECHANISMS AND RESEARCH METHODS IN THE STUDY OF HUMAN MILK AND LACTATION
Edited by Margarett K. Davis, Charles E. Isaacs, Lars Å. Hanson, and Anne L. Wright

Volume 504
MYCOTOXINS AND FOOD SAFETY
Edited by Jonathan W. DeVries, Mary W. Trucksess, and Lauren S. Jackson

Volume 505
FLAVONOIDS IN CELL FUNCTION
Edited by Béla A. Buslig and John A. Manthey

Volume 506
LACRIMAL GLAND, TEAR FILM, AND DRY EYE SYNDROMES 3: Basic Science and Clinical Relevance
Edited by David A. Sullivan, Michael E. Stern, Kazuo Tsubota, Darlene A. Dartt, Rose M. Sullivan, and B. Britt Bromberg

Volume 507
EICOSANOIDS AND OTHER BIOACTIVE LIPIDS IN CANCER, INFLAMMATION, AND RADIATION INJURY, 5
Edited by Kenneth V. Honn, Lawrence J. Marnett, Santosh Nigam, and Charles Serhan

Volume 508
SENSORIMOTOR CONTROL OF MOVEMENT AND POSTURE
Edited by Simon C. Gandevia, Uwe Proske, and Douglas G. Stuart

Volume 509
IRON CHELATION THERAPY
Edited by Chaim Hershko

A Continuation Order Plan is available for this series. A continuation order will bring delivery of each new volume immediately upon publication. Volumes are billed only upon actual shipment. For further information please contact the publisher.

Iron Chelation Therapy

Edited by

Chaim Hershko
Hebrew University School of Medicine
Jerusalem, Israel

Kluwer Academic / Plenum Publishers
New York, Boston, Dordrecht, London, Moscow

Library of Congress Cataloging-in-Publication Data

Iron chelation therapy/edited by Chaim Hershko.
    p.   cm. — (Advances in experimental medicine and biology; v. 509)
    Includes bibliographical references and index.
    ISBN 0-306-46785-2
    1. Chelation therapy.   2. Iron—Metabolism—Disorders.   I. Hershko, Chaim.

RC632.I7 I75 2002
616.3'96—dc21

2002022131

ISBN 0-306-46785-2

©2002 Kluwer Academic/Plenum Publishers, New York
233 Spring Street, New York, New York 10013

http://www.wkap.nl/

10 9 8 7 6 5 4 3 2 1

A C.I.P. record for this book is available from the Library of Congress

All rights reserved

No part of this book may be reproduced, stored in a retrieval system, or transmitted in any form or by any means, electronic, mechanical, photocopying, microfilming, recording, or otherwise, without written permission from the Publisher, with the exception of any material supplied specifically for the purpose of being entered and executed on a computer system, for exclusive use by the purchaser of the work.

Printed in the United States of America

PREFACE

Within the last few years, iron research has yielded exciting new insights into the understanding of normal iron homeostasis. However, normal iron physiology offers little protection from the toxic effects of pathological iron accumulation, because nature did not equip us with effective mechanisms of iron excretion. Excess iron may be effectively removed by phlebotomy in hereditary hemochromatosis, but this method cannot be applied to chronic anemias associated with iron overload. In these diseases, iron chelating therapy is the only method available for preventing early death caused mainly by myocardial and hepatic iron toxicity. Iron chelating therapy has changed the quality of life and life expectancy of thalassemic patients. However, the high cost and rigorous requirements of deferoxamine therapy, and the significant toxicity of deferiprone underline the need for the continued development of new and improved orally effective iron chelators. Such development, and the evolution of improved strategies of iron chelating therapy require better understanding of the pathophysiology of iron toxicity and the mechanism of action of iron chelating drugs.

The timeliness of the present volume is underlined by several significant developments in recent years. New insights have been gained into the molecular basis of aberrant iron handling in hereditary disorders and the pathophysiology of iron overload (Chapters 1–5). We shall review the impact of long-term iron chelating therapy using deferoxamine or the new, but controversial oral iron chelator deferiprone based on experience gained by multicenter trials, with special emphasis on survival, morbidity and drug toxicity (Chapters 6–7); review the development of new and improved orally effective chelators that may be suitable for clinical use in the near future (Chapters 8–10) and; examine novel strategies of iron chelating treatment for the control of cell proliferation in malignant disease or malaria (Chapters 12–13).

All of the designated contributors of this volume are leading authorities in the field of iron pathophysiology and chelation research and had major personal contributions to the advancement of basic and clinical research in this field.

Chaim Hershko

CONTENTS

PATHOPHYSIOLOGY OF IRON OVERLOAD

1. Animal Models of Hereditary Iron Transport Disorders 1
   Nancy C. Andrews

2. Mechanism of Iron Toxicity 19
   Antonello Pietrangelo

3. Role of Non-Transferrin-Bound Iron in the Pathogenesis of Iron Overload and Toxicity 45
   Pierre Brissot and Olivier Loréal

4. Intracellular and Extracellular Labile Iron Pools 55
   Ioav Cabantchik, Or Kakhlon, Silvina Epsztejn, Giulianna Zanninelli, and William Breuer

5. Cardioprotective Effect of Iron Chelators 77
   Chaim Hershko, Gabriela Link and Abraham M. Konijn

CHELATION THERAPY IN TRANSFUSIONAL SIDEROSIS

6. Results of Long Term Iron Chelation Treatment with Deferoxamine 91
   Bernard A. Davis and John B. Porter

7. Long Term Deferiprone Chelation Therapy 127
   Victor Hoffbrand and Beatrix Wonke

DEVELOPMENT OF NEW IRON CHELATORS

8. Iron Chelator Chemistry 141
   Zu D. Liu, Ding Y. Liu, and Robert C. Hider

9. Structure-Activity Relationships Among Desazadesferrithiocin Analogues .. 167
   Raymond J. Bergeron, Jan Wiegand, James S. McManis, William R. Weimar, and Guangfei Huang

10. ICL670A: Preclinical Profile 185
    Hanspeter Nick, Agnes Wong, Pierre Acklin, Bernard Faller, Yi Jin, René Lattmann, Thomas Sergejew, Suzanne Hauffe, Helmut Thomas, and Hans Peter Schnebli

11. Pyridoxal Isonicotinoyl Hydrazone and Its Analogues 205
    Joan L. Buss, Marcelo Hermes-Lima, and Prem Ponka

NOVEL STRATEGIES IN IRON CHELATION TREATMENT

12. Therapeutic Potential of Iron Chelators in Cancer Therapy 231
    Des Richardson

13. Antimalarial Effect of Iron Chelators 251
    Victor R. Gordeuk and Mark Loyevsky

Index 273

ANIMAL MODELS OF HEREDITARY IRON TRANSPORT DISORDERS

Nancy C. Andrews[*]

1. INTRODUCTION

Animal models have long been recognized to be useful in understanding disorders of human iron homeostasis. Although the advent of molecular biology has vastly improved our understanding of the biochemistry and cellular biology of iron transport, studies in testtubes and tissue culture cells cannot lead to a total understanding of human iron disorders. This is because human iron disorders are invariably abnormalities of iron balance, resulting from defects in regulation of intestinal iron absorption or inappropriate distribution of iron among the tissues of the body. Thus, molecular studies cannot replace animal models in this area.

Several different types of organisms, ranging from yeast to plants to mammals, have been used to gain insight into iron metabolism. However, only mammals are similar enough to human patients to allow the study of iron balance. Mice and rats offer valid models for understanding human iron absorption, distribution, utilization and storage. Mice have been particularly valuable for two reasons. First, there are several spontaneous mouse mutations, recognized in inbred mouse colonies, which affect iron metabolism and have been characterized in great detail. These will be discussed extensively below. Second, it is feasible to manipulate the mouse genome, through gene targeting technology exploiting the potential for homologous recombination in totipotential embryonic stem cells, and through the production of transgenic animals expressing engineered genes. This chapter will discuss the existing animal models, with emphasis on mice, to show how we have taken advantage of vertebrate genetics to understand human iron disorders.

---

[*] Nancy C. Andrews, Howard Hughes Medical Institute, Children's Hospital and Harvard Medical School, Boston, MA 02115.
Iron Chelation Therapy.
Edited by Chaim Hershko, Kluwer Academic/Plenum Publishers, 2002

2. IRON DEFICIENCY DISORDERS

Human iron deficiency is not usually attributable to genetic disease. Most iron deficient patients enter negative iron balance because of iron losses, in the form of blood, which cannot be compensated for by increased iron absorption, or because dietary intake is inadequate to meet iron needs. A small subset of patients with marked iron deficiency, perhaps 1-5% of that total group, probably have inherited defects in genes involved in iron metabolism, but those defects have not yet been identified on a molecular level. Nonetheless, animals with inherited iron deficiency have been invaluable for learning about iron transport processes. For that reason, they will be discussed in detail here. It is likely that, in the future, human patients will be described who carry mutations similar to those that have been observed in mouse, rat and zebrafish.

2. 1. Mouse mutations

As in human patients, iron deficiency in mice leads to insufficient iron for hemoglobin synthesis and consequent anemia. Over the past 80 years, mouse fanciers and mouse breeders have frequently noticed pale mice within inbred colonies, and demonstrated that their pallor was due to heritable iron deficiency anemia resulting from spontaneous mutations. Through careful and elegant physiological investigations, the phenotypes of the mutant animals have been characterized in detail. Many of them are summarized in reviews by Bannerman and Russell (Bannerman, 1976; Russell, 1984).

2.1.1. Microcytic anemia

Microcytic anemia (*mk*) mice were first described in 1964, after they were discovered in a colony maintained at the Jackson Laboratory (Nash et al., 1964). The mutation was shown to be autosomal recessive, resulting in poor viability, skin lesions, runting, and severe hypochromic, microcytic anemia beginning during fetal development (Russell et al., 1970). Intestinal iron absorption was shown to be defective, with an apparent block in apical uptake of dietary iron that results in diminished body iron stores (Bannerman et al., 1972; Edwards and Hoke, 1972). However, the mice also have a defect in uptake of iron by erythroid precursors, as demonstrated by bone marrow transplantation experiments (Harrison, 1972) and analysis of reticulocytes in vitro (Edwards and Hoke, 1975).

The nature of the *mk* defect remained mysterious until 1997, when it was shown that *mk* mice carry a missense mutation in the gene encoding divalent metal transporter 1 (DMT1, formerly Nramp2, also known as DCT1) (Fleming et al., 1997). Almost simultaneously, DMT1 was shown to function as a proton-coupled metal cation transporter when expressed in *Xenopus* oocytes (Gunshin et al., 1997). Taken together, these results strongly suggested that DMT1 is the major, if not only, transmembrane transporter molecule functioning to bring dietary non-heme iron across the apical membrane of absorptive cells in the intestine. DMT1 can also transport other metals, including manganese, cobalt, copper, zinc and even lead (Gunshin et al., 1997). It was later shown that DMT1 is expressed in absorptive enterocytes (Canonne-Hergaux et al., 1999) and that the *m k* mutation severely impairs DMT1 localization and activity (Canonne-Hergaux et al., 2000; Su et al., 1998). Other studies showed that DMT1 is also expressed in transferrin cycle endosomes, where it also functions in transmembrane iron transport, explaining the erythroid defect in *mk* mice (see below; (Fleming et al., 1998; Gruenheid et al., 1999; Su et al., 1998)).

2.1.2 Sex-linked anemia

Sex-linked anemia (*sla*) mice appeared in an X-irradiated colony around the same time that *mk* mice were discovered (Falconer and Isaacson, 1962). Early in life, hemizygous *sla*/Y males and homozygous *sla/sla* females are pale, and have hypochromic, microcytic erythrocytes and minimal body iron stores (Bannerman and Cooper, 1966; Grewal, 1962). However, the *sla* defect only disrupts placental and intestinal iron absorption; when the mice are given parenteral iron their erythropoiesis is normal (Bennett et al., 1968; Kingston et al., 1978; Pinkerton et al., 1970). Over time the anemia improves, but body iron stores remain subnormal. Using everted duodenal sacs, it was shown that apical iron uptake is normal, but basolateral iron transfer is greatly reduced in *sla* mutants (Edwards and Bannerman, 1970; Manis, 1971). Accordingly, abundant non-heme iron is apparent when *sla* enterocytes are stained with Perls Prussian blue (Levy et al., 2000; Pinkerton, 1968). The stainable iron is not only derived from uptake of dietary metal; radiolabeled iron that is administered parenterally also shows up within the developing enterocytes (Bedard et al., 1976). The defect appears to be relatively specific for iron; *sla* animals show no abnormality in the uptake of zinc (Flanagan et al., 1984).

The molecular identity of *sla* was worked out by Vulpe and colleagues (Vulpe et al., 1999). They mapped the chromosomal location of *sla*, and found a plausible candidate gene, which they designated hephaestin after the Greek god Hephaestus, who forged iron. Hephaestin has suffered a large deletion in *sla* mice, presumably as a result of X-ray mutagenesis. It is not yet clear whether the deletion completely blocks protein expression, or leads to production of an abnormal hephaestin protein. Hephaestin is highly homologous to ceruloplasmin, which has been known for almost half a century to play a role in iron metabolism (see below). Both hephaestin and ceruloplasmin are multicopper proteins; they differ in that hephaestin includes a membrane anchor sequence. Ceruloplasmin functions as a ferroxidase (Holmberg and Laurell, 1948) hephaestin presumably does the same. A ferroxidase activity appears to be important for efficient export of iron from cells. The working model is that hephaestin serves this function for placental syncytiotrophoblasts and enterocytes, and ceruloplasmin operates in iron export from macrophages and hepatocytes. However, the fact that both *sla* mutant mice and ceruloplasmin knockout mice (see below) are viable suggests that the ferroxidase activity, while contributing to iron export, is not absolutely essential. Proof of this hypothesis awaits the generation of animals lacking both hephaestin and ceruloplasmin.

2.1.3. Hemoglobin deficit

Hemoglobin deficit (*hbd*; also called hemoglobin deficient) appeared spontaneously in an inbred mouse colony in the former East Germany (Scheufler, 1969). This is also an autosomal recessive mutation, resulting in a profound hypochromic, microcytic anemia with abundant target forms. Interestingly, though the anemia is comparable in severity to that of *mk* homozygotes, the viability and overall health of *hbd* mutants is much better. This is probably because the *hbd* defect is isolated to erythroid cells. Anemia begins prenatally, and continues throughout life. It is not corrected by parenteral administration of iron (Scheufler, 1969). The enzymes of heme biosynthesis appear to be normal, as do transferrin and the transferrin receptor, suggesting that the *hbd* mutation directly impairs erythroid iron procurement at a stage beyond the transferrin cycle (Garrick et al., 1987).

The *hbd* mutation maps to mouse chromosome 19 (Bloom et al., 1998). There are no known genes related to iron metabolism in the vicinity of *hbd* (J Lim and NCA, unpublished data) and the *hbd* gene has not yet been identified. Discovery of *hbd* is likely to add new insight into trafficking of iron in erythroid precursor cells.

2.2. Rat mutation

Although rat colonies have not been as rich a source of iron mutants as mouse colonies, and rat genomics is primitive compared to genomics of mice and humans, one rat mutant has been very useful in working out the molecular details of iron metabolism. Belgrade (*b*) rats were first described in the 1960s in the former Yugoslavia (Sladic-Simic et al., 1969). These animals have hypochromic, microcytic anemia, and deficiency of both iron and manganese (Chua and Morgan, 1997; Farcich and Morgan, 1992a; Oates and Morgan, 1996). Their defect is manifest at the level of the intestine and also in erythroid precursors (Bowen and Morgan, 1987; Edwards et al., 1986; Edwards et al., 1980; Farcich and Morgan, 1992a; Farcich and Morgan, 1992b; Garrick et al., 1993; Oates and Morgan, 1996). Erythroid iron uptake has been extensively characterized in these animals, and double labeling experiments showed that diferric transferrin is internalized appropriately, but iron does not exit from transferrin cycle endosomes to enter the cytoplasm. This suggests a defect in a component of the transmembrane iron transporter functioning at that site.

The *b* gene was identified through a gene mapping/candidate gene approach (Fleming et al., 1998). The *b* mutation maps to the proximal portion of rat chromosome 7. The gene encoding rat DMT1 maps to the same location. Perhaps not surprisingly, at least in retrospect, the *b* mutation alters DMT1. It is astonishing, however, that the mutation is identical to that found in *mk* mice, resulting in the substitution of arginine for a glycine residue in transmembrane domain 4 of the protein (Fleming et al., 1998). Rat DMT1 and mouse DMT1 are highly similar; it is clear that the mutation must lead to a severe functional defect in the rat protein just as it does in mouse. The double labeling experiments, implicating the *b* gene product in export of iron from transferrin cycle endosomes, support the conclusion that DMT1 functions at that site, as well as along the apical membrane of absorptive enterocytes. Accordingly, immunofluorescence and green fluorescent fusion protein experiments have demonstrated that DMT1 and transferrin co-localize to the same, endosomal, subcellular compartment (Gruenheid et al., 1999; Su et al., 1998).

2.3. Zebrafish mutations

Over the past decade, there has been increasing interest in zebrafish (*Danio rerio*) as a model genetic system. The advantages are that they are easily maintained in captivity, that they reproduce to yield large numbers of progeny, and that various techniques are available to facilitate genetic analysis. The disadvantage, at least for comparative physiology, is that they are significantly more distant from humans on an evolutionary scale than mammals. Nonetheless, interpreted cautiously, information relevant to mammalian iron biology can be gleaned from zebrafish experiments. The most fruitful results have come from characterization of blood mutants produced in large-scale chemical mutagenesis screens (Haffter et al., 1996). Although zebrafish erythrocytes retain their nuclei, hypochromia can be recognized by evaluation of cellular hemoglobin content. Two hypochromic mutants have been shown to have defects in iron transporter proteins.

2.3.1. Weissherbst

Weissherbst (weh) mutant fish have severe anemia (Donovan et al., 1999). Two independent alleles have been described. Although the total numbers of erythroid cells are approximately normal at 33 and 48 hours post fertilization, the cells are severely deficient in hemoglobin, and have immature features. Development of other tissues and organs appears to be intact. By 7 to 14 days of development, all animals die, apparently as a result of severe anemia. Total circulating red blood cell iron was decreased several fold in mutants as compared to wild type animals of the same age. Importantly, injection of iron dextran full rescued the mutant phenotype (Donovan et al., 2000).

The *weh* gene was localized within the zebrafish genome, and identified by positional cloning (Donovan et al., 2000). Designated ferroportin1, it encodes a predicted protein of 562 amino acids. It is disrupted by a missense mutation at codon 167 in one zebrafish mutant, and by a nonsense mutation at codon 361 in the other mutant. Injection of wild type ferroportin1 mRNA into mutant embryos leads to partial rescue of their phenotype. Two other groups identified mammalian orthologs simultaneously through other approaches, and used the alternative names IREG1 and MTP1 (Abboud and Haile, 2000; McKie et al., 2000).

In all species examined, ferroportin1/IREG1/MTP1 has approximately 10 predicted transmembrane domains, consistent with a transporter function. It localizes to the yolk syncytial layer in developing zebrafish embryos, where nutrients pass from yolk to embryo-proper (Donovan et al., 2000). In mammals, it localizes to the basal surface of placental syncytiotrophoblasts, adjacent to the fetal circulation, and to the basolateral surface of absorptive enterocytes in the small intestine (Donovan et al., 2000). Both locations are suggestive of a role in the export of iron from cells. Ferroportin1/IREG1/MTP1 is also highly expressed in mammalian hepatocytes and reticuloendothelial macrophages, suggesting that it may play a role in export of iron from those storage cell types, as well.

Iron export function has been formally demonstrated through forced expression of ferroportin1/IREG1/MTP1 mRNA in *Xenopus* oocytes (Donovan et al., 2000; McKie et al., 2000). When cells are pre-loaded with iron through DMT1, they release it more efficiently in the presence of ferroportin1/IREG1/MTP1. However, export assays are difficult, because oocytes are prone to spontaneous leakage. Unfortunately, a more robust assay for ferroportin1/IREG1/MTP1 function has not yet been designed.

2.3.2. Chardonnay

A second zebrafish mutant, *chardonnay (cdy)*, also has abnormal iron metabolism. Similar to *weh* mutants, *cdy* mutants develop a hypochromic anemia. This results from mutations in a zebrafish protein that is homologous to DMT1, and which may be the zebrafish ortholog (Donovan et al., 1999).

3. IRON OVERLOAD DISORDERS

In contrast to iron deficiency, iron overload does not result in a highly characteristic, readily apparent phenotype that can be easily recognized in large animal colonies. Consequently, no spontaneous mutations leading to iron overload have been identified in non-human models. However, it is now clear that there are multiple genes that, when mutated, lead to iron overload in human patients. The identification of human mutations has laid the groundwork for generating new, deliberately mutated models of human diseases. To date, several models of HFE-associated genetic hemochromatosis have been described. These have been used for several types of investigations to study the pathogenesis of human hemochromatosis, and of human porphyria cutanea tarda.

3.1. Hemochromatosis

Hereditary hemochromatosis encompasses a group of prevalent iron loading disorders that typically presents in middle aged individuals. It results from a small, chronic, increase in absorption of dietary iron. Over time, the excess iron is deposited in the liver, heart, endocrine tissues and other sites, leading to tissue damage. Classically, the disease appeared with a distinctive triad of hyperpigmentation, cirrhosis and diabetes ("bronze diabetes"). The genetic defect responsible for the most common European form of hemochromatosis was identified as a cysteine 282 to tyrosine (C282Y) mutation in a gene designated *HFE* (formerly *HLA-H*; (Feder et al., 1996)). A second, rare, form of hemochromatosis is attributable to mutations in the gene encoding a homolog of the transferrin receptor, transferrin receptor-2 (*TFR2*; (Camaschella et al., 2000)). It is clear that there are still other hemochromatosis genes to be discovered (Camaschella et al., 1997; Cazzola et al., 1983; Pietrangelo et al., 1999), as recently discussed in a review by Camaschella (2001).

3.1.1. Targeted disruption of the B2m gene

The first faithful mouse model of HFE-associated hemochromatosis was generated and characterized before *HFE* itself was identified. The gene encoding murine beta-2 microglobulin (*B2m*) was disrupted by targeted mutagenesis in an effort to study the importance of beta-2 microglobulin in immune function (Koller et al., 1990; Zijlstra et al., 1990). Beta-2 microglobulin forms a heterodimer with all class I major histocompatibility proteins, and with many related (atypical class I) proteins. Not surprisingly, global immune defects were evident in *B2m-/-* mice. Unexpectedly, however, they also developed tissue iron overload, in a pattern identical to that seen in human hemochromatosis (de Sousa et al., 1994; Rothenberg and Voland, 1996; Santos et al., 1996). In retrospect, this was a very important clue. It had been shown almost twenty years earlier that most Northern European patients with hemochromatosis had a particular human leukocyte antigen (HLA) class I haplotype pattern, indicating linkage to chromosome 6p, near the HLA-A gene (Simon et al., 1976). Considering that information, and the iron loading phenotype of *B2m-/-* mice, it was not surprising that the *HFE* gene was later found to encode a protein resembling HLA class I molecules (Feder et al., 1996). Similar to other class I-like molecules, HFE forms a heterodimer with beta-2 microglobulin. The C282Y mutation is predicted to disrupt an intramolecular disulfide bond within HFE, distorting its conformation (Feder et al., 1996; Lebron et al., 1998).

3.1.2. Targeted mutagenesis of the murine Hfe gene

Three groups independently generated mice carrying null mutations in the murine *Hfe* gene (Bahram et al., 1999; Levy et al., 1999b; Zhou et al., 1998). In addition, Levy and co-workers introduced the C282Y mutation into an otherwise intact *Hfe* gene to generate a mutant mouse with the same missense substitution as most human hemochromatosis patients (Levy et al., 1999b). All *Hfe* mutations result in increased iron absorption and abnormal iron accumulation in the liver, indicating that hemochromatosis results from a loss of HFE function (Bahram et al., 1999; Levy et al., 1999b; Zhou et al., 1998). Iron loading is less severe in C282Y mutant mice than in *Hfe*-null mice, making it likely that the human disease results from a subtotal loss of HFE activity (Levy et al., 1999b). Iron accumulation begins at birth; there is no apparent abnormality in placental iron transfer (J Levy, L Montross and NCA, unpublished data). Mice heterozygous for either null or C282Y mutations accumulate significantly more iron than wild type mice do, though the amounts are still lower than in mice homozygous for *Hfe* mutations. Iron loading occurs early in life; after approximately 3 months of age the amount of liver iron plateaus (J Levy, L Montross and NCA, unpublished data). Interestingly, mice carrying mutations in *Hfe* have decreased splenic iron accumulation, probably because there is less iron in splenic macrophages (Levy et al., 1999b). This is reminiscent of human hemochromatosis patients, who have decreased iron stored in bone marrow macrophages.

Although mice carrying mutations in *Hfe* have many phenotypic features indicating that they are a faithful model of the human disease, there is one significant difference. Unlike human hemochromatosis patients, mice with hemochromatosis do not become cirrhotic or diabetic (J Levy, L Montross and NCA, unpublished data). They do have a mild elevation of hepatic transaminases, and occasionally they develop liver tumors in the second year of life. It seems likely that the difference in organ damage results from a difference in the murine response to oxidative stress insults, rather than a difference in primary iron accumulation. This point is underscored by the fact that homozygous *hpx* mice develop iron overload exceeding that of *Hfe* mutant mice by more than an order of magnitude yet they do not develop cirrhosis either (Trenor et al., 2000).

3.1.3. Identification of genetic modifiers of the hemochromatosis phenotype in mice

It has become increasingly apparent that there is a wide range in phenotypic expression of iron loading in human patients who are homozygous for the *HFE* C282Y mutation (Beutler et al., 2000a; Bulaj et al., 2000; Sham et al., 2000). Some patients develop severe iron overload by the third decade of life, and others live into old age with no clinical evidence of disease. While there are undoubtedly environmental influences that affect disease penetrance, there are almost certainly other genes that modify the hemochromatosis phenotype. Now that *HFE* itself has been identified, much attention has turned to finding and characterizing modifier genes. Two approaches have been used thus far. First, genes that have known roles in intestinal iron absorption have been considered, by crossing mice carrying mutations in those genes with *Hfe* knockout mice (Levy et al., 2000). In that way, it was determined that iron loading in mouse hemochromatosis occurs through transport pathways involving DMT1 and hephaestin.

Mild mutations in either of those genes could affect the rate at which iron overload occurs, modifying the hemochromatosis phenotype. Mutations in genes encoding beta-2 microglobulin and transferrin receptor also affect iron loading (Levy et al., 2000).

As an alternate approach, investigators have looked for differences in iron loading among genetically homogeneous inbred mouse strains. There are major differences in tissue iron accumulation between strains (Fleming et al., 2001b; Sproule et al., 2001). Working from the assumption that some or all of the genetic determinants of iron loading are likely to be modifiers of the hemochromatosis phenotype, attempts are underway to map and identify the genes accounting for strain differences.

3.1.4. Pathogenesis of porphyria cutanea tarda in mouse models

Hfe knockout mice have also been useful for studying the pathogenesis of porphyria cutanea tarda (PCT), an acquired defect in an enzyme of heme biosynthesis (uroporphyrinogen decarboxylase) that results in a phototoxic dermatosis and progressive liver disease. It has long been known that hepatic iron accumulation contributes to the pathogenesis of this disorder, but the mechanism remains unclear. There are both familial forms, in which patients are heterozygous for a mutation in the gene encoding uroporphyrinogen decarboxylase, and sporadic forms, in which the disease appears to develop spontaneously. In recent studies, it has been shown that *Hfe* knockout mice are more prone to develop PCT than their normal littermates when bred to mice lacking one allele of the uroporphyrinogen decarboxylase (Phillips et al., 2001) or when treated with an agent known to induce the disease in animals (Sinclair et al., 2001). These experiments establish animal models for familial and sporadic PCT, respectively.

4. DISORDERS OF TISSUE IRON DISTRIBUTION

There are several disorders of iron metabolism that cannot simply be considered to be iron deficiency or iron overload. These are more appropriately categorized as disorders of iron distribution, resulting in excessive iron accumulation at some sites, and relative iron deficiency elsewhere in the body. The two best examples of this both involve deficiencies in plasma proteins.

4.1. Hypotransferrinemia

Hypotransferrinemia has been described in human patients (Beutler et al., 2000b; Goya et al., 1972; Hamill et al., 1991; Heilmeyer, 1966) and in mice (Bernstein, 1987). In both cases, transferrin is severely deficient or absent. The mouse mutation, *hpx*, occurred spontaneously in an inbred colony. It was shown to result from an abnormality of mRNA splicing (Huggenvik et al., 1989) attributable to a single nucleotide change within a splice donor signal (Trenor et al., 2000). As a result, approximately 1% of the normal amount of transferrin is produced in homozygous *hpx* animals. The transferrin that is made contains a nine amino acid internal deletion near its carboxyl terminus. Although it has not been examined experimentally, this deletion probably does not interfere with transferrin function, because it does not remove any of the known functional residues (Trenor et al., 2000).

In the absence of transferrin, most plasma iron circulates weakly bound to albumin and other abundant plasma constituents (Simpson et al., 1992). In that form it is not available for uptake through the transferrin cycle. Consequently, the erythron, which requires the transferrin cycle for iron uptake, becomes markedly iron deficient (Bernstein, 1987; Trenor et al., 2000). Animals generally die before birth or within the first two weeks of life if they are not treated with exogenous transferrin or red blood cell transfusions. However most or all non-hematopoietic tissues can assimilate iron efficiently via as yet unknown, non-transferrin cycle pathways, and they become massively iron overloaded in surviving animals (Simpson et al., 1993; Trenor et al., 2000). Findings are similar in human patients, who also require transferrin or repeated red blood cell transfusions (Beutler et al., 2000b; Goya et al., 1972; Hamill et al., 1991; Heilmeyer, 1966).

4.2. Aceruloplasminemia

Ceruloplasmin is another plasma metalloprotein, but it contains copper, rather than iron. As discussed earlier, it has a ferroxidase activity, and it appears to be important in the export of iron from absorptive cells and storage cells (Harris et al., 1999). Human patients with mutations disrupting the ceruloplasmin gene develop progressive retinopathy and neurodegenerative disease accompanied by liver iron accumulation, diabetes and mild microcytic anemia (Harris et al., 1998; Harris et al., 1995; Miyajima et al., 1996; Okamoto et al., 1996; Yoshida et al., 1995). To date, aceruloplasminemia has proved very difficult to treat, though desferrioxamine therapy may be of some benefit (Miyajima et al., 1997).

This disorder has been modeled by creation of a ceruloplasmin knockout mouse (Harris et al., 1999). Mice lacking ceruloplasmin demonstrate accumulation of iron in tissues that would normally export significant amounts of iron, including hepatocytes and reticuloendothelial macrophages. Their pathology may be explained by the fact that they accumulate too much iron in those sites and in the central nervous system, yet are unable to appropriately recycle iron from effete erythrocytes to make iron available for new erythropoiesis. The recycled iron is retained by macrophages because export is impaired. It will be interesting to breed these mice with *sla* mice lacking hephaestin to determine the extent of redundancy between the two homologous proteins, and to assess how important their ferroxidase function is for iron homeostasis.

5. DISORDERS OF MITOCHONDRIAL IRON METABOLISM

Abnormalities of intracellular iron distribution also cause human disease. To date, the intracellular iron disorders that have been described have been characterized by excessive accumulation of iron in mitochondria. This probably relates to the singular importance of mitochondria in iron metabolism. Two important types of iron-containing co-factors, heme and iron-sulfur clusters, are synthesized at least in part within the mitochondrion.

5.1. Friedreich Ataxia

Friedreich ataxia (FRDA) is a lethal, autosomal recessive, neurodegenerative disease that presents in childhood. It is characterized by degeneration of large sensory neurons and spinocerebellar tracts, cardiomyopathy and increased incidence of diabetes. There is no known, effective treatment for the disease. The gene responsible for FRDA, frataxin, was discovered by positional cloning (Campuzano et al., 1996), but several years passed before clues to the function of frataxin were revealed through studies of Yfh, its yeast homologue (Babcock et al., 1997). Yfh was discovered because it is important in mitochondrial iron homeostasis and respiratory function. Although the details remain to be worked out, loss of either frataxin, in patients, or Yfh, in yeast, leads to mitochondrial iron accumulation.

In accord with the fact that total loss of frataxin function has never been described in a human patient, a complete knockout of the frataxin gene results in an embryonic lethal phenotype in mice (Cossee et al., 2000). However, tissue-specific, selective knockout of the gene has led to the development of mouse models of FRDA that recapitulate features of the human disease. Two models have been generated: mice that lack frataxin in neurons and mice that lack frataxin in striated muscle (Puccio et al., 2001). These may be useful for working out the pathogenesis of FRDA, and for testing potential pharmacological treatments of the disease.

5.2. Sideroblastic/Siderocytic Anemia

Sideroblastic anemias comprise a second class of human diseases characterized by mitochondrial iron accumulation. Sideroblastic anemia may be inherited or acquired. There are two inherited forms, both due to mutations of genes on the X chromosome. Classical X-linked sideroblastic anemia results from missense mutations in the gene encoding erythroid aminolevulinic acid synthase (*ALAS2*) (Cotter et al., 1999; Cox et al., 1994). Nearly all patients are male. They have varying degrees of anemia, which may or may not be treatable with pyridoxine, a vitamin co-factor for ALAS2. The diagnosis is usually made by examination of the bone marrow and identification of characteristic ringed sideroblasts, which are erythroid precursor cells with iron-laden mitochondria forming a circle around the nucleus. The anemia is associated with ineffective erythropoiesis, and many patients develop clinically significant iron overload, even in the absence of transfusion therapy.

A second form of inherited sideroblastic anemia, termed X-linked sideroblastic anemia with ataxia, results from mutations in the ABC7 transporter protein (Allikmets et al., 1999; Bekri et al., 2000). The anemia is much less severe in this disorder, and ataxia is the prominent feature. The function of the ABC7 protein is not definitively known, but it is believed to function similarly to its yeast homolog, ATM1, a protein involved in biosynthesis of iron-sulfur clusters.

5.2.1. Anemia of flexed tail mice

There are no mouse models of the X-linked sideroblastic anemias. A knockout of murine *Alas2* has been done, but the homozygous phenotype is embryonic lethal (Nakajima et al., 1999). There are zebrafish carrying mutations in the fish erythroid aminolevulinic acid synthase gene that have been touted as animal models of sideroblastic anemia, but they do not have sideroblasts (Brownlie et al., 1998). Targeted mutagenesis of the murine *Abc7* gene has not yet been reported.

There is, however, a mutant mouse that develops a sidero*cytic* anemia. Mice homozygous for the flexed tail (*f*) mutation have normal erythroblasts (Fleming et al.,

2001a) but their mature erythrocytes contain iron-laden mitochondria from mid-gestation through the second week of life (Gruneberg, 1942). Thereafter, the siderocytic anemia resolves spontaneously and erythropoiesis appears normal unless it is stressed. The gene mutated in *f* animals was identified by positional cloning (Fleming et al., 2001a). It encodes a multi-transmembrane domain mitochondrial protein, designated sideroflexin, of unknown function. Its molecular structure suggests that it is a channel or carrier protein, but its cargo has not yet been identified. Elucidation of the function of sideroflexin is likely to shed more light on mitochondrial iron homeostasis. This will be an important area of inquiry over the next few years.

6. OTHER MOUSE MODELS

Not all iron gene targeting efforts have been directed towards creating mouse models of human diseases. Several genes have been disrupted because they were known or suspected to have important roles in iron metabolism, but their *in vivo* importance had not been fully worked out.

6.1. Iron regulatory protein 2 knockout mice

Iron regulatory proteins are regulatory molecules that recognize stem-loop structures in non-coding regions of mRNAs that encode several proteins of iron metabolism (reviewed in (Eisenstein, 2000)). They have been studied extensively in cultured cells, and shown to effect post-transcriptional regulation both by creating steric interference of translational initiation and by protecting mRNA from nucleolytic destruction. However, little is known about their importance or function *in vivo* in living animals. To address this, Rouault and colleagues have undertaken targeted disruption of genes coding for iron regulatory proteins 1 and 2. To date, they have only published their results with iron regulatory protein 2 (LaVaute et al., 2001).

Mice homozygous for null mutations in the murine iron regulatory protein 2 gene (*Ireb2*) are viable but abnormal (LaVaute et al., 2001). They have increased stainable iron present in the intestinal mucosa and the central nervous system. With increasing age, they develop noticeable neurologic abnormalities, including ataxia, bradykinesia and tremor. Although this phenotype does not closely resemble that of any inherited human disease, it has been compared to Parkinson's disease (LaVaute et al., 2001). While fascinating, this phenotype has not yet let to a complete understanding of the roles of IRPs in mammalian biology.

6.2. Transferrin receptor knockout mice

The phenotype of homozygous *hpx* mice (described above) provided important information about the major role of the transferrin cycle in erythropoiesis. However, it remained uncertain whether the phenotype of animals lacking the transferrin receptor would be identical. First, the small amount of circulating transferrin in *hpx* mice might be sufficient to meet the needs of cells that were dependent upon the transferrin cycle but had relatively modest iron requirements. Second, it was possible that transferrin, transferrin receptor or both proteins might have functions outside of their known ligand-receptor interactions. To investigate these possibilities, the gene encoding the murine transferrin receptor (*Trfr*) was disrupted by homologous recombination (Levy et al., 1999a).

Mice homozygous for a null *Trfr* allele invariably died *in utero* before embryonic day 12.5 (Levy et al., 1999a). The cause of death appeared to be severe anemia, due to an inability to produce red blood cells in adequate supply. In addition, the embryos had a characteristic kinking of their neural tubes, associated with increased apoptosis of neuroepithelial precursors at around embryonic day 9.5. Other tissues appeared to be normal, at least at that stage. These findings confirmed and extended information gleaned from studies of *hpx* mice, leading to the conclusions that the transferrin cycle is absolutely required for normal erythropoiesis, and that the transferrin receptor plays an important role in early development of the nervous system that is yet to be defined.

Animals that were heterozygous for the null *Trfr* allele appeared well, but had subtle abnormalities. They had a normal total red blood cell mass, but the cells were smaller than normal and increased in number (Levy et al., 1999a). This suggests that haploinsufficiency for *Trfr* results in fewer transferrin receptors on the surfaces of erythroid precursors, leading to decreased uptake of iron by each cell, and consequent microcytosis due to relative iron deficiency. Heterozygous loss of transferrin receptor gene expression should be considered in patients with unexplained microcytosis, particularly in the absence of anemia.

6.3. H-ferritin deficient mice

Ferritin is important for storage and sequestration of intracellular iron. In mammals, ferritin is a heteropolymer of heavy (H) and light (L) subunits. H-ferritin has a ferroxidase activity that is thought to be important for initial steps in iron incorporation; L-ferritin is thought to potentiate the formation of the inorganic iron core (reviewed in (Theil et al., 1999)).

The gene encoding the murine ferritin H subunit (*Fth*) was disrupted by homologous recombination (Ferreira et al., 2000). Mice homozygous for the null *Fth* allele die very early in gestation, probably between days 3.5 and 9.5, indicating that there is no functional redundancy between the two subunits. Heterozygous mice are grossly normal, though analysis of their iron status has not yet been reported in detail.

7. PERSPECTIVE

Animal models have proved to be very useful for understanding mammalian iron metabolism. The physiology of iron transport, storage and utilization appears to be very similar among mammalian species, and some parallels appear to extend as far down the vertebrate evolutionary scale as zebrafish. Mice have been particularly valuable for two reasons -- because of the existence of a series of well-characterized, spontaneous mutants, and because of our ability to manipulate the mouse genome through targeted mutagenesis. Although all predictions made from animal studies must ultimately be validated in humans, much recent progress has been made in this field.

8. REFERENCES

Abboud, S., and Haile, D. J., 2000, A novel mammalian iron-regulated protein involved in intracellular iron metabolism. *J Biol Chem* **275**, 19906-19912.

Allikmets, R., Raskind, W. H., Hutchinson, A., Schueck, N. D., Dean, M., and Koeller, D. M., 1999, Mutation of a putative mitochondrial iron transporter gene (ABC7) in X- linked sideroblastic anemia and ataxia (XLSA/A). *Hum Mol Genet* **8**, 743-749.

Babcock, M., De Silva, D., Oaks, R., Davis-Kaplan, S., Jiralerspong, S., Montermini, L., Pandolfo, M., and Kaplan, J., 1997, Regulation of mitochondrial iron accumulation by Yfh1p, a putative homolog of frataxin. *Science* **276**, 1709-1712.

Bahram, S., Gilfillan, S., Kuhn, L. C., Moret, R., Schulze, J. B., Lebeau, A., and Schumann, K., 1999, Experimental hemochromatosis due to MHC class I HFE deficiency: immune status and iron metabolism. *Proc Natl Acad Sci U S A* **96**, 13312-13317.

Bannerman, R. M., 1976, Genetic defects of iron transport. *Federation Proceedings* **35**, 2281.

Bannerman, R. M., and Cooper, R. G., 1966, Sex-linked anemia: a hypochromic anemia of mice. *Science* **151**, 581-582.

Bannerman, R. M., Edwards, J. A., Kreimer-Birnbaum, M., McFarland, E., and Russell, E. S., 1972, Hereditary microcytic anaemia in the mouse; studies in iron distribution and metabolism. *British Journal of Haematology* **23**, 235-245.

Bedard, Y. C., Pinkerton, P. H., and Simon, G. T., 1976, Uptake of circulating iron by the duodenum of normal mice and mice with altered iron stores, including sex-linked anemia. High resolution radioautographic study. *Laboratory Investigation* **34**, 611-615.

Bekri, S., Kispal, G., Lange, H., Fitzsimons, E., Tolmie, J., Lill, R., and Bishop, D. F., 2000, Human ABC7 transporter: gene structure and mutation causing X-linked sideroblastic anemia with ataxia with disruption of cytosolic iron- sulfur protein maturation. *Blood* **96**, 3256-3264.

Bennett, M., Pinkerton, P. H., Cudkowicz, G., and Bannerman, R. M., 1968, Hemopoietic progenitor cells in marrow and spleen of mice with hereditary iron deficiency anemia. *Blood* **32**, 908-921.

Bernstein, S. E., 1987, Hereditary hypotransferrinemia with hemosiderosis, a murine disorder resembling human atransferrinemia. *Journal of Laboratory and Clinical Medicine* **110**, 690-705.

Beutler, E., Felitti, V., Gelbart, T., and Ho, N., 2000a, The effect of HFE genotypes on measurements of iron overload in patients attending a health appraisal clinic. *Ann Intern Med* **133**, 329-337.

Beutler, E., Gelbart, T., Lee, P., Trevino, R., Fernandez, M. A., and Fairbanks, V. F., 2000b, Molecular characterization of a case of atransferrinemia. *Blood* **96**, 4071-4074.

Bloom, M. L., Simon-Stoos, K. L., and Mabon, M. E., 1998, The hemoglobin-deficit mutation is located on mouse chromosome 19. *Mamm Genome* **9**, 666-667.

Bowen, B. J., and Morgan, E. H., 1987, Anemia of the Belgrade rat: evidence for defective membrane transport of iron. *Blood* **70**, 38-44.

Brownlie, A., Donovan, A., Pratt, S., Paw, B., Oates, A., Brugnara, C., Witkowska, H., Sassa, S., and Zon, L., 1998, Positional cloning of the zebrafish sauternes gene: a model for congenital sideroblastic anaemia. *Nat Genet* **20**, 244-250.

Bulaj, Z. J., Ajioka, R. S., Phillips, J. D., LaSalle, B. A., Jorde, L. B., Griffen, L. M., Edwards, C. Q., and Kushner, J. P., 2000, Disease-related conditions in relatives of patients with hemochromatosis. *N Engl J Med* **343**, 1529-1535.

Camaschella, C., De Gobbi, M., and Roetto, A., 2001, Hereditary hemochromatosis: progress and perspectives. *Rev Clin Exp Hematol* **4**, 302-321.

Camaschella, C., Roetto, A., Cali, A., De Gobbi, M., Garozzo, G., Carella, M., Majorano, N., Totaro, A., and Gasparini, P., 2000, The gene TFR2 is mutated in a new type of haemochromatosis mapping to 7q22. *Nat Genet* **25**, 14-15.

Camaschella, C., Roetto, A., Cicilano, M., Pasquero, P., Bosio, S., Gubetta, L., Di Vito, F., Girelli, D., Totaro, A., Carella, M., Grifa, A., and Gasparini, P., 1997, Juvenile and adult hemochromatosis are distinct genetic disorders. *Eur J Hum Genet* **5**, 371-375.

Campuzano, V., Montermini, L., Molto, M. D., Pianese, L., Cossee, M., Cavalcanti, F., Monros, E., Rodius, F., Duclos, F., Monticelli, A., and et al., 1996, Friedreich's ataxia: autosomal recessive disease caused by an intronic GAA triplet repeat expansion. *Science* **271**, 1423-1427.

Canonne-Hergaux, F., Fleming, M. D., Levy, J. E., Gauthier, S., Ralph, T., Picard, V., Andrews, N. C., and Gros, P., 2000, The Nramp2/DMT1 iron transporter is induced in the duodenum of microcytic anemia mk mice but is not properly targeted to the intestinal brush border. *Blood* **96**, 3964-3970.

Canonne-Hergaux, F., Gruenheid, S., Ponka, P., and Gros, P., 1999, Cellular and subcellular localization of the Nramp2 iron transporter in the intestinal brush border and regulation by dietary iron. *Blood* **93**, 4406-4417.

Cazzola, M., Ascari, E., Barosi, G., Claudiani, G., Dacco, M., Kaltwasser, J. P., Panaiotopoulos, N., Schalk, K. P., and Werner, E. E., 1983, Juvenile idiopathic haemochromatosis: a life-threatening disorder presenting as hypogonadotropic hypogonadism. *Hum Genet* **65**, 149-154.

Chua, A. C., and Morgan, E. H., 1997, Manganese metabolism is impaired in the Belgrade laboratory rat. *J Comp Physiol* **167**, 361-369.

Cossee, M., Puccio, H., Gansmuller, A., Koutnikova, H., Dierich, A., LeMeur, M., Fischbeck, K., Dolle, P., and Koenig, M., 2000, Inactivation of the friedreich ataxia mouse gene leads to early embryonic lethality without iron accumulation. *Hum Mol Genet* **9**, 1219-1226.

Cotter, P. D., May, A., Li, L., Al-Sabah, A. I., Fitzsimons, E. J., Cazzola, M., and Bishop, D. F., 1999, Four new mutations in the erythroid-specific 5-aminolevulinate synthase (ALAS2) gene causing X-linked sideroblastic anemia: increased pyridoxine responsiveness after removal of iron overload by phlebotomy and coinheritance of hereditary hemochromatosis. *Blood* **93**, 1757-1769.

Cox, T. C., Bottomley, S. S., Wiley, J. S., Bawden, M. J., Matthews, C. S., and May, B. K., 1994, X-linked pyridoxine-responsive sideroblastic anemia due to a Thr388-to- Ser substitution in erythroid 5-aminolevulinate synthase. *N Engl J Med* **330**, 675-679.

de Sousa, M., Reimao, R., Lacerda, R., Hugo, P., Kaufmann, S. H. E., and Porto, G., 1994, Iron overload in beta2-microglobulin deficient mice. *Immunol Lett* **39**, 105-111.

Donovan, A., Brownlie, A., Pratt, S., Shepard, J., Paw, B., Dorschner, M. O., Fleming, M. D., Thisse, B., Andrews, N. C., and Zon, L. (1999). Positional cloning of two zebrafish mutant genes involved in hypochromic anemia: the iron transporter DMT1 (Nramp2) and a novel integral membrane protein. Paper presented at: BioIron 1999 (Sorrento, Italy).

Donovan, A., Brownlie, A., Zhou, Y., Shepard, J., Pratt, S. J., Moynihan, J., Paw, B. H., Drejer, A., Barut, B., Zapata, A., Law, T. C., Brugnara, C., Lux, S. E., Pinkus, G. S., Pinkus, J. L., Kingsley, P. D., Palis, J., Fleming, M. D., Andrews, N. C., and Zon, L. I., 2000, Positional cloning of zebrafish ferroportin1 identifies a conserved vertebrate iron exporter. *Nature* **403**, 776-781.

Edwards, J., Huebers, H., Kunzler, C., and Finch, C., 1986, Iron metabolism in the Belgrade rat. *Blood* **67**, 623-628.

Edwards, J. A., and Bannerman, R. M., 1970, Hereditary defect of intestinal iron transport in mice with sex-linked anemia. *J Clin Invest* **49**, 1869-1871.

Edwards, J. A., and Hoke, J. E., 1972, Defect of intestinal mucosal iron uptake in mice with hereditary microcytic anemia. *Proc Soc Exp Biol Med* **141**, 81-84.

Edwards, J. A., and Hoke, J. E., 1975, Red cell iron uptake in hereditary microcytic anemia. *Blood* **46**, 381-388.

Edwards, J. A., Sullivan, A. L., and Hoke, J. E., 1980, Defective delivery of iron to the developing red cell of the Belgrade laboratory rat. *Blood* **55**, 645-648.

Eisenstein, R. S., 2000, Iron regulatory proteins and the molecular control of mammalian iron metabolism. *Annu Rev Nutr* **20**, 627-662.

Falconer, D. S., and Isaacson, J. H., 1962, The genetics of sex-linked anaemia in the mouse. *Genet Res* **3**, 248-250.

Farcich, E. A., and Morgan, E. H., 1992a, Diminished iron acquisition by cells and tissues of Belgrade laboratory rats. *Am J Physiol* **262**, R220-224.

Farcich, E. A., and Morgan, E. H., 1992b, Uptake of transferrin-bound and nontransferrin-bound iron by reticulocytes from the Belgrade laboratory rat: comparison with Wistar rat transferrin and reticulocytes. *American Journal of Hematology* **39**, 9-14.

Feder, J. N., Gnirke, A., Thomas, W., Tsuchihashi, Z., Ruddy, D. A., Basava, A., Dormishian, F., Domingo, R., Ellis, M. C., Fullan, A., Hinton, L. M., Jones, N. L., Kimmel, B. E., Kronmal, G. S., Lauer, P., Lee, V. K., Loeb, D. B., Mapa, F. A., McClelland, E., Meyer, N. C., Mintier, G. A., Moeller, N., Moore, T., Morikang, E., Prass, C. E., Quintana, L., Starnes, S. M., Schatzman, R. C., Brunke, K. J., Drayna, D. T., Risch, N. J., Bacon, B. R., and Wolff, R. K., 1996, A novel MHC class I-like gene is mutated in patients with hereditary haemochromatosis. *Nature Genetics* **13**, 399-408.

Ferreira, C., Bucchini, D., Martin, M. E., Levi, S., Arosio, P., Grandchamp, B., and Beaumont, C., 2000, Early embryonic lethality of H ferritin gene deletion in mice. *J Biol Chem* **275**, 3021-3024.

Flanagan, P. R., Haist, J., MacKenzie, I., and Valberg, L. S., 1984, Intestinal absorption of zinc: competitive interactions with iron, cobalt and copper in mice with sex-linked anemia (sla). *Canadian Journal of Physiology and Pharmacology* **62**, 1124-1128.

Fleming, M. D., Campagna, D. R., Haslett, J. N., Trenor, C. C., 3rd, and Andrews, N. C., 2001a, A mutation in a mitochondrial transmembrane protein is responsible for the pleiotropic hematological and skeletal phenotype of flexed-tail (f/f) mice. *Genes Dev* **15**, 652-657.

Fleming, M. D., Romano, M. A., Su, M. A., Garrick, L. M., Garrick, M. D., and Andrews, N. C., 1998, Nramp2 is mutated in the anemic Belgrade (b) rat: evidence of a role for Nramp2 in endosomal iron transport. *Proc Natl Acad Sci USA* **95**, 1148-1153.

Fleming, M. D., Trenor, C. C. I., Su, M. A., Foernzler, D., Beier, D. R., Dietrich, W. F., and Andrews, N. C., 1997, Microcytic anemia mice have a mutation in Nramp2, a candidate iron transporter gene. *Nature Genetics* **16**, 383-386.

Fleming, R. E., Holden, C. C., Tomatsu, S., Waheed, A., Brunt, E. M., Britton, R. S., Bacon, B. R., Roopenian, D. C., and Sly, W. S., 2001b, Mouse strain differences determine severity of iron accumulation in Hfe knockout model of hereditary hemochromatosis. *Proc Natl Acad Sci U S A* **98**, 2707-2711.

Garrick, L. M., Edwards, J. A., Hoke, J. E., and Bannerman, R. M., 1987, Diminished acquisition of iron by reticulocytes from mice with hemoglobin deficit. *Experimental Hematology* **15**, 671-675.

Garrick, M. D., Gniecko, K., Liu, L., Cohan, D. S., and Garrick, L. M., 1993, Transferrin and the transferrin cycle in Belgrade rat reticulocytes. *Journal of Biological Chemistry* **268**, 14867.

Goya, N., Miyazaki, S., Kodate, S., and Ushio, B., 1972, A family of congenital atransferrinemia. *Blood* **40**, 239-245.

Grewal, M. D., 1962, A sex-linked anaemia in the mouse. *Genet Res* **3**, 238-247.

Gruenheid, S., Canonne-Hergaux, F., Gauthier, S., Hackam, D. J., Grinstein, S., and Gros, P., 1999, The iron transport protein NRAMP2 is an integral membrane glycoprotein that colocalizes with transferrin in recycling endosomes. *J Exp Med* **189**, 831-841.

Gruneberg, H., 1942, The anaemia of flexed-tail mice (Mus musculus L.). II. Siderocytes. *J Genet* **44**, 246-271.

Gunshin, H., Mackenzie, B., Berger, U. V., Gunshin, Y., Romero, M. F., Boron, W. F., Nussberger, S., Gollan, J. L., and Hediger, M. A., 1997, Cloning and characterization of a mammalian proton-coupled metal-ion transporter. *Nature* **388**, 482-488.

Haffter, P., Granato, M., Brand, M., Mullins, M. C., Hammerschmidt, M., Kane, D. A., Odenthal, J., van Eeden, F. J., Jiang, Y. J., Heisenberg, C. P., Kelsh, R. N., Furutani-Seiki, M., Vogelsang, E., Beuchle, D., Schach, U., Fabian, C., and Nusslein-Volhard, C., 1996, The identification of genes with unique and essential functions in the development of the zebrafish, Danio rerio. *Development* **123**, 1-36.

Hamill, R. L., Woods, J. C., and Cook, B. A., 1991, Congenital atransferrinemia: a case report and review of the literature. *American Journal of Clinical Pathology* **96**, 215-218.

Harris, Z. L., Durley, A. P., Man, T. K., and Gitlin, J. D., 1999, Targeted gene disruption reveals an essential role for ceruloplasmin in cellular iron efflux. *Proc Natl Acad Sci U S A* **96**, 10812-10817.

Harris, Z. L., Klomp, L. W., and Gitlin, J. D., 1998, Aceruloplasminemia: an inherited neurodegenerative disease with impairment of iron homeostasis. *Am J Clin Nutr* **67**, 972S-977S.

Harris, Z. L., Takahashi, Y., Miyajima, H., Serizawa, M., MacGillivray, R. T. A., and Gitlin, J. D., 1995, Aceruloplasminemia: Molecular characterization of this disorder of iron metabolism. *Proceedings of the National Academy of Sciences (USA)* **92**, 2539-2543.

Harrison, D. E., 1972, Marrow Transplantation and Iron Therapy in Mouse Hereditary Microcytic Anemia. *Blood* **40**, 893-901.

Heilmeyer, L., 1966, Atransferrinemias [German]. *Acta Haematologica* **36**, 40-49.

Holmberg, C. G., and Laurell, C. B., 1948, *Acta Chem Scand* **2**, 550-556.

Huggenvik, J. I., Craven, C. M., Idzerda, R. L., Bernstein, S., Kaplan, J., and McKnight, G. S., 1989, A splicing defect in the mouse transferrin gene leads to congenital atransferrinemia. *Blood* **74**, 482-486.

Kingston, P. J., Bannerman, C. E., and Bannerman, R. M., 1978, Iron deficiency anaemia in newborn sla mice: a genetic defect of placental iron transport. *Br J Haematol* **40**, 265-276.

Koller, B. H., Marrack, P., Kappler, J. W., and Smithies, O., 1990, Normal development of mice deficient in beta 2M, MHC class I proteins, and CD8+ T cells. *Science* **248**, 1227-1230.

LaVaute, T., Smith, S., Cooperman, S., Iwai, K., Land, W., Meyron-Holtz, E., Drake, S. K., Miller, G., Abu-Asab, M., Tsokos, M., Switzer, R., Grinberg, A., Love, P., Tresser, N., and Rouault, T. A., 2001, Targeted deletion of the gene encoding iron regulatory protein-2 causes misregulation of iron metabolism and neurodegenerative disease in mice. *Nat Genet* **27**, 209-214.

Lebron, J. A., Bennett, M. J., Vaughn, D. E., Chirino, A. J., Snow, P. M., Mintier, G. A., Feder, J. N., and Bjorkman, P. J., 1998, Crystal structure of the hemochromatosis protein HFE and characterization of its interaction with transferrin receptor. *Cell* **93**, 111-123.

Levy, J. E., Jin, O., Fujiwara, Y., Kuo, F., and Andrews, N. C., 1999a, Transferrin receptor is necessary for development of erythrocytes and the nervous system. *Nature Genet* **21**, 396-399.

Levy, J. E., Montross, L. K., and Andrews, N. C., 2000, Genes that modify the hemochromatosis phenotype in mice. *J Clin Invest* **105**, 1209-1216.

Levy, J. E., Montross, L. K., Cohen, D. E., Fleming, M. D., and Andrews, N. C., 1999b, The C282Y mutation causing hereditary hemochromatosis does not produce a null allele. *Blood* **94**, 9-11.

Manis, J., 1971, Intestinal iron-transport defect in the mouse with sex-linked anemia. *American Journal of Physiology* **220**, 135-139.

McKie, A. T., Marciani, P., Rolfs, A., Brennan, K., Wehr, K., Barrow, D., Miret, S., Bomford, A., Peters, T. J., Farzaneh, F., Hediger, M. A., Hentze, M. W., and Simpson, R. J., 2000, A novel duodenal iron-regulated transporter, IREG1, implicated in the basolateral transfer of iron to the circulation. *Mol Cell* **5**, 299-309.

Miyajima, H., Takahashi, Y., Kamata, T., Shimizu, H., Sakai, N., and Gitlin, J. D., 1997, Use of desferrioxamine in the treatment of aceruloplasminemia. *Ann Neurol* **41**, 404-407.

Miyajima, H., Takahashi, Y., Shimizu, H., Sakai, N., Kamata, T., and Kaneko, E., 1996, Late onset diabetes mellitus in patients with hereditary aceruloplasminemia. *Intern Med* **35**, 641-645.

Nakajima, O., Takahashi, S., Harigae, H., Furuyama, K., Hayashi, N., Sassa, S., and Yamamoto, M., 1999, Heme deficiency in erythroid lineage causes differentiation arrest and cytoplasmic iron overload. *Embo J* **18**, 6282-6289.

Nash, D. J., Kent, E., Dickie, M. M., and Russell, E. S., 1964, The inheritance of "mick," a new anemia in the house mouse [abstract]. *American Zoologist* **4**, 404-405.

Oates, P. S., and Morgan, E. H., 1996, Defective iron uptake by the duodenum of Belgrade rats fed diets of different iron contents. *Am J Physiol* **270**, G826-832.

Okamoto, N., Wada, S., Oga, T., Kawabata, Y., Baba, Y., Habu, D., Takeda, Z., and Wada, Y., 1996, Hereditary ceruloplasmin deficiency with hemosiderosis. *Hum Genet* **97**, 755-758.

Phillips, J. D., Jackson, L. K., Bunting, M., Franklin, M. R., Thomas, K. R., Levy, J. E., Andrews, N. C., and Kushner, J. P., 2001, A mouse model of familial porphyria cutanea tarda. *Proc Natl Acad Sci U S A* **98**, 259-264.

Pietrangelo, A., Montosi, G., Totaro, A., Garuti, C., Conte, D., Cassanelli, S., Fraquelli, M., Sardini, C., Vasta, F., and Gasparini, P., 1999, Hereditary Hemochromatosis in Adults without Pathogenic Mutations in the Hemochromatosis Gene. *N Engl J Med* **341**, 725-732.

Pinkerton, P. H., 1968, Histological evidence of disordered iron transport in the x-linked hypochromic anaemia of mice. *J Pathol Bacteriol* **95**, 155-165.

Pinkerton, P. H., Bannerman, R. M., Doeblin, T. D., Benisch, B. M., and Edwards, J. A., 1970, Iron metabolism and absorption studies in the X-linked anaemia of mice. *Br J Haematol* **18**, 211-228.

Puccio, H., Simon, D., Cossee, M., Criqui-Filipe, P., Tiziano, F., Melki, J., Hindelang, C., Matyas, R., Rustin, P., and Koenig, M., 2001, Mouse models for Friedreich ataxia exhibit cardiomyopathy, sensory nerve defect and Fe-S enzyme deficiency followed by intramitochondrial iron deposits. *Nat Genet* **27**, 181-186.

Rothenberg, B. E., and Voland, J. R., 1996, Beta 2 knockout mice deveop parenchymal iron overload: a putative role for class I genes of the major histocompatibility complex in iron metabolism. *Proceedings of the National Academy of Sciences (USA)* **93**, 1529-1534.

Russell, E. S., 1984, Developmental studies of mouse hereditary anemias. *Am J Med Genet* **18**, 621-641.

Russell, E. S., McFarland, E. C., and Kent, E. L., 1970, Low viability, skin lesions, and reduced fertility associated with microcytic anemia in the mouse. *Transpl Proc* **2**, 144-151.

Santos, M., Schilham, M. W., Rademakers, L. H. P. M., Marx, J. J. M., deSousa, M., and Clevers, M., 1996, Defective iron homeostasis in beta2-microglobulin knockout mice recapitulates hereditary hemochromatosis in man. *J Exp Med* **184**, 1975-1985.

Scheufler, V. H., 1969, Eine weitere Mutante der Hausmaus mit Anamie (hbd). *Z Versuchstierk* **11**, 348-353.

Sham, R. L., Raubertas, R. F., Braggins, C., Cappuccio, J., Gallagher, M., and Phatak, P. D., 2000, Asymptomatic hemochromatosis subjects: genotypic and phenotypic profiles. *Blood* **96**, 3707-3711.

Simon, M., Bourel, M., Fauchet, R., and Genetet, B., 1976, Association of HLA-A3 and HLA-B14 antigens with idiopathic haemochromatosis. *Gut* **17**, 332-334.

Simpson, R. J., Cooper, C. E., Raja, K. B., Halliwell, B., Evans, P. J., Aruoma, O. I., Singh, S., and Konijn, A. M., 1992, Non-transferrin-bound iron species in the serum of hypotransferrinaemic mice. *Biochim Biophys Acta* **1156**, 19-26.

Simpson, R. J., Konijn, A. M., Lombard, M., Raja, K. B., Salisbury, J. R., and Peters, T. J., 1993, Tissue iron loading and histopathological changes in hypotransferrinaemic mice. *J Pathol* **171**, 237-244.

Sinclair, P. R., Gorman, N., Walton, H. S., Bement, W. J., Sinclair, J. F., Gerhard, G. S., Szakacs, J. G., Andrews, N. C., and Levy, J. E., 2001, Uroporphyria in Hfe mutant mice given 5-aminolevulinate: A new model of Fe-mediated porphyria cutanea tarda. *Hepatology* **33**, 406-412.

Sladic-Simic, D., Martinovich, P. N., Zivkovic, N., Pavic, D., Martinovic, J., Kahn, M., and Ranney, H. M., 1969, A thalassemia-like disorder in Belgrade laboratory rats. *Annals of the New York Academy of Sciences* **165**, 93-99.

Sproule, T. J., Jazwinska, E. C., Britton, R. S., Bacon, B. R., Fleming, R. E., Sly, W. S., and Roopenian, D. C., 2001, Naturally variant autosomal and sex-linked loci determine the severity of iron overload in beta 2-microglobulin-deficient mice. *Proc Natl Acad Sci U S A* **98**, 5170-5174.

Su, M. A., Trenor, C. C., Fleming, J. C., Fleming, M. D., and Andrews, N. C., 1998, The G185R mutation disrupts function of iron transporter Nramp2. *Blood* **92**, 2157-2163.

Theil, E. C., Takagi, H., Small, G. W., He, L., Tipton, A. R., and Danger, D., 1999, The ferritin iron entry and exit problem. *Inorganica Clinica Acta* **297**, 242-251.

Trenor, C. C. I., Campagna, D. R., Sellers, V. M., Andrews, N. C., and Fleming, M. D., 2000, The molecular defect in hypotransferrinemic mice. *Blood* **96**, 1113-1118.

Vulpe, C. D., Kuo, Y. M., Murphy, T. L., Cowley, L., Askwith, C., Libina, N., Gitschier, J., and Anderson, G. J., 1999, Hephaestin, a ceruloplasmin homologue implicated in intestinal iron transport, is defective in the sla mouse. *Nature Genet* **21**, 195-199.

Yoshida, K., Furihata, K., Takeda, S., Nakamura, A., Yamamoto, K., Morita, H., Hiyamuta, S., Ikeda, S., Shimizu, N., and Yanagisawa, N., 1995, A mutation in the ceruloplasmin gene is associated with systemic hemosiderosis in humans. *Nat Genet* **9**, 267-272.

Zhou, X. Y., Tomatsu, S., Fleming, R. E., Parkkila, S., Waheed, A., Jiang, J., Fei, Y., Brunt, E. M., Ruddy, D. A., Prass, C. E., Schatzman, R. C., O'Neill, R., Britton, R. S., Bacon, B. R., and Sly, W. S., 1998, HFE gene knockout produces mouse model of hereditary hemochromatosis. *Proc Natl Acad Sci USA* **95**, 2492-2497.

Zijlstra, M., Bix, M., Simister, N. E., Loring, J. M., Raulet, D. H., and Jaenisch, R., 1990, Beta 2-microglobulin deficient mice lack CD4-8+ cytolytic T cells. *Nature* **344**, 742-746.

MECHANISM OF IRON TOXICITY

Antonello Pietrangelo

As iron is essential for fundamental vital activities, iron deprivation threatens cell survival, thus making iron deficiency in humans a public health problem throughout the world and iron supplementation the only therapeutical option. On the other hand, a number of disease states are pathogenetically linked to excess body iron stores and iron removal therapy is an effective life-saving strategy in many circumstancies.

It is the iron-oxygen connection thatsets the basis, paradoxically, for both life and death in living cells. In fact, the capacity of readily exchanging electrons in aerobic conditions not only makes iron essential for fundamental cell functions, but also a potential catalyst for chemical reactions involving free radical formation, oxidative stress and cell damage. To avoid this latter event, cells have developed systems to adjust intracellular iron concentration to levels that are adequate for their metabolic needs but below the toxicity threshold.

The rapid progress of iron research in the field of biochemistry, genetics, cell and molecular biology is also changing our concept of iron toxicity. At the chemical and biochemical level, within the traditional conceptual framework of iron as catalyst of free radical reactions, we are now recognize the pathogenetic role of perturbations in the genetic control of the catalitically active iron pool, the so-called "free iron pool" or "labile iron pool" (LIP). In addition, at the pathological and clinical level iron is emerging as a leading pathogenetic factor not only in conditions of overt "iron overload" (such as genetic or acquired iron overload states) but also in those disease states where total iron burden may not change but intracellular iron is "delocalised" and the LIP expands (e.g. inflammation, atherosclerosis, neurodegeneration etc.).

This chapter attempts to reconcile the new findings in iron metabolism research with the established concepts of oxidative-stress-driven iron toxicity, using the liver disease during iron overload as exemplification of "iron overload disease".

1. IRON AND OXIDATIVE STRESS. BASIC ASPECTS.

1.1 Iron, Reactive oxygen species and oxidative stress

In every cell during its normal life in aerobic conditions, a small amount of the consumed oygen is reduced in a specific way, yielding a variety of highly reactive chemical entities. These are collectively called reactive oxygen species (ROS) or reactive oxidative intermediates (ROIs).

ROS include a variety of molecular species such as hydrogen peroxide (H_2O_2), singlet molecular oxygen ($^1O_2$), hydroxyl (OH·), superoxide (O_2^-), alkoxyl (RO·), peroxyl (ROO·) and nitric oxide (NO·) radicals, highly heterogeneous in terms of reactivity against cellular targets, diffusion capability and half-lives.[1] Although molecular O_2 contains an even number of electrons, it has two unpaired electrons in its molecular orbitals with the same spin quantum number. Thus, when accepting a pair of electrons from a molecule, new electrons must be of parallel spin to fit in the vacant spaces of the orbitals. This imposes one electron transfer reaction at a time. These reactions may lead to generation of reactive intermediates, such as superoxide radicals, peroxyl radicals or singlet molecular oxygen. Transition metal ions as iron, having frequently unpaired electrons, are excellent catalysts and play a decisive role in the generation of the very reactive species from the less reactive ones, for instance by catalyzing the formation of hydroxyl radicals, from reduced forms of O_2 (Figure 1A). Under normal conditions, cellular organelles, including mitochondria, microsomes, peroxisomes, produces a significant amount of ROS. The concentration of ROS in the cell is kept fairly constant by enzymic and nonenzymic activities that are able to dispose unwanted ROS and generate nontoxic by-products (Table 1). In addition to cellular or plasma antioxidants, the modulation of iron availability is the main means by which cells keep ROS levels under strict control because the appropriate sequestration of iron may allow the physiologic roles of the relatively safe O_2^-, and H_2O_2 to take place without the production of the highly reactive OH·, by Fenton chemistry [2] (Figure 1A). ROS are capable of causing oxidative damage to macromolecules leading to lipid peroxidation (Figure 1B), oxidation of amino acid side chains (especially cysteine), formation of protein-protein cross-links, oxidation of polypeptide backbones resulting in protein fragmentation, DNA damage, and DNA strand breaks.[1] High doses of ROS, which may be generated during chronic and acute inflammatory diseases or on environmental stresses, are cytotoxic.

A

$$Fe^{3+} + O_2^- \rightarrow Fe^{2+} + O_2$$
$$H_2O_2 + Fe^{2+} \rightarrow OH· + OH^- + Fe^{3+}$$

B

$$LH + R· \rightarrow L· + RH$$
$$L· + O_2 \rightarrow LOO·$$
$$LOO· + LH \rightarrow LOOH + L·$$
$$LOOH\ (+ Fe^{2+}/Fe^{3+}) \Longrightarrow alkanes$$
$$alkanals$$
$$alkenals$$
$$4\text{-hydroxyalkenals}$$

Figure 1. A. Hydroxyl radical generation through classical H_2O_2/Fe^{2+} interaction (Fenton reaction). The Haber-Weiis reaction where O_2^- reduces Fe^{3+} to Fe^{2+} is also reported. B. Lipid peroxidation of biological membranes. LH= polyunsaturated fatty acids; R· = free radical; L· = lipid free radical; LOO· = lipid peroxyl radical; LOOH = lipid peroxide.

Table 1. Antioxidants

Non-enzymic	Enzymic
α-tocopherol (vitamin E)	Superoxide dismutases
β-carotene	Glutathione peroxidase
glutathione	Catalase
urate	Glutathione and oxidized glutathione reductase
bilirubin	Glutathione-S-transferase
flavonoids	UDP-glucuronosyl-transferase
plasma proteins (albumin, ceruloplasmin, transferrin)	NADPH-quinone oxidoreductas
	NADPH supply transport system

Small amounts of ROS, produced as a consequence of electron transfer reactions in mitochondria, peroxisomes, and microsomes play a role in physiological pathways, such as signal transduction, cell signalling and redox regulation of cell proliferation and apoptosis. In fact, in view of the short half-life, small size and high diffusibility, ROS are excellent candidates for second messenger function (see below).

A state of increased levels of intracellular ROS leading to cell toxicity is referred to as oxidative stress. Cells respond to these adverse conditions by modulation of their antioxidant levels, induction of new gene expression, and protein modification[3]. The homeostatic modulation of oxidant levels is a highly efficient mechanism that appeared early in evolution, allowing all cells to tightly control their redox status within a very narrow range. Critical steps in the signal transduction cascade are sensitive to oxidants and antioxidants.[4,5] In fact, many basic events of cell regulation such as protein phosphorylation and binding of transcription factors to consensus sites on DNA are driven by physiological oxidant-antioxidant homeostasis, especially by the thiol disulfide balance (Table 2).

As to the regulation of gene expression by intracellular reduction-oxidation (redox) state, at least two well-defined transcription factors, nuclear factor (NF) kB and activator protein (AP)-1 have been identified to be regulated by the intracellular redox state. The Rel-NFκB family of transcriptional factors regulate expression of numerous cellular and viral genes and play important roles in immune and stress responses, inflammation, and apoptosis.[6] Binding sites of the redox-regulated transcription factors NF-kB and AP-1 are located in the promoter region of a large variety of genes (cytokine, adesion molecules, chemokines, major histocompatibility class proteins and others) that are directly involved in the pathogenesis of diseases, e.g., AIDS, cancer, atherosclerosis and diabetic complications.[6,7]

In this context, iron, as a master regulator of ROS production is pathogenetically involved in a variety of pathophysiological states (e.g. apoptosis, inflammation, cancer etc.) where unwanted ROS generation takes place.

Table 2. Signalling molecules and transcription factors controlled by the redox state

Signalling molecules	Transcription factors
Protein tyrosine kinase: epidermal growth factor receptor, insulin receptor, platelet-derived growth factor receptor	Activator protein-1 (c-Fos/c-Jun)
	Nuclear Factor kB (p50)
	Upstream stimulatory factor
Protein tyrosine phosphatase	cAMP responsive element binding protein
Protein serine/threonine kinase: MAP kinase, c-Jun N terminal kinase, p38	Thyroid transcription factor 1
	Nuclear factor 1
Protein serine/threonine phosphatase: protein poshatase 1 and 2A, calcineurin	Activating transcription factor
Small G protein: Ras	
Lipid signaling: phospholipase L, D and A_2, phosphatidylinositol 3-kinase	
Ca++ signal: inositol(1,4,5) triphosphate, Ca^{2+}-ATPse, Ca^{2+}/Na^+ exchanger	

1.2 Sensing iron and oxygen through iron-sulfur clusters: the iron regulatory proteins

As discussed above, fluctuations of iron (and oxygen) levels in the cell are critical for cell survival or death. Consequently, "sensing" iron and oxygen and controlling their levels is vital for living organisms.

Very early in earth's history, in the absence of oxygen, elemental iron and sulfur could freely interact and assemble spontaneously. The iron-sulfur proteins, in which the iron is at least partially coordinated by sulfur, are found in a wide range of organisms from bacteria to man.[8] In many of these proteins, the role of the cluster is to allow electron transfer, catalyze functions, or maintain structural integrity. Thus, their chemical versatility and flexibility in catalytic and electron transfer reactions make them suitable for many fundamental functions.[9] These clusters, particularly the [4Fe-4S] clusters, can function as strong reductants. Indeed, iron-sulfur proteins are responsible for cellular generation of superoxide, a by-product of respiration formed by reducing agents that are powerful enough to perform a single electron reduction of O_2. Because these clusters have the capacity to accept or donate single electrons, they are frequently found in enzymes in which single electrons must be supplied or removed to catalyse transformations of substrate, as is the case in a number of mammalian respiratory chain proteins, such as succinate dehydrogenase, NADH dehydrogenase and the cytochrome bc1 complex. The Fe-S cluster can be very unstable in the presence of oxidants.[10] The disintegration may originate intermediate products, but the specific pathways have not been elucidated yet. In a few cases the cluster is resistant to oxidation: the cluster is bound in a region of the protein that is inaccessible to solvent and oxidants. This may explain why these clusters are commonly found not only in anaerobic organisms, but also in aerobic organisms, where many of them perform key functions. In conditions in which oxidants are overproduced or are able to reach the core pocket, it is reasonable to speculate that loss of electrons, beyond a certain point, results in partial free radical formation at bridging sulfurs and cysteinyl ligands. This is followed by coupling reactions between free radicals and hydrolytic degradation of the oxidized center by solvent. The disintegration of iron-sulfur clusters in certain settings can lead to loss of activity of the associated protein. In E. coli, target enzymes such as aconitase, dihydroxyacid dehydratase and S phosphogluconate dehydratase are inactivated by the superoxide anion, and the loss of function of these enzymes could contribute significantly to the toxicity observed under conditions of oxidative stress. [10,11]

One major function of the Fe-S cluster, due to their chemistry, is "sensing" oxygen and iron. In fact, in these instances, the instability of clusters might be advantageous. This is the case of the fumarate nitrate reduction protein (FNR) of E. coli, an oxygen sensor, in which the loss of integrity of the cluster in the presence of oxygen is the key to sensing. In response to lowered oxygen tension, the metabolism of E coli switches from use of oxygen to use of alternative terminal electron acceptors such as fumarate and nitrate, a change that requires the simultaneous transcriptional activation of over 50 genes of anaerobic metabolism. [12] Oxygen levels are sensed by the E. coli transcription factor FNR. In the absence of oxygen, FNR binds to promoters of the genes of anaerobic metabolism and activates transcription. The ability of FNR to activate transcription was observed to be impaired by iron deprivation, and the relationship between nutritional iron status and oxygen sensing led to speculation that an iron co-factor was involved in sensing of oxygen levels. Sensing of oxygen depends on the sensitivity of a [4Fe-4S] iron-sulfur cluster to oxygen. The cluster is destroyed within seconds upon exposure to oxygen and an apoprotein devoid of the cluster is the product of the reaction. In the presence of oxygen, the cluster disintegrates, and the protein no longer specifically binds DNA. Thus, FNR provides a clear example in which a reactive iron-sulfur cluster is used to sense concentrations of the destabilizing agent, oxygen.

What about sensing of iron? As mentioned, iron is essential for cell division and growth, but tight regulation of iron uptake and distribution is necessary because excess iron can be toxic, particularly because iron species and oxygen can interact to form ROS, including superoxide and hydroxyl radicals. Thus, iron uptake and storage within most cells is carefully regulated by a set of genes that are highly conserved in mammalian cells to maintain an adequate substrate while also minimising the pool of potentially toxic "free iron". Particularly, uptake, storage and utilization of iron must be carefully controlled to maintain a suitable level of catalytically active iron pool. Proper "sensing" of iron and oxygen, in the aerobic environment, and the need for a tight control of iron and oxidants are also crucial for the cell to adapt to unfavorable and hazardous situations. Recent investigations suggest that a single genetic regulatory system, based on iron-sulfur interactions, "senses" iron and "orchestrates" all these aspects at the postranscriptional level: the iron regulatory proteins, IRPs.[13] The IRPs are the key players in this complex interaction insofar as they represent the *sensors* of cytoplasmic iron and the *controllers* of ferritin (Ft) and transferrin receptor (TfR). These proteins are used by cells to adjust intracellular iron concentration to levels that are adequate for their metabolic needs but below the toxicity threshold.

Iron homeostasis is controlled through several genes (e.g. Ft and TfR) that have been found to contain non coding sequences (i.e. the iron responsive elements, IRE) which are recognized at mRNA level by two cytoplasmic iron regulatory proteins (IRP-1 and IRP-2). The IRP belong to the aconitase superfamily: by means of a Fe-S cluster-dependent switch, IRP-1 can function as an mRNA binding protein or as an enzyme that converts citrate to isocitrate.[14,15] Although structurally and functionally similar to IRP-1, IRP-2 does not seem to assemble a cluster nor to possess aconitase activity; moreover, it has a distinct pattern of tissue expression and is modulated by means of proteasome-mediated degradation. In response to fluctuations in the level of the LIP, IRP act as key regulators of cellular iron homeostasis as a result of the translational control of the expression of a number of iron-genes. Conversely, various agents and conditions may affect IRP activity, thereby modulating iron and oxygen radical levels in different pathobiological settings. [16]

Mammalian cells react to iron deficiency by presenting a higher number of TfR at the cell surface in order to internalize iron-laden transferrin. At the same time, the synthesis of the iron storage protein Ft is halted to enhance metal availability. The opposite is true when iron overload occurs: TfR is down-regulated in order to stop iron uptake and Ft synthesis is increased to sequester excess iron in newly formed Ft shells. Although iron-controlled transcriptional regulation of the Ft H and L subunits and the TfR gene has been described,[16] intracellular iron homeostasis is mainly controlled at post-transcriptional level. When iron is scarce in the LIP, Ft and TfR mRNAs are specifically recognized and bound by the active form of IRP that modulate Ft and TfR mRNAs translation and stability, respectively. On the contrary, when iron is abundant, IRP are devoid of mRNA binding activity and target transcripts are freely accessible to translation complexes or nucleases. The close inverse relationship between cellular iron levels and changes in IRP activity has been recently confirmed by studies directly measuring variations in the level of the LIP. [17] IRP therefore control cell iron status by means of divergent but coordinated regulation of Ft and TfR levels.

The IRP model of iron homeostasis should ideally be able to protect the cell from iron overload under any circumstances. Unfortunately, during experimental and human chronic iron overload, this does not occur. In the bloodstream, in these conditions, non-transferrin bound forms of iron appear, that are clearly pathogenetically linked to iron toxicity in specific human pathological settings such as thalassemia and hemochromatosis. [18] Moreover, many cells, such as, liver parenchymal cells, may take up iron from the bloodstream through different TfR-independent systems, that are perfectly functional also during iron overload. Also the recently described transferrin receptor-2, [19] which is capable of delivering transferring iron, appears to be normally expressed during liver iron overload. [20] Therefore, the cell, when facing high iron level in the bloodstream, may still be exposed to a continuous influx of iron in spite of down-regulation of the TfR. The iron storage capacity is mainly devoted to ferritin: within the ferritin shells iron is kept in a safe state. Nevertheless this is not an inert state, and redox changes in the cytoplasm, xenobiotics or other conditions may rapidly mobilize this iron and make it catalitically active.

In the example of FNR, cluster disassembly results from exposure to oxygen, and although iron sufficiency is required in sensing, the level of oxygen in the cell is the primary stimulus to which the regulatory response is oriented. In the case of IRP1, however, the major regulatory response is in sensing of iron levels and in the control of synthesis of proteins of iron metabolism. The state of the cluster in cytosolic aconitase might reflect a balance between cluster disassembly, which is likely to occur under conditions of normal aerobic growth, and cluster reassembly, which would require sufficient iron and sulfur, along with assembly enzymes. The relative efficiencies and sensitivities of the various components would determine whether the sensing is primarily for oxygen or for iron, although clearly the sensing of each reactant would require the availability of the other. [13] In this fashion, IRP1 and perhaps other proteins, such as FNR, simultaneously sense levels of oxidants and iron. Although levels of iron might have little impact on the process of cluster disassembly, iron levels determine whether the cluster can be reassembled and thereby control the transition from apoprotein to holoprotein. The holoprotein can function as a sensor of oxidants, while the apoprotein can function as a sensor of iron levels. The mechanisms of assembly and disassembly of iron-sulfur clusters are the key to understanding how iron-sulfur proteins can serve as sensors of both oxidants and iron.

1.3 Perturbations of the IRP system and iron toxicity

An important aspect to be evaluated when considering the biochemical basis for iron toxicity is that the IRP system itself may be a target of inciting insults, including oxidants, nitric oxide and xenobiotics (Figure 2), which may eventually modify, through the IRP, iron homeostasis and cell survival.

It is now clear that the activity of IRP is altered under conditions of oxidative stress. Given the role of Ft and TfR in iron chelation and uptake, one would expect their respective up- and down-regulation under conditions of oxidative stress in order to limit iron availability. Indeed, it has been demonstrated that various cell types react to a variety of stressful conditions by increasing Ft synthesis [21-24] and oxidative stress response has been found to be reduced by Ft H subunit overexpression. [25] However, this effect is not consistent with the enhanced activity of IRP in H_2O_2-treated cells [26,27] because this would expand the LIP, and thus exacerbate ROS toxicity. A thorough analysis of the mechanisms and signals underlying IRP activation under these conditions has made it evident that this prompt effect occurs more as a response to phosphorylation-dependent signalling pathways, possibly related to the growth-promoting effects of H_2O_2, [28] than to oxidative stress itself. In fact, neither raising intracellular H_2O_2 levels [29] nor treating cytosolic extracts with H_2O_2 [26,27,30,31] led to any stimulatory effect and the response of IRP, which requires ATP, GTP and phosphorylation events, was restored by the addition of membranes.[32]

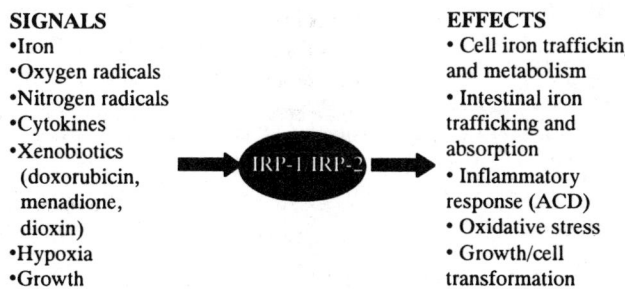

Figure 2. IRPs as target of pathobiological signals and effector of pathobiological events. ACD=Anemia of Chronic Disease.

Furthermore, it has been shown that the simultaneous generation of H_2O_2 plus O_2^-, which is more likely to occur under oxidative stress conditions than the production of H_2O_2 alone, inhibits IRP activity.[30] IRP inactivation has also been found following the treatment of cells with menadione, a redox cycling quinone that yields H_2O_2 plus O_2^- [33] and it has been shown that the *in vivo* ROS production obtained by treating with dioxin,[34] or glutathione-depleting drugs [35] or by means of post-ischemic rat liver reperfusion [36] downregulates IRP activity.

All of these considerations mainly refer to the 4Fe-4S-endowed IRP-1, whereas there are fewer data concerning the effect of ROS on IRP-2. The mechanism of IRP-2 regulation, i.e. ubiquitin-dependent proteasome degradation after direct oxidative modification of residues in the degradation domain,[37] suggests that the molecule is highly susceptible to changes in the redox status of the cell. Indeed, IRP-2, which is not activated by extracellular H_2O_2,[38, 39] is downregulated in rat liver after exposure to oxidative stress [35,36] and in cell cultures treated with menadione.[33]

In brief, recent results have offered new perspectives concerning the role of IRP in oxidative stress that have allowed the reconciliation of its function in a coherent framework aimed at limiting the availability of catalytically active iron (Figure 2). The interactions between IRP activation and ROS establish a direct link between iron metabolism and oxidative stress.

In this context, even more complex interactions may take place during inflammatory processes, in which profound perturbations of iron homeostasis occur, and inflammatory mediators (such as nitric oxide - NO - and cytokines) on one side affect iron metabolism, on the other, are affected by iron availability and IRP activation state. The fact that iron sulfur clusters are preferential targets of NO and that mitochondrial aconitase inhibition is involved in the cytotoxic effect of NO (reviewed in [40]) has prompted a number of investigators to address whether this molecule, which is known to be involved in a variety of physiological and pathological processes, also plays a role in the IRP-mediated regulation of iron metabolism. A close link between iron homeostasis and NO has been clearly established, also in consideration of the effect of intracellular iron levels on iNOS gene transcription.[41] However, depending on the investigated cells, the applied stimulus and the involved redox-related NO species, it has been reported that NO has a number of effects on both IRP-1 and IRP-2. Modulation of IRP by NO may be of pathophysiological relevance in macrophages particularly in conditions such as inflammation and anaemia of chronic disease (ACD).

In agreement with the first reports describing the NO-mediated activation of IRP binding activity,[42,43] increased IRP-1 activity has been repeatedly found after treatment of murine macrophagic cell lines with cytokines - leading to NO production - or with NO donors (reviewed in [16]). It is still unclear whether the activating effect of NO depends on a direct interaction with the cluster, which would imply disassembly beyond the 3Fe-4S stage and acquisition of the RNA-binding apoform, or a slow effect of NO on the LIP.

Conversely, a NO-dependent downregulation of IRP-2 by LPS/IFNγ has been reported [44-46] possibly through degradation, at least in macrophages.[47, 48]

In summary, a number of uncertainties remain concerning the effect of NO on IRP partially because the intrinsic chemistry of the molecule means that its reactivity also depends on the redox state of the environment in which it is generated. This is exemplified by the fact that IRP-1 treated with peroxynitrite, a not uncommon product of the reaction of NO with constantly generated O_2^-, loses its aconitase activity without gaining RNA-binding activity.[49] Furthermore, most of the information relating to the NO-

IRP interaction has been obtained by treating cell extracts with NO donors. Although this is a convenient means of exploring the molecular mechanisms underlying the effect of NO on IRP, it hardly recreates *in vivo* conditions. In fact, full *in vitro* response to NO requires the addition of a cellular component: i.e. thioredoxin.[50] However, to summarize the physiological implications of the stimulation of the NO pathway on IRP, it does seem possible to conclude that, in spite of IRP-1 activation, the loss of IRP-2, which is highly expressed in macrophages and has greater affinity for target IRE,[51] may be pathophysiologically more relevant.

These findings provide novel insights into the molecular mechanisms underlying the enhanced iron retention in macrophages that is characteristic of inflammation. A human pathological correlate is anemia of chronic disease, ACD, in which the supply of iron for hemoglobin synthesis to erythroid precursors is restricted as a result of its sequestration in the reticuloendothelial system. As in the case of IRP-2 down-regulation in mouse macrophage cell lines, IRP (IRP-1 plus IRP-2) activity in human monocytes/macrophages is markedly repressed and accompanied by enhanced Ft content.[52] To this regard, the elucidation of iron metabolism in activated macrophages and further information about the interaction between IRP and the NO pathway may provide unexpected explanations for clinical phenomena in patients with inflammatory states and the ensuing ACD.

The IRP-IRE machinery is also a possible target of the xenobiotics that undergo redox cycling. A loss of both IRP functions has been reported in cells exposed to a quinone [33] and in the liver of mice treated with dioxin.[34] Doxorubicin (Dox) is an anthracycline whose clinical use as an anticancer drug is limited by its severe cardiotoxicity. Reconstitution of the alcohol metabolite of Dox with human myocardial cytoplasmic fractions has been shown to impair both the enzymatic and IRE-binding activities of IRP-1[54] in a way that is independent of the presence of free-radicals. The effect of Dox has highlighted the fact that exceptions to the model of mutually exclusive IRP functions can exist insofar as Doxol mobilizes iron from the cluster, leading to a consequent loss of aconitase activity, but cluster disassembly is not accompanied by the acquisition of IRE-binding activity. This damage is mediated by Dox-Fe(II) complexes and reflects oxidative modifications of the –SH residues that are essential for both cluster formation and mRNA binding. Interestingly, the irreversible damage of the homeostatic mechanism of aconitase/IRP-1 may disrupt cardiac iron homeostasis and play a significant role in Dox cardiotoxicity.[54] Thus, different signals, including iron, ROS, NO, and xenobiotics appear to affect IRP activity thereby regulating IRE-containing mRNAs (Figure 2). It can be easily predicted that new target genes for the IRP-IRE regulatory system will be soon discovered. It is thus possible that the induction of IRP by the descibed signals may lead to the regulation of as yet unidentified IRE-containing mRNAs involved in the oxidative stress response, energy production or other cellular signalling or fundamental metabolic pathways which rely on a finely tuned balance between iron and oxidants. This may set the basis for important cell activities but also for pathobiological states. In this context, particular attention should be paid to malignant cell growth. In fact, while the necessity of iron for cell replication is well recognized, the specific mechanisms by which iron-binding proteins modulate cell growth have not been elucidated. Recent evidence indicated that activated oncogenes might favour unremitting growth by coordinately altering a gene expression program controlling iron homeostasis.[55] Thus, further clarification of the interconnection between modulation of IRP activity and cell proliferation may offer insights into the alterations of iron metabolism that accompany cancer development and progression. In conclusion, the recent scientific discoveries on the genetic control of intracellular and body iron homeostasis, have greatly broadened the concept of iron toxicity. Nevertheless, several *in vitro* findings on IRP and ROS have not been confirmed when tested in *in vivo* systems, where other signals and interfering factors may play complex roles.[16] There is still uncertainty on whether the same pathways operating in test tubes can be extrapolated to paythophysioloigical settings where, for instance, iron overload or deficiency, inflammation and degeneration may coexist, and the IRP control may be overruled by other regulatory signals.

2. IRON OVERLOAD DISEASE

2.1 Liver toxicity during iron overload as a paradigm for organ disease due to iron

There are several inherited and acquired disorders that can result in chronic iron overload in humans, and whose major clinical outcome include hepatic fibrosis, cirrhosis, hepatocellular cancer, cardiac disease, and diabetes (Figure 3). The liver is the main organ for iron storage and also the main site of iron toxicity during iron overload. Therefore, we can consider the hepatic damage due to iron as the prototype of organ toxicity during iron overload, and many basic aspects of such toxicity can be extrapolated to other organ diseases due to iron.

1. (Hereditary) haemochromatosis
   Haemochromatosis, HFE-related
       C282Y homozygosity
       C282Y/H63D compound heterozygosity
       Other mutations
   Haemochromatosis, non-HFE related
       • Autosomal recessive : transferrin receptor-2
       •Autosomal dominant: ferroportin-1
   Juvenile haemochromatosis
2. Acquired Iron Overload
   Iron-loading anaemias
   Dietary iron overload
   Chronic liver disease
       Hepatitis C
       Alcoholic cirrhosis, especially when advanced
       Nonalcoholic steatohepatitis
       Porphyria cutanea tarda
       Dysmetabolic iron overload syndrome
       Post portacaval shunting
3. Miscellaneous
       Iron overload in sub-Sahara Africa
       Neonatal iron overload
       Aceruloplasminaemia
       Congenital atransferrinaemia

Figure 3. Classification of human iron overload disorders

Clinical evidence for toxicity caused by excess iron is well established and has been provided by studies in patients with hereditary hemochromatosis (HC),[56-58] and secondary hemochromatosis caused by thalassemia.[59,60] In these conditions a correlation has been demonstrated between the hepatic iron concentration and the occurrence of liver injury, and therapeutic reduction of hepatic iron by either phlebotomy or chelation therapy has resulted in clinical improvement. In addition to the known association of chronic iron overload and the development of hepatic fibrosis and cirrhosis, there is sufficient clinical evidence to indicate an increased risk of hepatocellular carcinoma in hemochromatotic patients with cirrhosis.[56] In recent years many aspects of the pathophysiological mechanisms for hepatocellular injury during iron overload have been clarified. Others, mainly related to the key molecular events leading to fibrosis (see below) and hepatocarcinogenesis, are still illdefined.

2.2 Organelle dysfunctions and iron toxicity

Several mechanisms by which excess hepatic iron causes cellular injury have been proposed (see below). However, oxidative injury in cell organelles is possibly the unifying mechanism underlying the several theories of cellular injury during iron overload (Figure 4).

Since iron accumulation largely occurs in lysosomes, the hypothesis has been put forward that the excess of ferritin and/or hemosiderin in lysosomes physically disrupts these organelles with intracellular release of their hydrolytic enzymes leading to cell death. This hypothesis does not rule out the possibility that oxidant damage of lysosome contributes to their disruption during iron overload. Peters and colleagues [61-63] first proposed that hepatocellular injury in iron overload is mediated by increased lysosomal fragility, resulting in the release of cell-damaging hydrolytic enzymes into the cytosol, and have shown an increase in lysosomal enzyme activity in homogenates prepared from liver biopsy specimens from patients with various types of hemochromatosis. In patients with HC who were successfully treated by phlebotomy, the increased fragility returned to normal, thus demonstrating the iron dependency of this change. In addition they demonstrated a strong correlation between increased lysosomal fragility and hepatic hemosiderin content.[63] Antioxidant levels were not found to be dereased in in liver biopsy specimens from patients with iron overload, except a reduction in glutathione reductase activity noted in the livers of patients with thalassemia and secondary hemochromatosis.[64]

Experimentally, using different methods for producing iron overload (e.g. parenteral administration of iron salts; dietary iron enrichment) it has been found reductions in latent lysosomal enzyme activity, suggesting an increase in lysosomal fragility,[65,66] similar to the changes in liver biopsy specimens from iron-loaded patients reported before. In addition, using the carbonyl iron model, others have shown evidence of increased lipid peroxidation in lysosomal membranes (MDA levels) with an associated decrease in membrane fluidity and a defect in lysosomal acidification (increased lysosomal pH).[67]

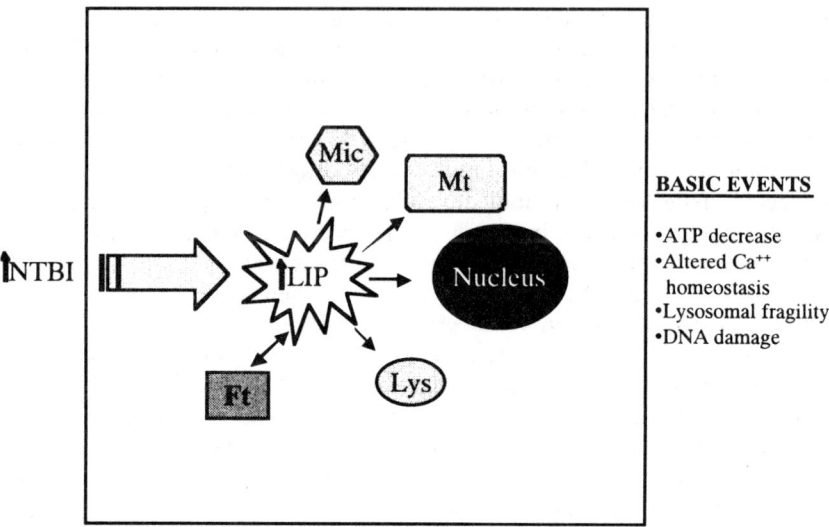

Figure 4. Mechanisms of iron toxicity. NTBI= Non-Transferrin Bound Iron; LIP= Labile Iron Pool; Mic= microsome; Mt=mitochondrion; Lys=lysosome; Ft=ferritin

An additional explanation is that pathological accumulation of iron initiates membrane lipid peroxidation in cellular organelles, a process which is implicated in tissue damage in a wide variety of disorders. Several investigators have reported decreases in hepatic cytochrome P-450 concentration and in some of the microsomal drug metabolizing enzyme systems using models of either acute or chronic parenteral iron loading. [68-71] When dietary carbonyl iron supplementation was used as a method for producing experimental iron overload, evidence of hepatic microsomal lipid peroxidation was found at high hepatic iron concentrations in association with a decrease in the concentrations of cytochromes and in aminopyrine demethylase activity.[72] From these experiments it was concluded that iron-induced lipid peroxidation to microsomal membranes was probably responsible for the decrease in cytochrome P-450 although a cause and effect relationship was unproven.

2.3 The central role of mitochondria

Mitochondria occupy an unique position within the cell both in the handling of iron and in energy production by oxidative phosphorylation process. They represent a primary target of oxidative injury, but at the same time, are a major source for ROS. In fact, ROS are produced through the respiratory chain at a rate which is dependent on the metabolic state. It is reasonable to assume that the destabilization of one or more components of the electron transport chain, occurring after the rise in the level of chelatable iron, enhances the potential for autoxidation and increases the production of O^{2-} and other ROS. [73-76]

Hanstein and colleagues [71,76] were the first to study mitochondrial oxidative metabolism in experimental iron overload and to find a decrease in the mitochondrial respiratory control ratio. In studies by Tangeras [77] and Masini et al. [78-80] minimal or no changes in mitochondrial lipid peroxidation or substrate oxidation were found while the mitochondrial transmembrane potential was lower. In enterally iron overloaded rats, a significant decreases in the state 3 (ADP-stimulated) respiratory rates and in the mitochondrial respiratory control ratio was reported [81] concomitant with a rise of hepatic conjugated diene formation, indicative of mitochondrial lipid peroxidation. Experiments have also been performed in which the various mitochondrial oxidases and reductases have been examined in dietary carbonyl iron overload. At moderate degrees of hepatic iron overload, significant decreases in succinate-cytochrome-c-reductase activity (complex II to III) and marked (70%) decreases in cytochrome oxidase activity were found (complex IV). Cytochrome oxidase is functionally dependent on intact cardiolipin, a phospholipid unique to mitochondria, which contains a high percentage of polyunsaturated fatty acids and which is possibly susceptible to iron-induced peroxidative damage. Finally, at high hepatic iron concentrations at which mitochondrial substrate oxidation was significantly reduced, there was a 60-70% reduction in hepatic ATP concentration with a corresponding decrease in the hepatic energy charge and mass action ratio.

Therefore, experimental evidence has been accumulating that chronic iron overload may result in *in vivo* lipid peroxidation of mitochondrial membranes. This process may modify the fluidity state of the mitochondrial membrane with alterations in the transport of solutes and ions, may lead to swelling and lysis of isolated mitochondria associated with the enhancement of lipid peroxidation.[82] Furthermore, hydroperoxidic products, derived from the free radical oxidation of mitochondrial polyunsaturated fatty acids (PUFAs), and H_2O_2 may result in more subtle and earlier modifications in the functional properties of liver mitochondria. Indeed, a hydroperoxide-induced release of Ca^{2+} from mitochondria via a separate release pathway, has been reported. [83-86] This event was shown to be elicited via the oxidation of pyridine nucleotides as originally suggested by Lehninger et al. [87] Furthermore, hydroperoxydes and Ca^{2+} may lead to the opening of Ca^{2+} pore in mitochondrial inner membrane. [88,89] Finally, under conditions of oxidative stress, the oxidation of PUFAs may give rise to products acting as specific Ca^{2+} ionophores. [91] A perturbation in mitochondrial Ca^{2+} transport results in energy loss and it leads to serious mitochondrial damage. [83,84,91,92] In rats fed carbonyl iron the induction of lipoperoxidative reaction appeared to be associated with the activation of Ca^{2+} release from mitochondria. This was shown to occur as a consequence of rather subtle modifications in the inner membrane structure via a specific efflux route, which appeared to be linked to the oxidation level of mitochondrial pyridine nucleotides. The induction of this Ca^{2+} release from iron-treated mitochondria resulted in enhancement of $Ca2+$ cycling, a process which dissipates energy to reaccumulate into mitochondria the released Ca^{2+}.

Although some report does not support this view [77] and a clear correlation between lipid peroxidation and total iron level does not appear to exist from all these studies, the accepted mechanism by which cellular toxicity may occur in response to excess iron involves oxidant stress, including lipid peroxidation of biological membranes, resulting in functional impairment of hepatic subcellular organelles. The form of intracellular iron responsible for the induction of oxidant stress is still unknown. A likely possibility is that an increase in the intracellular transit pool of iron may account for the induction of lipid peroxidation. In fact, this pool of low-molecular weight iron chelatable by desferrioxamine (DFO-chelatable iron)-i.e. the LIP- is catalytically active in initiating free-radical reactions and lipid peroxidation [2].

It is reasonable to assume that the destabilization of one or more components of the electron transport chain, occurring after the rise in the level of chelatable iron, enhances the potential for autoxidation and increases the production of O_2^- and other ROS,[93] so exacerbating the effects of iron toxicity. The *in vivo* administration of silybin, a flavonoid with ROS scavenger activity and iron chelating properties, in combination with dietary iron overload in rats, has been recently reported to fully counteract hepatic mitochondrial oxidative damage and ATP decrease.[94] A desferrioxamine chelatable iron pool may stimulate lipid peroxidation of hepatic organelles in siderotic rats. Recently, in the gerbil model of iron overload, lipid peroxidation was paralleled by an increase in the chelatable-iron pool in mitochondria,[95] a derangement in the mitochondria energy-transducing capability, resulting from a reduction in the respiratory chain enzyme, and followed by a dramatic drop in tissue ATP level. Sylibin significantly reduced both mitochondria functional anomalies and the associate tissue fibrosis by blocking the mitochondrial free iron pool. Many mechanisms may be responsible for the reduction in mitochondrial enzyme activity. First of all, a reduction in the synthesis of respiratory chain enzymes may be involved, but in this study it was excluded due to the lack of activation by Lubrol, a non-ionic detergent, which facilitates the substrate accessibility.[96] Post-translational events due to iron-induced oxidative stress may also play a role. Iron induced peroxidative damage to the polyunsaturated fatty acids of inner membrane phospholipids, especially cardiolipin, may account for the reduced cytochrome c oxidase activity.[97] The lack of lubrol activation may indicate that the inner membrane organization has been altered. Alternatively, reactive aldehydic products of lipid peroxidation, may form adducts with proteins, thereby partially inactivating them. In fact, in this report, it was found a significant accumulation of MDA-adducts in nonparenchymal-inflammatory cells of iron-treated livers. In this vein, it has been reported that the presence of 4-hydroxy-2-nonenal,[98] within a micromolar range, strongly impairs mitochondrial oxidative metabolism. In addition, iron induced free radicals may directly attack the protein components of the enzyme complexes and partially inhibit them. Mitochondrial oxidative phosphorylation is responsible for supplying over 95% of the total ATP requirement in eukaryotic cells.[99] The depressed mitochondrial oxidative metabolism may thus reasonably account for the large decrease up to 61% in the hepatic ATP concentration observed in iron treated gerbils.

In conclusion, the expansion of the catalytically active iron pool in the mitochondria due either to cell iron overload (e.g. during hemochromatosis) or to iron decompartmentalization (e.g. in response to pathological insults: cytokines, oxidants, xenobiotics) may exert a central role in iron toxicity. Mitochondria are key to energy metabolism, they are highly enrichment in membranes and membrane-associated carriers and activities -the targets of free radical attack- they posses several iron-sulfur proteins sensitive to redox changes and important in cell respiration and energy production. Overall, a decrease in ATP production and in microsomal activities, and alteration of Ca^{2+} homeostasis are key to cell damage and necrosis due to iron.

2.4 Iron overload and liver fibrosis

Progressive fibrosis of organs, such as the liver, the kidney or the lung, includes a variety of mechanistically related disorders that are major causes of morbidity and mortality. As keloids and hypertrophic dermal scarring can result in disfigurement and disability, scarring in visceral organs, such as the liver, leads to disruption of normal lobular architecture, nodular regeneration and loss of function. As conditions of liver toxicity, iron overload diseases are invariably associated to liver fibrosis. Whenever is a dying hepatocyte, monocyte/macrophage-type cells that may be acutely extravasated from the circulation or be normally resident in the organ, as in the case of Kupffer cells in the liver, and other "inflammatory" cells will be activated. The resulting inflammatory cascade will lead to recruitment, activation and proliferation of fibroblasts or fibroblast-like cells, such as the hepatic stellate cells (HSC) (formerly known as Ito cells or lipocytes) and to a marked increase in their production of extracellular matrix (ECM).

Hepatic fibrosis is a dynamic process from chronic liver damage to cirrhosis. It is predominantly characterized by excessive accumulation of ECM components in the liver caused by both markedly increased production and unbalanced degradation.[101] Extracellular matrices consist of pericellular insoluble macromolecules organized in either interstitial or basement membrane structures that interact with most cells in multicellular organisms.[101] They provide mechanical cohesiveness for cells and, consequently, play a major role in tissue architecture. However, the old concept that ECM is inert and merely functions as a scaffold for liver cells has been abandoned. It is now clear that ECM components modulate parenchymal cell function and differentiation. They have multiple domains wich allow direct contact with cell membrane receptors and other ECM components. Signaling to cells is also achieved through cytokines, growth hormones and other biological peptides normally stored in the ECM. Cross-talk between different cells is allowed by these interactions as well as by direct communications through formation of gap junctions. Overall, a finely tuned biological system made of a network of physical entities and biochemical interactions will result. For these reasons, ECM plays a crucial role in physiological and pathophysiological states, such as liver development, regeneration, fibrosis and carcinogenesis. Acute insults, regardless of their intrinsic nature, will disrupt this equilibrium causing release of bioactive molecules from ECM, altered signalling to epithelial cells and activation of local or incoming mesenchymal cells. Eventually, tissue repair and epithelial cell regeneration will follow. Under normal conditions, ECM undergoes a continuous remodeling where fibrogenesis is always associated to fibrolysis. When the former event will exceed following chronic insults, fibrosis will result.

That inflammation has a key role in the activation of HSC and liver fibrogenesis is beyond any despute. In fact, most forms of chronic liver injury in humans, such as viral and alcoholic liver disease, have a prominent inflammatory component. However, in other types of hepatic fibrosis, such as in HC, there is little or no inflammation, raising the possibility that additional factors may activate lipocytes. In this context, the role of oxidative stress in regulating collagen synthesis has recently attracted the interest of many investigators. It is generally accepted that mediators of increased oxidative stress may also lead to increased collagen synthesis *in vivo*. Thus, chronic inflammation, redox cycling drugs, hyperbaric oxygen therapy, and other oxidants may increase collagen synthesis and deposition (see below). Indeed, that collagen should be regulated by the oxidative state of the cell is not unrealistic. Relative collagen synthetic rates in wound microenvironments are extremely sensitive to oxygen concentration and oxygen per se exertes a regulatory role on prolyl hydroxylase as a source of oxygen atoms.[102] In addition, *in vitro*, oxygen tension above physiologic levels has been reported to increase collagen gene expression.[103] Yet, increased fibroblast migration and proliferation are induced by oxygen free radicals *in vitro*.[104] It is thus probable that ROS play a role in expression of collagen synthesis *in vivo,* and that when uncontrolled, conditions of fibrosis may occur.

2.5 ROS and fibrogenesis

Although parenchymal cells are the major site for oxygen utilization, both parenchymal and nonparenchymal cells are source of ROS. Hepatocytes release significant amounts of ROS under normoxic control conditions. Usually hepatocytes can deal with them without any overt sign of injury by virtue of efficient antioxidant defenses (Table 1).[101] Under certain conditions, however, this capability may be impaired. In fact, increased intracellular formation of ROS, as may be induced by xenobiotics, or by additional confrontation with extra-hepatocellularly generated ROS (e.g. from activated Kupffer cells) may overcome their defence systems, thus leading to cell injury. In addition, alterations in the hepatocytes may lead to an increased sensitivity to the damaging potential of ROS.

Besides hepatocytes, all non-parenchymal cells may be capable of releasing ROS and may thereby contribute to parenchymal cell necrosis. A major source of ROS are Kupffer cells, fibroblasts and invading leucocytes. Kupffer cells are activated by a variety of stimuli to release ROS such as opsonized zymosan, phorbol esters, lipopolysaccharides and Ca^{2+} ionophores.[105] In Kupffer cells, O_2^- is produced during the oxidative burst of phagocytic activity. A membrane respiratory burst oxidase catalyzes one electron reduction of O_2 to O_2^- at expenses of NADPH. A further radical species, $NO^.$, is produced in macrophages from L-arginine by catalysis of NO synthase following a different set of stimuli including lipopolysaccharides and the combined addition of prostaglandin E_2 and tumor necrosis factor-α. Both O_2^- and $NO^.$ possess a significant cytotoxic potential. Basically, under control conditions, Kupffer cells have a limided producing ability of ROS, signifcantly lower than granulocytes. However, after a "priming" event (such as stimulation by cytokines or binding by other ligands) they increase ROS production which becomes maximal during phagocytic activity. Likewise, release of ROS by fibroblasts can be stimulated by opsonized zymosan, phorbol ester, interleukin-1 and tumor necrosis factor-α.[106] Granulocytes, which in addition to O_2^-/H_2O_2 generated by NADPH oxidase, may form ROS through myeloperoxidase, significantly contribute to ROS production during inflammation.[107] Both Kupffer cells and activated leucocytes adhere to sinusoidal endothelial cells, without direct interactions with hepatocytes. Thus sinusoidal endothelial cells may represent the first target of ROS-mediated toxicity. After the sinusoidal lining damage, plasma membrane of hepatocytes may be directly exposed to toxicity. Moreover, endothelial cells themselves may produce ROS.[108] In conditions where liver fibrosis occur, even due to preexisting injuries, an increased intracellular formation of ROS may result from hypoxia due to perisinusoidal fibrosis. In this condition, connective tissue deposition in the perisinusoidal space (perisinusoidal fibrosis) may severely impair substrate supply to, and product clearance from, the hepatocytes and non-parenchymal cells, due to increased diffusion barriers and narrowing of sinusoidal blood flow. Deprivation of an adequate O_2 supply to the hepatocytes results in hepatocellular hypoxia. Since this may not be permanent, or at least the degree of hypoxia may vary with time in the various parts of the liver, a situation very similar to the one defined in experimental systems as hypoxia-reoxygenation may exist.[109] It is known from several experimental evidences that under these conditions hepatocytes may respond with an increased generation of ROS leading to cell damage and necrosis (reoxygenation injury). Release of ROS by hepatocytes does not cease when the cells lose their viability, and for a certain period it may even increase. Hence, non-viable hepatocytes are capable of releasing significant amounts of reactive oxygen species for several hours. In this way they may contribute to injury of still viable neighboring parenchymal cells.

As mentioned above, ROS released by hepatocytes, resident nonparenchymal cells or invading mononuclear cells, may be involved in hepatic fibrogenesis as inducers of parenchymal cell necrosis and/or as activators of cells which are effectors (e.g. HSC, (myo)fibroblasts) or key-mediators (e.g. infiltrating mononuclear cells and resident macrophages) of the fibrogenic process. In fact, besides acting as direct cytotoxic agents, ROS may trigger an increased synthesis of collagen in myofibroblast like cells or activate granulocytes and Kupffer cells, resulting in an increased formation of cytokines and eicosanoids, and further ROS. This may constitute a cascade of amplifying loops which self-perpetuate the fibrogenic process (Figure 5).

Lipid peroxidation of biological membrane is associated with experimental iron overload *in vivo*. By attacking PUFAs of membrane phospholipids free radicals of these lipids are formed as well as respective peroxyl radicals (Figure 1B). Following initiation, in the propagation step, transition metal catalysis, such as iron and copper, are essential. Main consequence of lipid peroxidation of biological membranes are the functional and structural modification of the membranes and generation of by-products, such as highly reactive aldehydes, from oxidative breakdown of PUFAs.[110] ROS posses tremendous intrinsic reactivity which lead to interaction with cell targets close by the site of generation. For instance, OH· can only diffuse 5-10 molecular diameters from its site of formation before it reacts. Unlike highly reactive free radicals, lipid peroxidation by-products such as aldehydes are rather long-lived and can diffuse from the site of formation and reach intracellular and extracellular targets. In fact, these by-products may form covalent bonds (adducts) with proteins thus inactivating them and damaging the cell. *In vitro*, lipid-peroxidation by-products have been found to be fibrogenic for HSCs.[111]

2.6 Iron overload and liver fibrogenesis

In humans with iron overload, due to either genetic or acquired causes, fibrosis and cirrhosis are common findings. Early studies reported signs of fibrosis in young individuals with genetic HC in the absence of necro-inflammatory events.[112-113] Occasionally, HC patients with extensive cirrhosis or signs of portal hypertension have been found without significant hepatic biochemical abnormalities. These findings emphasized the possible role of iron as a profibrogenic agent itself. In recent years, investigators have tried to give experimental support to the idea that iron is a potent fibrogenic agent and to understand the cellular pathways mediating the fibrogenic effect of iron.

A major difficulty in approaching this issue experimentally has been the lack of an animal model producing the full pathological picture of the cirrhotic liver seen in HC. In fact, in animal models of iron overload, in spite of significant increase in liver iron content, occasionally approaching that seen in human disease, necrotic events are limited and fibrotic response minimal. Even in human HC, in the absence of other inciting insults, the hepatic disease is rather "mild" and usually slowly progresses over a long-time period to the micronodular cirrhosis. This is quite dissimilar from post-necrotic cirrhosis following, for instance, chronic hepatitis. Thus, it seems that iron alone represents a "mild" hepatotoxin and very efficient protective systems operate in the liver to quench high influx of iron during iron overload. In this vein, HC Knock-out mice show clearly signs of iron overload but very little fibrosis and no cirrhosis.[114] Some explanations can be put forth. Rodents may have a natural resistance to iron load and intrinsic differences in iron metabolism may exist between humans and rodents. Moreover, other "modifying" factors (genes?) may play a role in human HC as compared to mouse HC.

Some of the models used, however, such as the carbonyl-iron-fed rat model,[115] have offered the opportunity of studying important aspects of liver cell biology during iron overload and early fibrogenesis. Using this model, it was originally shown that enteral iron overload in the absence of significant necro-inflammatory activates collagen type I gene expression [116] and that HSC are the major source of collagen expression. [117] The main features of HSC activation during fibrogenesis have been also found during *in vivo* iron overload, namely, increased rough endoplasmic reticulum, loss of vitamin A droplets, enhanced collagen mRNA expression, active proliferation and smooth muscle (-actin expression. [117,118] Thus, it is beyond any dispute that HSC are the main effector of iron induced liver fibrogenesis, while the contributing role of periportal fibroblasts and true myofibroblasts, more difficult to identify on histological grounds, should not be underscored. In the gerbil model of iron overload, in which the fibrogenic response is most dramatic (see below), all the events critical to liver fibrosis, are completely abolished by antioxidant treatment which prevents liver cirrhosis and preserve a normal lobular architecture in animals with a dramatic burden of iron in the liver.[118] Thus, this strongly suggests that these "early" fibrogenic phenomena induced by iron are mediated by an oxidative stress (Figure 5). As ideal catalyst of free radical reactions, iron accumulation into HSC might directly mediate collagen gene activation. However, several observations based on ultrastructural, immunohistochemical and molecular analyses, ruled out the possibility of a significant increase of iron [119] and/or products of lipid peroxidation [120,121] into HSC at the time of enhanced collagen expression. An alternative pathway may include "paracrine" activation of HSC by oxidative-stress induced cytokines. [121] A microenvironment rich in free radicals may stimulate resident or invading nonparenchymal cells to release fibrogenic mediators (such as transforming growth factor-β_1 -TGFβ_1- or platelet-derived growth factor) ultimately responsible for activation of HSC (Figure 5). Interestingly, during iron overload, a significant enhancement of TGFβ_1 into iron-filled cells where oxidative stress is maximal has been described, and this phenomenon is counteracted by antioxidant treatment.[118] This strongly suggests a causal relationship between iron-induced oxidative stress and cytokine release by iron-loaded cells. Thus, at early stages fibrogenesis might be sustained by an iron-induced oxidant-stress and by the release of fibrogenic factors from iron-loaded hepatocytes, before reaching the stage of overt necrosis (Figure 5). At advanced stages, in HC, the drive to fibrogenesis may be due to iron-induced hepatocellular necrosis. At these stages, nonparenchymal cell iron overload appears as a mere consequence of phagocitic activity of sideronecrotic hepatocytes.

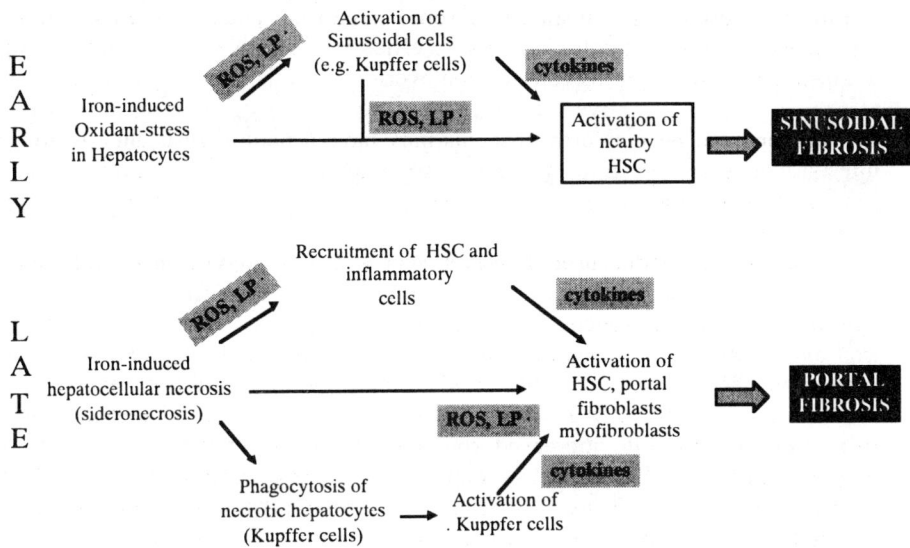

Figure 5. Mechanisms of iron-induced fibrogenesis in hemochromatosis.

Indeed, selective targeting of considerable amount of iron to Kupffer cells in parenterally iron-treated rats was unable to induce activation of nearby HSC in the absence of hepatocellualr necrosis, whereas comparable hepatocellular iron loading in iron-fed rats led to HSC activation.[122]

Of greater clinical impact is the potential "fibrogenicity" of iron when acting in concert with other hepatotoxins (e.g. viruses, alcohol, and other xenobiotics). In this instances, even at low level of tissue iron burden, dramatic toxicity may arise, with rapid acceleration of the liver disease. This is probably due to the fact that in the presence of a damaging event, regardless of its nature, the prooxidant action of iron may amplify and propagate the initial toxic effect or contribute to its cytotoxicity. Even a slight increase in the highly reactive "free iron pool" may favour these events by offering an optimal catalyst for free radical reactions and cell toxicity. Of course, besides a direct toxic effect on hepatocytes, a "free radical" microenvironment favoured by iron excess would contribute to activation of Kupffer cells and exert a chemotactic effect on mononuclear cells toward the area of primary damage. The fact that the only experimental models in which liver cirrhosis has been obtained following iron treatment include iron-loaded animals treated with ethionine or low protein diet,[123] iron-loaded gerbils with pre-existing micro-inflammatory foci due to gut endotoxins,[124] iron-treated rats with CCl$_4$ alone[125] or in combination with alcohol[126] and iron-treated rats with alcohol intoxication and high-fat diet,[127] strongly support these concepts. Common to many of these experimental models of hepatic fibrosis is a necro-inflammatory event which is pre-existing or concomitant to iron load. When iron hits, even in the absence on direct sideronecrosis -provided that Kupffer cells have been "primed" by a necrotic event or by other stimuli- the fibrogenic process explodes. In summary, as indicated in Figure 5, we can envisage different pathways leading to fibrosis during iron-overload associated liver disease, all having the hepatocytes as primary target, the Kupffer cell as key mediator and the HSC as effectors. In conclusion, iron has the potential of inducing all the biochemical and morphological modifications which underlie the hepatic wound healing process directly in those liver cells which are effectors and mediators of fibrogenesis. Sideronecrosis and local inflammatory (macrophagic) response appear to be required as mediators of these effects in the genuine iron overload liver disease. However, even in other pathological settings non primarily related to iron metabolism disturbancies, the cytotoxic and fibrogenic potential of iron may be dramatic, even at low level of hepatic iron burden, wherever a pre-existing necrogenic event has activated resident Kupffer cells.

REFERENCES

1. E. Cadenas, Biochemistry of oxygen toxicity, Annu. Rev. Biochem. **58**, 79-110, (1989).
2. B. Halliwell, and J.M. Gutteridge, Biologically relevant metal ion-dependent hydroxyl radical generation. An update, FEBS Lett. **307**,108-112, (1992).
3. H. Sies, Oxidative stress: from basic research to clinical application, Am. J. Med. **91**, 31S-38S (1991).
4. C.K. Sen, and L. Packer, Antioxidant and redox regulation of gene transcription, FASEB J. **10**, 709-720 (1996).
5. R.G. Allen, and M. Tresini, Oxidative stress and gene regulation, Free Radic. Biol. Med. **28**:463-499 (2000).
6. N. Li, and M. Karin, Is NF-kappaB the sensor of oxidative stress? FASEB J. **13**:1137-1143 (1999).
7. M. Karin, Z.G. Liu, and E. Zandi, AP-1 function and regulation, Curr. Opin. Cell. Biol. **9**, 240-246 (1997).
8. H. Beinert, Recent developments in the field of iron-sulfur proteins, FASEB J. **4**, 2483-2491 (1990).
9. H. Beinert, and M.C. Kennedy, 19th Sir Hans Krebs Lecture. Engineering of protein bound iron-sulfur clusters. A tool for the study of protein and cluster chemistry and mechanism of iron-sulfur enzymes, Eur. J. Biochem. **186**, 5-15 (1989).
10. D.H. Flint, E. Smyk-Randall, J.F. Tuminello, B. Draczynska-Lusiak, and O.R. Brown, The inactivation of dihydroxy-acid dehydratase in Escherichia coli treated with hyperbaric oxygen occurs because of the destruction of its Fe-S cluster, but the enzyme remains in the cell in a form that can be reactivated, J. Biol. Chem. **268**, 25547-25552 (1993).
11. P.R. Gardner, and I. Fridovich, Inactivation-reactivation of aconitase in Escherichia coli. A sensitive measure of superoxide radical, J. Biol. Chem. **267**: 8757-63, 1992.
12. R.P. Gunsalus, and S.J. Park. Aerobic-anaerobic gene regulation in Escherichia coli: control by the ArcAB and FNR regulons, Res. Microbiol. **145**, 437-50 (1994).
13. R.D. Klausner, and J.B. Harford. Cis-trans models for post-transcriptional gene regulation, Science **246**, 870-872 (1989).
14. T.A. Rouault, and R.D. Klausner. Iron-sulfur clusters as biosensors of oxidants and iron. Trends in Biochem. Sci. **21**, 174-177 (1996).
15. Hentze MW, Translational regulation: versatile mechanisms for metabolic and developmental control. Curr. Opin. Cell Biol. **7**, 393-398 (1995).
16. G. Cairo, and A. Pietrangelo, Iron Regulatory proteins in pathobiology, Biochem J. **352**, 241-250 (2000).
17. A. M. Konijn, H. Glickstein, B. Vaisman, E. G. Meyron-Holtz, I. N. Slotki, and Z. I. Cabantchik, The cellular labile iron pool and intracellular ferritin in K562 cells, Blood **94**, 2128-2134 (1999).
18. W. Breuer, C. Hershko, and Z. I. Cabantchik, The importance of non-transferrin bound iron in disorders of iron metabolism, Trasf. Sci. **23**, 185-192, (2000).
19. Kawabata H, Yang R, Hirama T, Vuong PT, Kawano S, Gombart AF, Koeffler HP. Molecular cloning of transferrin receptor 2. A new member of the transferrin receptor-like family. J. Biol. Chem. **274**, 20826-20832 (1999).
20. R.E. Fleming, M.C. Migas, C.C. Holden, A. Waheed, R.S. Britton, S. Tomatsu, B.R. Bacon, W.S. Sly. Transferrin receptor 2: continued expression in mouse liver in the face of iron overload and in hereditary hemochromatosis, Proc. Natl. Acad. Sci. U.S.A **97**, 2214-9 (2000).

21. G. Balla, H. S. Jacob, J. Balla, M. Rosenberg, K. A. Nath, F. Apple, J. W. Eaton and G. M. Vercellotti, Ferritin: a cytoprotective antioxidant statagem of endothelium, J. Biol. Chem. **267**, 18148-18153 (1992).
22. G. F. Vile, and R. M. Tyrrell, Oxidative stress resulting from ultraviolet A irradiation of human skin fibroblasts leads to a heme oxygenase-dependent increase in ferritin, J. Biol. Chem. **268**, 14678-14681 (1993)
23. E. L. Kwak, D. A. Larochelle, C. Beaumont, S. V. Torti and F. M. Torti, Role for NF-kB in the regulation of ferritin H by tumor necrosis factor-alpha, J. Biol. Chem. **270**, 15285-15293 (1995)
24. S. Lobreaux, S. Thoiron, and J. F. Briat, Induction of ferritin synthesis in maize leaves by an iron-mediated oxidative stress, Plant J. **8**, 443-449 (1995).
25. S. Epsztejn, H. Glickstein, V. Picard, I. N. Slotki, W. Breuer, C. Beaumont, and Z. I. Cabantchik, H-ferritin subunit overexpression in erythroid cells reduces the oxidative stress response and induces resistance properties, Blood **94**, 3593-3603 (1999)
26. E. A. L. Martins, R. L. Robalinho, and R. Meneghini, Oxidative stress induces activation of a cytosolic protein responsible for control of iron uptake, Arch. Biochem. Biophys. **316**, 128-134 (1995)
27. K. Pantopoulos, and M. W. Hentze, Rapid responses to oxidative stress mediated by iron regulatory protein. EMBO J. **14**, 2917-2924 (1995).
28. R. H. Burdon, Control of cell proliferation by reactive oxygen species, Biochem. Soc. T. **24**, 1028-1032 (1996).
29. K. Pantopoulos, S. Mueller, A. Atzberger, W. Ansorge, W. Stremmel, and M. W. Hentze, Differences in the regulation of iron regulatory protein-1 (IRP-1) by extra- and intracellular oxidative stress, J. Biol. Chem. **272**, 9802-9808 (1997).
30. G. Cairo, E. Castrusini, G. Minotti, and A. Bernelli-Zazzera, Superoxide and hydrogen peroxide-dependent inhibition of iron regulatory protein activity: a protective stratagem against oxidative injury. FASEB J. **10**, 1326-1335 (1996).
31. C. Bouton, M. Raveau, and J. C. Drapier, Modulation of iron regulatory protein functions. Further insights into the role of nitrogen- and oxygen-derived reactive species, J. Biol. Chem. **271**, 2300-2306 (1996)
32. K. Pantopoulos, and M. W. Hentze, Activation of iron regulatory protein-1 by oxidative stress in vitro. Proc. Natl. Acad. Sci. U.S.A. **95**, 10559-10563 (1998).
33. N. H. Gehring, M. W. Hentze, and K. Pantopoulos, Inactivation of both RNA binding and aconitase activities of iron regulatory protein-1 by quinone-induced oxidative stress, J. Biol. Chem. **274**, 6219-6225 (1999).
34. A. G. Smith, B. Clothier, S. Robinson, M. J. Scullion, P. Carthew, R. Edwards, J. Luo, C. K. Lim, and M. Toledano, Interaction between iron metabolism and 2,3,7,8-tetrachlorodibenzo-p-dioxin in mice with variants of the Ahr gene: a hepatic oxidative mechanism, Mol. Pharmacol. **53**, 52-61 (1998)
35. G. Cairo, L. Tacchini, G. Pogliaghi, E. Anzon, A. Tomasi, and A. Bernelli-Zazzera, Induction of ferritin synthesis by oxidative stress. Transcriptional and post-transcriptional regulation by expansion of the "free" iron pool, J. Biol. Chem. **270**, 700-703 (1995)
36. L. Tacchini, S. Recalcati, A. Bernelli-Zazzera, and G. Cairo, Induction of ferritin synthesis in ischemic-reperfused rat liver: analysis of the molecular mechanisms, Gastroenterology **113**, 946-953 (1997)
37. K. Iwai, S. K. Drake, N. B. Wehr, A. M. Weissman, T. LaVaute, N. Minato, R. D. Klausner, R. L. Levine, and T. A. Rouault, Iron-dependent oxidation, ubiquitination, and degradation of iron regulatory protein 2: implications for degradation of oxidized proteins, Proc. Natl. Acad. Sci. U.S.A. **95**, 4924-4928 (1998)
38. E. Menotti, B. R. Henderson, and L. C. Kuhn, Translational regulation of mRNAs with distinct IRE sequences by iron regulatory proteins 1 and 2, J. Biol. Chem. **273**, 1821-1824 (1998)
39. K. Pantopoulos, G. Weiss, and M. W. Hentze, Nitric oxide and oxidative stress (H2O2) control mammalian iron metabolism by different pathways, Mol. Cell. Biol. **16**, 3781-3788 (1996)
40. J. C. Drapier, and C. Bouton, Modulation by nitric oxide of metalloprotein regulatory activities, BioEssays **18**, 549-556 (1996)
41. G. Weiss, G. Werner-Felmayer, E. R. Werner, K. Grunewald, H. Wachter, and M. W. Hentze, Iron regulates nitric oxide synthase activity by controlling nuclear transcription, J. Exp. Med. **180**, 969-976 (1994)
42. J. C. Drapier, H. Hirling, J. Wietzerbin, P. Kaldy, and L. C. Kuhn, Biosynthesis of nitric oxide activates iron regulatory factor in macrophages, EMBO J. **12**, 3643-3649 (1993)
43. G. Weiss, B. Goossen, W. Doppler, D. Fuchs, K. Pantopoulos, G. Werner-Felmayer, H. Wachter, and M. W. Hentze, Translational regulation via iron-responsive elements by the nitric oxide/NO-synthase pathway, EMBO J. **12**, 3651-3657, (1993)
44. S. Recalcati, D. Taramelli, D. Conte, and G. Cairo, Nitric oxide-mediated induction of ferritin synthesis in J774 macrophages by inflammatory cytokines: role of selective iron regulatory protein-2

downregulation. Blood **91**, 1059-1066, (1998)
45. V. Mulero, and J. H. Brock, Regulation of iron metabolism in murine J774 macrophages: role of nitric oxide-dependent and -independent pathways following activation with gamma interferon and lipopolysaccharide, Blood **94**, 2383-2389, (1999)
46. S. Kim, and P. Ponka, Control of transferrin receptor expression via nitric oxide-mediated modulation of iron regulatory protein 2, J. Biol. Chem. **274**, 33035-33042 (1999)
47. C. Bouton, L. Oliveira, and J. C. Drapier, Converse modulation of IRP1 and IRP2 by immunological stimuli in murine RAW 264.7 macrophages. J. Biol. Chem. **273**, 9403-9408, (1998)
48. S. Kim, and P. Ponka, Effects of interferon-gamma and lipopolysaccharide on macrophage iron metabolism are mediated by nitric oxide-induced degradation of iron regulatory protein 2, J. Biol. Chem. **275**, 6220-6226, (2000)
49. C. Bouton, H. Hirling, and J. C. Drapier, Redox modulation of iron regulatory proteins by peroxynitrite. J. Biol. Chem. **272**, 19969-19975 (1997)
50. L. Oliveira, C. Bouton, and J. C. Drapier, Thioredoxin activation of iron regulatory proteins. Redox regulation of RNA binding after exposure to nitric oxide, J. Biol. Chem. **274**, 516-521, (1999)
51. E. C. Theil, Targeting mRNA to regulate iron and oxygen metabolism, Biochem. Pharmacol. **59**, 87-93 (2000)
52. S. Recalcati, R. Pometta, S. Levi, D. Conte, and G. Cairo, Response of monocyte iron regulatory protein activity to inflammation: abnormal behavior in genetic hemochromatosis, Blood **91**, 2565-2572, (1998)
53. G. Minotti, S. Recalcati, A. Mordente, G. Liberi, A. M. Calafiore, C. Mancuso, P. Preziosi, and G. Cairo, The secondary alcohol metabolite of doxorubicin irreversibly inactivates aconitase/iron regulatory protein-1 in cytosolic fractions from human myocardium, FASEB J. **12**, 541-552, (1998)
54. G. Minotti, G. Cairo, and E. Monti Role of iron in anthracycline cardiotoxicity: new tunes for an old song? FASEB J. **13**, 199-212 (1999)
55. K. J. Wu, A. Polack, and R. Dalla-Favera, Coordinated regulation of iron-controlling genes, H-ferritin and IRP2, by c-MYC. Science **283**, 676-679, (1999)
56. C. Niederau, R. Fisher, A. Sonnenberg, W. Stremmel, H.J. Trampisch, and G. Strohmeyer, Survival and causes of death in cirrhotic and noncirrhotic patients with primary hemochromatosis, N. Engl. J. Med. **313**, 1256-1262. (1985)
57. M. Basset, J. Halliday, and L.W. Powell, Value of hepatic iron measurements in early hemochromatosis and determination of the critical level associated with fibrosis, Hepatology **6**,24-29 (1986)
58. Y.M. Deugnier, O. Loréal and B. Turlin, Liver pathology in genetic hemochromatosis: A review of 135 homozygous cases and their bioclinical correlations, Gastroenterology, **102**:2050-2059 (1992)
59. M. Barry, D.M. Flynn, E.A. Letsky, and R.A. Risdon, Long-term chelation therapy in thalassaemia major: effect on liver iron concentration, liver histology, and clinical progress, Br. Med. J. **2**, 16-20 (1974)
60. E. Angelucci, D. Baronciani, G. Lucarelli, C. Giardini, M. Galimberti, P. Polchi, F. Martinelli, M. Baldassarri, and P. Muretto, Liver iron overload and liver fibrosis in thalassemia, Bone Marrow Transplant. **12**, 29-31 1993
61. C.A. Seymour, and T.J. Peters, Organelle pathology in primary and secondary haemochromatosis with special reference to lysosomal changes, Br. J. Haematol. **40**, 239-53 (1978).
62. T.J. Peters, and C.A. Seymour, Acid hydrolase activities and lysosomal integrity in liver biopsies from patients with iron overload, Clin. Sci. Mol. Med. **50** :75-8 (1976).
63. C. Selden, M. Owen, J.M. Hopkins, and T.J. Peters. Studies on the concentration and intracellular localization of iron proteins in liver biopsy specimens from patients with iron overload with special reference to their role in lysosomal disruption, Br. J. Haematol. **44**, 593-603 (1980).
64. C. Selden, C.A. Seymour, T.J. Peters, Activities of some free-radical scavenging enzymes and glutathione concentrations in human and rat liver and their relationship to the pathogenesis of tissue damage in iron overload. Clin Sci. **58**, 211-219 (1980)
65. R. Hultcrantz, J. Ahlberg, and H. Glaumann, Isolation of two lysosomal populations from iron-overloaded rat liver with different iron concentration and proteolytic activity, Virchows Arch. B Cell. Pathol. Incl. Mol. Pathol. **47**, 55-65 (1984)
66. B.M. Myers, F.G. Prendergast, R. Holman, S.M. Kuntz, and N.F. LaRusso, Alterations in the structure, physicochemical properties, and pH of hepatocyte lysosomes in experimental iron overload, J. Clin. Invest, **88**, 1207-1215 (1991).
67. B.R. Bacon, R.S. and Britton, The pathology of hepatic iron overload: a free radical--mediated process? Hepatology **11**, 127-37 (1990).
68. F. De Matteis, R.G. Sparks. Iron-dependent loss of liver cytochrome P-450 haem *in vivo* and in vitro, FEBS Lett. **15**, 141-144 (1973)
69. M. Louw, A.C. Neethling, V.A. Percy, M. Carstens, and B.C. Shanley, Effects of hexachlorobenzene feeding and iron overload on enzymes of haem biosynthesis and cytochrome P 450 in rat liver, Clin Sci

Mol Med. ;**53**, 111-5. (1977).
70. H.L. Bonkowsky, J.F. Healey, P.R. Sinclair, J.F. Sinclair, and J. S. Pomeroy, Iron and the liver. Acute and long-term effects of iron-loading on hepatic haem metabolism, Biochem. J. **15**, 57-64 (1981).
71. W.G. Hanstein, T.D. Heitmann, A. Sandy, H.L. Biesterfeldt, H.H. Liem, U. Muller-Eberhard. Effects of hexachlorobenzene and iron loading on rat liver mitochondria. Biochim. Biophys. Acta Dec **18**, 293-299 (1981).
72. B.R. Bacon, J.F. Healey, G.M. Brittenham, C.H. Park, J. Nunnari, A.S. Tavill, H.L. Bonkovsky. Hepatic microsomal function in rats with chronic dietary iron overload. Gastroenterology **90**, 1844-1853. (1986).
73. A. Boveris, and B. Chance, The mitochondrial generation of hydrogen peroxide, Biochem. J. **134**, 707-716 (1973).
74. E. Cadenas, and A. Boveris, C.I. Ragan, and A.O.M. Stoppani, Production of superoxide radicals and hydrogen peroxide by NADH-ubiquinone reductase and ubiquinol-cytochrome c reductase from beef-heart mitochondria, Arch. Biochem. Biophys. **180**, 248-257 (1977).
75. H. Nohl, and D. Hegner. Do mitochondria produce oxygen radicals *in vivo*? Eur. J. Biochem. **82**, 563-567 (1978).
76. W.G. Hanstein, P.V. Sacks, and U. Muller-Eberhard. Properties of liver mitochondria from iron-loaded rats. Biochem Biophys. Res. Commun. **1**, 1175-1184 (1975).
77. A. Tangeras, Iron content and degree of lipid peroxidation in liver mitochondria isolated from iron loaded rats, Biochim. Biophys. Acta **757**, 59-68 (1983).
78. A. Masini, T. Trenti, E. Ventura E, D. Ceccarelli-Stanzani, U. Muscatello. Functional efficiency of mitochondrial membrane of rats with hepatic chronic iron overload. Biochem. Biophys. Res. Commun. (1984)
79. A. Masini, D. Ceccarelli-Stanzani, T. Trenti, E. Rocchi, and E. Ventura. Structural and functional properties of rat liver mitochondria in hexachlorobenzene induced experimental porphyria. Biochem. Biophys. Res. Commun. **118**, 356-363 (1984)
80. A. Masini, D. Ceccarelli-Stanzani, T. Trenti, and E. Ventura, Transmembrane potential of liver mitochondria from hexachlorobenzene-and iron-treated rats, Biochim. Biophys. Acta **802**, 253-258 (1984).
81. B.R. Bacon, C.H. Park, G.M. Brittenham, R. O'Neill,a and A.S. Tavill A, Hepatic mitochondrial oxidative metabolism in rats with chronic dietary iron overload, Hepatology **5**, 789-797 (1985).
82. R.C. McKnight, and F.E. Hunter, Mitochondrial membrane ghosts produced by lipid peroxidation induced by ferrous ion. II. Composition and enzymatic activity, J. Biol. Chem. **241**, 2757-2765 (1966).
83. G. Bellomo, A. Martino, P. Richelmi, G.A. Moore, S.A. Jewell, and S. Orrenius, Pyridine-nucleotide oxidation, Ca2+ cycling and membrane damage during tert-butyl hydroperoxide metabolism by rat-liver mitochondria, Eur. J. Biochem. **140**, 1-6 (1984).
84. B. Frei , K.H. Winterhalter, and C. Richter, Quantitative and mechanistic aspects of the hydroperoxide-induced release of Ca2+ from rat liver mitochondria, Eur. J. Biochem. **149**, 633-9 (1985)
85. H.R. Lotscher, K.H. Winterhalter, E. Carafoli, and C. Richter, Hydroperoxides can modulate the redox state of pyridine nucleotides and the calcium balance in rat liver mitochondria. Proc. Natl. Acad. Sci. U. S. A. **76**, 4340-4344. (1979).
86. H.R. Lotscher, K.H. Winterhalter, E. Carafoli, and C. Richter, Hydroperoxide-induced loss of pyridine nucleotides and release of calcium from rat liver mitochondria, J. Biol. Chem.**255**, :9325-3930 (1980).
87. A.L. Lehninger, A. Vercesi, E.A. Bababunmi, Regulation of Ca2+ release from mitochondria by the oxidation-reduction state of pyridine nucleotides, Proc. Natl. Acad. Sci. U. S.A. **75**, 1690-16944 (1978).
88. I. Al-Nasser, and M. Crompton, The reversible Ca2+-induced permeabilization of rat liver mitochondria. Biochem. J **239**, 19-29 (1986).
89. M. Crompton M, and A. Costi, Kinetic evidence for a heart mitochondrial pore activated by Ca2+, inorganic phosphate and oxidative stress. A potential mechanism for mitochondrial dysfunction during cellular Ca2+ overload. Eur J Biochem. **178**, 489-501 (1988).
90. C. Serhan, P. Anderson, E. Goodman, P. Dunham, G. Weissmann, Phosphatidate and oxidized fatty acids are calcium ionophores. Studies employing arsenazo III in liposomes, J. Biol. Chem. **256,** 2736-2741 (1981).
91. G. Bellomo, S.A. Jewell, and S. Orrenius, The metabolism of menadione impairs the ability of rat liver mitochondria to take up and retain calcium, J. Biol. Chem. **257**, 11558-11562 (1982).
92. B. Frei, K.H. Winterhalter, and C.Richter. Mechanism of alloxan-induced calcium release from rat liver mitochondria. J. Biol. Chem. **260,** 7394-7401 (1985).
93. S. E. Dryer, R. L. Dryer, and A. P. Autor, Enhancement of mitochondrial, cyanide-resistant superoxide dismutase in the livers of rats treated with 2,4-dinitrophenol, J. Biol. Chem. **10**, 1054-1057(1980).

94. A. Pietrangelo, F. Borella, G. Casalgrandi, G. Montosi, D. Ceccarelli, D. Gallesi, F. Giovannini, and A. Masini, Antioxidant activity of silybin in vivo during chronic iron overload in rats, Gastroenterology **109**, 1941-1949 (1995).
95. A. Masini, D.Ceccarelli, F. Giovannini, G. Montosi, C. Garuti, A. Pietrangelo, Iron-induced oxidant stress leads to irreversible mitochondrial dysfunctions and fibrosis in the liver of chronic iron-dosed gerbils. The effect of silybin, J. Bioe. Biom. **32**, 175-182 (2000).
96. U. Muscatello, and E. CarafoliThe oxidation of exogenous and endogenous cytochrome c in mitochondria. A Biochemical and Ultrastructural Study, J. Cell. Biol. **40**, 602-621 (1969).
97. M. Fry, and D. E. Green, Cardiolipin requirement by cytochrome oxidase and the catalytic role of phospholipid, Biochem. Biophys. Res. Commun. **93**, 1238-1246 (1980).
98. M.U. Dianzani, Biochemical effect of saturated and unsaturated aldehydes. In: McBrien DCH. Slater TF eds. Free radicals, lipid peroxidation and cancer. Academic press, 1982:129-151
99. M. Inoue, Protective mechanisms against reactive oxygen species, in *The Liver. Biology and Pathobiology*, edited by I.M. Arias, J.L. Bojer, N. Fausto, W.B. Jacoby, D. Schachter, and D.A. Shafritz (Raven Press New York) 1994 pp. 443-459,
100. S.L. Friedman, The cellular basis of hepatic fibrosis. Mechanisms and treatment strategies, N. Engl. J. Med. **328**,1828-1835, (1993).
101. D.M. Bissell, Cell-matrix interaction and hepatic fibrosis, Prog. Liv. Dis. **9**,143-155 (1990).
102. C.I. Levene, C.J. Bates, The effect of hypoxia on collagen synthesis in cultured 3T6 fibroblasts and its relationship to the mode of action of ascorbate, Biochim. Biophys. Acta **444**, 446-452 (1976).
103. V. Falanga, T.A. Martin, H. Takagi, R.S. Kirsner, T. Helfman, J. Pardes, and M.S. Ochoa, Low oxygen tension increases mRNA levels of alpha 1 (I) procollagen in human dermal fibroblasts, J. Cell. Physiol. **157**, 408-12 (1993).
104. G.A.C. Murrel, M.J.O. Francis, and L. Bromley, Modulation of fibroblast proliferation by oxygen free radicals, Biochem. J. **265**, 659-665 (1990).
105. J. Kuiper, A. Brouwer, D.L. Knook, et al: Kupffer and sinusoidal endothelial cells. in *The Liver. Biology and Pathobiology*, edited by I.M. Arias, J.L. Bojer, N. Fausto, W.B. Jacoby, D. Schachter, and D.A. Shafritz (Raven Press New York1994) pp. 791-818.
106. B. Meier, H.H. Radeke, S. Seller, M. Younes, H. Sies, K. Resch, and G. G. Habermehl, Human fibroblast release reactive oxygen species in response to interleukin-1 or tumor necrosis factor-α. Biochem. J. **263**, 539-545, (1989).
107. S.J. Weiss, Tissue destruction by neutrophils, N. Engl. J. Med. **320**, 365-376 (1989).
108. T. Matsubara, M. Ziff, Superoxide anion release by human endothelial cells: synergism between phorbol ester and calcium ionophore, J. Cell. Physiol. **127**, 207-210 (1986).
109. H. de Groot, A. Littauer, Hypoxia, reactive oxygen, and cell injury, Free. Rad. Biol. Med. **6**, 541-551 (1989).
110. H. Estebauer, Cytotoxicity and genotoxicity of lipid-oxidation products, Am. J. Clin. Nutr. **57**, 779-786 (1993).
111. M. Parola, M. Pinzani, A. Casini, E. Albano, G. Poli, A. Gentilini, P. Gentilini, and M. U.Dianzani, Stimulation of lipid peroxidation or 4-hydroxynonenal treatment increases procollagen-a_l (I) gene expression in human liver fat-storing cells. Biochem Biophys Res Commun 194: 1044-1050, 1993.
112. S.C. Finch, and C.A. Finch, Idiopathic hemochromatosis: an iron storage disease, Medicine **34**, 381-430, (1955).
113. M. Block, G. Moore, P. Wasi, Histogenesis of the hepatic lesion in primary hemochromatosis: with consideration of the pseudo iron deficient state produced by phlebotomies, Am. J. Pathol. **47**, 89-123, (1965).
114. X.Y Zhou, S.Tomatsu, R. E. Fleming, S. Parkkila, A. Waheed, J. Jiang, Y. Fei, E. M. Brunt, D. A. Ruddy, C. E. Prass, R. C. Schatzman, R. O'Neill, R. S. Britton, B. R. Bacon, and W. S. Sly, HFE gene knockout produces mouse model of hereditary hemochromatosis. Proc. Natl. Acad. Sci. U.S.A. **95**, 2492-2497 (1998).
115. C.H. Park, B.R. Bacon, G.M. Brittenham, and A.S Tavill, Pathology of dietary carbonyl iron overload in rats, Lab. Invest. **57**, 555-563, (1987).
116. A. Pietrangelo, E. Rocchi, L. Schiaffonati, Liver gene expression during chronic dietary iron overload in rats, Hepatology **11**, 798-804 (1990).
117. A. Pietrangelo, R. Gualdi, G. Casalgrandi, Enhanced hepatic collagen type I mRNA expression into fat-storing cells in a rodent model of hemochromatosis, Hepatology **19**, 714-721 (1994).
118. A. Pietrangelo, Gualdi, G. Casalgrandi, Molecular and cellular aspects of iron-induced hepatic cirrhosis in rodents, J. Clin. Invest. **95**,1824-1831 (1995).
119. G. Montosi, C. Garuti, A. Iannone, and A. Pietrangelo. Spatial and temporal dynamics of hepatic stellate cell activation during oxidant-stress-induced fibrogenesis, Am. J. Pathol. **152**, 1319-1326 (1998).

120. G. Montosi, C. Garuti, S. Martinelli, and A. Pietrangelo. Hepatic stellate cells are not subjected to oxidant stress during iron-induced fibrogenesis in rodents, Hepatology **27**, 1611-1622 (1998).
121. A. Pietrangelo, Metals, oxidative stress, and hepatic fibrogenesis, Semin. Liver Dis. **16,**13-30 (1996).
122. R. Gualdi, G. Casalgrandi, G. Montosi, and A. Pietrangelo, Excess iron into hepatocytes is required for activation of collagen type I gene during experimental siderosis, Gastroenterology **107**, 1118-1124. (1994).
123. L. Goldberg, and J.P. Smith, Iron overloading and hepatic vulnerability, Am. J. Clin. Pathol. **25,** 514-542 (1955).
124. P. Carthew, R.E. Edwards, A.G. Smith, Rapid induction of hepatic fibrosis in the gerbil after the parenteral administration of iron-dextran complex, Hepatology **13,** 534-539 (1991).
125. E. Orfei, F. I. Volini, F. Madera-Orsini, O. T. Minick, and G. Kent, Effect of iron loading upon the formation of collagen in the hepatic injury in duced by carbon tetrachloride, Am. J. Pathol. **45,**129-155 (1964).
126. M. Mackinnon, C. Clayton, J. Plummer, M. Ahern, P. Cmielewski, A. Ilsley, and P. Hall, Iron overload facilitates hepatic fibrosis in the rat alcohol/low-dose carbon tetrachloride model, Hepatology **21**, 1083-1088 (1995).
127. H. Tsukamoto, W. Horne, S. Kamimura, O. Niemela, S. Parkkila, S. Yla-Herttuala, G. M. Brittenham, Experimental liver cirrhosis induced by alcohol and iron. J. Clin. Invest. **96**, 620-630 (1995).

ROLE OF NON-TRANSFERRIN-BOUND IRON IN THE PATHOGENESIS OF IRON OVERLOAD AND TOXICITY

Pierre Brissot, Olivier Loréal

Mammalian cells accumulate iron from two main circulating sources. The first one, which is the classical source, consists of iron bound to transferrin. The second one, identified by Hershko and colleagues[1], is called Non-Transferrin-Bound Iron (NTBI). The latter source is increasingly acknowledged as being of primary importance in iron overload situations, due to its high uptake by parenchymal cells and its potential toxicity.

1. DEFINITION, NATURE AND DOSAGE OF NTBI

1.1. DEFINITION

Four main forms of circulating iron can be individualized :

1.1.1. Transferrin iron. Transferrin is a glyco-protein consisting of a single polypeptide chain with two binding sites (each capable of binding one atom of ferric iron), and of two branched carbohydrate chains (glycans). Plasma transferrin can exist under four molecular forms (apotransferrin, monoferricA, monoferric B, diferric transferrin) but can be considered, physiologically, as a single homogeneous pool. The normal plasma concentrations are 20 µmol/L for iron and 30 µmol/L for transferrin (which corresponds to a transferrin saturation rate of approximately 30%).

1.1.2. Haem iron forms. Haem iron is carried as haemoglobin bound to haptoglobin and as haem bound to haemopexin.

1.1.3. Ferritin-iron is likely to represent only a minute amount since the iron content of circulating ferritin is usually low. Serum ferritin levels are normally less than 200 ng/ml in men and 100 in women.

1.1.4. Non-Transferrin-Bound Iron (NTBI).
i) Physiologically: This iron species is either undetectable or at very low concentration (<1µmol/l) in the normal plasma. It represents, however, a significant iron form in the normal cerebrospinal fluid (CSF). Indeed, likely due to an efficient blood-brain barrier against plasma iron crossing, iron concentration in normal CSF is low close to 0.8

- Pierre Brissot, Service des Maladies du Foie et INSERM U-522, Pontchaillou University Hospital, Rennes, France

Iron Chelation Therapy.
Edited by Chaim Hershko, Kluwer Academic/Plenum Publishers, 2002.

μmol/L[2]. Transferrin levels are estimated[2-5] between 0.1 and 0.4 μmol/L, and it is established that, in most normal CSFs, transferrin is fully saturated, and therefore that other forms of iron must be present. This is supported by Gutteridge's data[6] showing normal mean levels of non-transferrin bound iron of 0.55 ± 0.27 μmol/L in CSF.
ii) In case of iron overload, either acute or chronic, plasma NTBI becomes a significant fraction of circulating iron, its concentration reaching several micromoles/l. It should be noted that despite its imperfection the term NTBI is confined to this special iron entity and does not apply to ferritin-iron, haptoglobin-iron or hemopexin-iron, which, strictly thinking, correspond also to iron species which are transferrin-bound.
iii) In various non iron-overloaded diseases, the presence of circulating NTBI has been reported. These conditions include fulminant hepatic failure[7], hematological diseases[8] especially under chemotherapy[9-12] and bone marrow transplantation[13-14], adult respiratory distress syndrome[15], and cardiopulmonary bypass surgery[16].

1.2. NATURE

The chemical form of NTBI remains poorly defined but is likely to be heterogeneous, involving both non-protein and protein-bound forms. The non-protein ligands appear to correspond to low molecular weight organic compounds such as ascorbate, phosphate, carbonate, organic acids and aminoacids. In the absence of iron overload a low-molecular-weight polypeptide has been isolated and purified from normal human cord and adult sera by gel filtration and HPLC. Its molecular weight has been estimated to be 2.5 KDa[17,18]. In genetic hemochromatosis, Grootveld et al[19] characterized NTBI by high performance liquid chromatography and nuclear resonance spectroscopy and proposed that NTBI in the plasma of iron-overloaded patients exists largely as complexes with citrate and possibly also as ternary iron-citrate-acetate complexes. Besides these non protein-bound forms, NTBI could be complexed to proteins such as albumin which offers nucleation sites for iron aggregation[20]. The investigation of hypotransferrinemic mouse serum[21] revealed a mixture of iron species as demonstrated by differing reactivity to acid extraction, differing elution from sephadex G200 and differing reactivity to NTBI assays. These species could not be identified with existing iron complexes or proteins.

1.3. METHOD OF QUANTIFICATION

A number of assays have been proposed, since the original description by Hershko and colleagues[1]. They, schematically, correspond to three main types of methodological approaches.

1.3.1. The chelation-ultrafiltration-detection approach. It is based on the prior mobilization of serum NTBI by agents such as EDTA[1], oxalate[22] and, mostly, nitrilotriacetate (NTA)[20]. After ultrafiltration, which separates the chelated NTBI from transferrin-iron, NTBI detection is performed according to colorimetric methods[23], HPLC[20,24-26] techniques. or graphite furnace atomic absorption spectrometry[13,27]. When transferrin is incompletely saturated, a significant improvement of the method consists to add a cobalt salt which blocks the free iron binding sites in order to avoid in vitro donation of iron from NTA onto the vacant sites of transferrin and therefore an underestimation of NTBI values[28,29].

1.3.2. The bleomycin approach. It is based upon the ability of the antitumor antibiotic bleomycin to degrade DNA via a free radical reaction in the presence of ferrous ions.

These ferrous irons are produced by adding a reducing agent (such as ascorbate) to the tested serum and the free radical reaction is detected by the production of thiobarbituric acid-reactive substances which include malondialdehyde[30]. However, some of these products can be generated in the high temperature phase of the method and therefore correspond to artifacts[31]. Applying the ethidium-binding assay of DNA damage to the measurement of bleomycin-detectable iron has been reported to improve the reliability of the method[32].

1.3.3. The fluorescent probe approach. Breuer et al[33] have introduced an assay for monitoring a component of serum non-transferrin-bound iron, termed "desferrioxamine-chelatable iron (DCI)". In this assay, serum DCI is measured with the probe fluorescein-desferrioxamine whose fluorescence is stoichiometrically quenched by iron.

In summary, the detection of serum NTBI remains technically difficult, probably due to the heterogeneity of the chemical forms of circulating NTBI, depending especially on the degree of transferrin saturation and on the etiology of the iron excess.

2 ROLE OF NTBI IN THE PATHOGENESIS OF IRON OVERLOAD

A series of arguments, emerging from experimental and clinical situations, support the important role of NTBI in the development of visceral iron excess.

2.1. EXPERIMENTAL DATA

2.1.1. NTBI AND HEPATIC IRON OVERLOAD. Several types of data illustrate the role of NTBI. i) <u>NTBI is very efficiently taken up by the liver</u>. The acute saturation, by intravenous infusion of iron, of plasma iron-binding capacity in rats produced hepatic deposition of a large fraction of intestinally absorbed iron[34]. Examining the hepatic removal process for iron in the single-pass perfused rat liver we found[35] that NTBI uptake was highly efficient (extraction of 1µM iron, 58-75%), contrasting with the previously reported very low hepatic uptake from the plasma of transferrin iron (<1%)[36]. Using normal rats and mice whose transferrin had been saturated by an intravenous injection of nonradiolabeled iron, Craven et al[37] showed that injected ^{59}Fe, which behaved like NTBI, was removed from the serum with a half-life of <30s, versus 50 min for transferrin-iron. Furthermore, these authors showed, in agreement with Bradbury et al[38] and with Bernstein[39], that, in the mice with congenital hypotransferrinemia, the newly absorbed radioactive iron was almost completely deposited in the liver, versus <1% in the erythrocytes. ii) <u>NTBI uptake by the liver is targeted mainly towards the hepatocytes</u>. Electron microscopy of rat livers perfused for 30 min with 1 µM ferrous iron showed accumulation of electron dense particles compatible with ferritin cores in the cytoplasm of parenchymal cells with preferential localization in the lysosomes[40]. Rat hepatocytes in primary cultures have been shown to exert a high capacity to take-up NTBI in the form of Fe-citrate[41]. Evidence for a low Km transporter for NTBI has been reported in isolated rat hepatocytes[42]. In hypotransferrinemic mice, Iancu et al[43], using electron microscopy and laser microprobe mass analysis, nicely showed preferential iron deposition within the hepatocytes, under the forms of ferritin particles and clusters as well as ferritin and haemosiderin in lysosomes (siderosomes) ii) <u>NTBI uptake by the liver, in striking contrast with transferrin-iron, is not down-regulated by hepatocytic iron overload</u>. This has been shown in the perfused rat liver[41] as well as in primary cultures of rat hepatocytes[42]. It has been proposed, from experiments using various transformed cell lines (fibroblasts[44]; HepG2cells[45,46]), that intracellular iron leads to an increase in the rate of NTBI uptake. Such data however have not been shown when studying possibly more relevant hepatic models such as the perfused rat liver[41] or rat hepatocytes in cultures[41].

2.1.2. NTBI AND EXTRA-HEPATIC IRON OVERLOAD.
Beside the liver, several other organs can be overloaded via NTBI.
• Pancreas. The exocrine pancreas is particularly overloaded in the hypotransferrinemic mouse[47], the involvement of centroacinar and intercalated duct cells being massive and generalized whereas macrophages contained only isolated siderosomes. In islets, B cells rarely showed siderosomes and no ferritin particles were seen in the cytosol[43].
• Heart. Significant iron excess is present in heart myocytes of hypotransferrinemic mice[43]. In heart cell cultures the rate of NTBI uptake is over 300-times greater than that of transferrin iron[48], and is increased by high tissue iron content[49].
• Brain. 59Fe uptake into brain was 80-95% times greater in hypotransferrinemic mice than in normal[50]. Dickinson et al[51] reported, in the same animal model, that distribution of ^{59}Fe to the brain was not affected by the absence of transferrin, indicating the existence of a NTBI delivery system.
• Erythroid cells. NTBI uptake by rabbit reticulocytes has been reported by Egyed[52] and Hodgson et al[53]. Studying NTBI uptake in erythroid cells of Belgrade rats (animals which develop a microcytic anemia related to a mutation in the transmembrane iron transporter DMT1) versus normal rats, Garrick et al[54] found two types of NTBI uptake processes : i) a high affinity NTBI uptake which provides iron for heme synthesis, occurs in reticulocytes but not erythrocytes and is affected by the Belgrade mutation, indicating that it requires DMT1, and ii) a low affinity NTBI uptake which does not provide iron for heme synthesis, is operative in both erythrocytes and reticulocytes and does not require DMT1.

2.2. CLINICAL DATA

2.2.1. NTBI AND GENETIC HEMOCHROMATOSIS
Using a triple radio-iron isotope technique, Fawwaz et al[55] demonstrated, in untreated hemochromatosis (with low plasma latent iron-binding capacity), a high first-pass hepatic deposition of intestinally absorbed iron : significant hepatic iron disposition was observed within 2 hours, and at 6 hours four-fifths of the total iron absorbed from the gastrointestinal tract was deposited in the liver. Several studies have reported the direct presence of plasma NTBI in hemochromatosis[56-58,29]. Using a method which includes a presaturation step of transferrin with cobalt, Loréal et al[29] reported that NTBI levels: i) reached values up to 5 µM and could be detected even when transferrin was not fully saturated ; ii) persisted almost until completion of phlebotomy treatment ; iii) were well correlated with transferrin saturation in a given patient. iv) were absent when transferrin saturation values were lower than 35%. de Valk et al[59], using the same method, reported that non-transferrin bound iron (with values usually lower than 2 µM) was present in serum of hemochromatosis heterozygotes. The proportion of compound heterozygotes in this series was not indicated.

2.2.2. NTBI AND OTHER TYPES OF CHRONIC IRON OVERLOAD.
Thalassemia. Serum NTBI was found in patients with beta-thalassemia trait and chronic active hepatitis[60]. Studying 37 patients with Hb H disease and 104 patients with ß-

thalassemia/Hb E disease, NTBI was found to be associated with higher transferrin saturation and higher plasma ferritin levels, and to be higher in splenectomized patients[61]. Congenital atransferrinemia[62-64]. This exceptional condition is obviously a NTBI related disease, characterized by the contrast between a microcytic anemia and systemic iron overload.
African iron overload. Studying 195 subjects, consisting of 25 black African individuals with iron overload documented by liver biopsy and 170 relatives and neighbours, NTBI (> 2µmol/L) was present in 43 people, 22 of patients of whom underwent liver biopsy and 21 relatives or neighbours[26].

3. ROLE OF NTBI IN IRON TOXICITY

A number of data document the toxic role of NTBI towards the cells[65].

3.1. EXPERIMENTALLY, iron promotes the formation of free hydroxyl radicals via the Haber-Weiss reaction. Reactive oxygen species are in turn able to generate increased lipid peroxidation[66-68]. Iron-induced peroxidative damage has been demonstrated on different cells and cellular targets : rat hepatocytes[69,70], hepatic lysosomal membranes[71], hepatic mitochondria in liver cells[72,73] and heart cells mitochondria[74]. Iron can also damage the nuclei[75,76].
Beside this peroxidative mechanism, NTBI damages cells by increasing the intracellular iron content through both massive NTBI entry and, for hepatocytes, decreased biliary excretion of plasma NBTI[77,78]. Enterohepatic recycling of NTBI has been shown in rats[79] but the pathophysiological importance of this mechanism remains to be evaluated.

3.2. IN HUMAN STUDIES, several arguments favor the clinical toxicity of NTBI. In genetic hemochromatosis, Gutteridge et al[57] reported that low-molecular-weight iron complexes stimulated both the peroxidation of membrane lipids and the formation of hydroxyl radicals. Peters et al[80] confirmed these data and extended them to transfusional iron overload. Studying 52 patients with beta-thalassemia major, Al-Refaie et al[25] found a significant difference between the level of serum NTBI and whether or not the patients had complications of iron overload. However, in this disease, no correlation was found, in the serum, between NTBI and either malondialdehyde or vitamin E[81]. Moreover, no relationship was observed between oxidative damage of ß-thalassaemia intermedia erythrocytes and extracellular NTBI[82]. In their study on African dietary iron overload, McNamara et al[26] found that, among all 195 subjects, the presence of NTBI in serum was independently related to elevations in alanine and aspartate aminotransferase activity and bilirubin concentration, and this relationship between serum NTBI and hepatic dysfunction was confirmed in the subgroup of 25 subjects with iron overload documented by liver biopsy. It cannot be excluded that NTBI originated from damaged hepatocytes ; however, this could not be the sole explanation since serum NTBI was also detected in patients with normal liver function tests. Acute iron intoxication[83], which is an important cause of poisoning in young children, is a life-threatening condition, involving severe gastroenteropathy, hepatic necrosis, shock syndrome, and plasmatic coagulation defect. It has been proposed that this coagulopathy is the consequence of the susceptibility of serine proteases to NTBI[84]. The finding of stainable myocardial iron supports the view that there is at least a factor of primary myocardial failure in acute iron poisoning[85].

In conclusion, serum NTBI plays an important role in iron overload and damage. Numerous informations, however, remain to be obtained, especially : i) The real levels of circulating NTBI (using reliable detection procedures) in various iron overload conditions ; ii) The precise biochemical nature of serum NTBI in these situations ; iii) The transmembrane and intracytosolic pathways used by this iron species, and particularly the exact link between the levels of serum NTBI and the cytosolic labile iron pool; iii) The mechanisms whereby serum NTBI exerts its cellular damaging effects. Upon these findings depends the possible use of new therapeutic approaches specifically targeted on the serum and cellular decrease of this intriguing form of iron.

REFERENCES

1. Hershko C, Graham G, Bates GW, Rachmilewitz EA. Non-specific serum iron in thalassemia: an abnormal serum iron fraction of potential toxicity. Br J Haematol 1978; **40**: 255-263.
2. Bleijenberg BG, Van Eijk HG, Leijnse B. The determination of nonheme iron and transferrin in cerebrospinal fluid. Clin Chim Acta 1979; **31**: 277-281.
3. Felgenhauer K. Protein size and cerebrospinal fluid. Klin Wochenschr. 1974; **52**: 1158-1164.
4. Lamoureux G, Jolicoeur R, Giard N, St-Hilaire M, Duplantis F. Cerebrospinal fluid proteins in multiple sclerosis. Neurology 1975; **25**: 537-552.
5. Del Principe D, Menichelli A, Colistra C. The ceruloplasmin and transferrin systems in cerebrospinal fluid in acute leukaemia patients. Acta Paediatr scand 1989; **78**: 327-328.
6. Gutteridge JMC. Hydroxyl radicals, iron, oxidative stress and neurodegeneration. Ann NY Acad Sci 1994; **738**: 201-213.
7. Evans PJ, Evans RW, Bomford RW, Williams R, Halliwell B. Metal irons catalytic for free radical reactions in the plasma of patients with fulminant hepatic failure. 1994; **20**: 139-144.
8. Nomdedeu J, Gimferrer NJ, Ubeda J, Marco N, Brunet S, Sureda A, Martino R, Altés A, Royo MT, Marigo GJ, Domingo-Albos A. Non-transferrin plasma iron in leukemia, lymphoma and myelodysplactic syndromes. Biol Clin Hematol 1995; **17**: 135-139.
9. Halliwell B, Aruoma OI, Mufti G, Bomford A. Bleomycin-detectable iron in serum from leukaemic patients before and after chemotherapy. Therapeutic implications for treatment with oxidant-generating drugs. FEBS Letters 1988; **241**: 202-204.
10. Harrison P, Marwah SS, Hughes RT, Bareford D. Non-transferrin bound iron and neutropenia after cytotoxic chemotherapy. J Clin Pathol 1994; **47**: 350-352.
11. Carmine TC, Evans P, Bruchelt G, Evans R, Handgretinger R, Niethammer D, Halliwell B. Presence of iron catalytic for free radical reactions in patients undergoing chemotherapy: implications for therapeutic management. Cancer Letters 1995; **94**: 219-226.
12. Bradley SJ, Gosriwatana I, Srichairatanakool S, Hider RC, Porter JB. Non-transferrin-bound iron induced by myeloablative chemotherapy. Br J Haematol 1997; **99**: 337-343.
13. Dürken M, Nielsen P, Knobel S, Finckh B, Herrning C, Dresow B, Kohlschütter B, Stockschläder M, Krüger WH, Kohlschütter A, Zander AR. Nontransferrin-bound iron in serum of patients receiving bone marrow transplants. Free Biol Med 1997; **22**: 1159-1163.
14. Sahlstedt L, Ebeling F, von Bonsdorff L, Parkkinen J, Ruutu T. Non-transferrin-bound iron during allogeneic stem cell transplantation. Br J Haematol 2001; **113**: 836-838.
15. Gutteridge JMC, Quinlan GJ, Mumby S, Heath A, Evans TW. Primary plasma antioxidants in adult respiratory distress syndrome patients: changes in iro-oxidizing, iron-binding, and free radical-scavenging proteins. J Lab Clin Med 1994; **124**: 263-273.
16. Pepper JR, Mumby S, Gutteridge JMC. Transient iron-overload with bleomycin-detectable iron present during cardiopumonary bypass surgery. Free Rad Res 1994; **21**: 53-58.
17. Lau S, Sarkar B. Comparative studies of manganese(II), nickel(II), zinc(II), copper(II), cadmium(II) and iron(III) binding components in human cord and adult sera. Can J Biochem Cell Biol 1984; **62**: 449-455.
18. Stojkovski S, Goumakos W, Sarkar B. Iron(III)-binding polypeptide in human cord and adult serum: isolation, purification and partial characterization? Biochim Biophys Acta. 1992; **1137**: 155-161.

19. Grootveld M, Bell JD, Halliwell B, Aruoma OI, Bomford A, Sadler PJ. Non-transferrin-bound iron in plasma or serum from patients with idiopathic hemochromatosis. Characterization by high performance liquid chromatography and nuclear resonance spectroscopy. J Biol Chem. 1989 ; **264** : 4417-4422.
20. Singh S, Hider RC, Porter JB. A direct method for quantification of non-transferrin-bound iron. Anal Biochem. 1990; **186**: 320-323.
21. Simpson RJ, Cooper CE, Raja KB, Halliwell B, Evans PJ, Aruoma OI, Singh S, Konijn AM. Non-transferrin-bound iron species in the serum of hypotransferrinaemic mice. Biochim Biophys Acta 1992; **1156**: 19-26.
22. Breuer W, Ronson A, Abramov A, Slotki I, Hershko H, Cabantchik ZI. The assessment of serum non-transferrin-bound iron (NTBI) in chelation therapy and iron supplementation. Blood 2000; **95**: 2975-2982.
23. Zang D, Okada S, Kawabata T, Yasuda T. An improved simple colorimetric method for quantitation of non-transferrin-bound iron in serum. Biochem Mol Biol Int 1995; **35**: 635-641.
24. Porter JB, Abeysinghe RD, Marshall L, Hider RC, Singh S. Kinetics of removal and reappearance of non-transferrin-bound plasma iron with deferoxamine therapy. Blood 1996; **88**: 705-713.
25. Al-Refaie FN, Wickens DG, Wonke B, Kontoghiorghes GJ, Hoffbrand AV. Serum non-transferrin-bound iron in beta-thalassaemia major patients treated with desferrioxamine and L1. BrJ Haematol 1992; **82**: 431-436.
26. McNamara L, Macphail AP, Mandishona E, Bloom P, Paterson AC, Rouault TA, Gordeuk VR. Non-transferrin-bound iron and hepatic dysfunction in African dietary iron overload. J Gastroenterol Hepatol 1999; **14**: 126-132.
27. Jakeman A, Thompson T, McHattie J, Lehotay DC. Sensitive method for nontransferrin-bound iron quantification by furnace atomic absorption spectrometry. Clin Biochem 2001; **34**: 43-47.
28. Gosriwatana I, Loréal O, Shuli L, Brissot P, Porter J, Hider RC. Quantification of non-transferrin-bound iron in the presence of unsaturated transferrin. Anal Biochem 1999; **273**: 212-220.
29. Loréal O, Gosriwatana I, Guyader D, Porter J, Brissot P, Hider RC. Determination of non-transferrin bound iron in genetic hemochromatosis using a new HPLC-based method. J Hepatol 2000; **32**: 727-733.
30. Gutteridge JMC, Rowley DA, Halliwell B. Superoxide-dependent formation of hydroxyl radicals in the presence of iron salts. Biochem J 1981; **199**: 263-265.
31. Richardson DR, Dean RT. Does free extracellular iron exist in haemochromatosis and other pathologies and is it redox active ? Clin Sci 2001; **100**: 237-238.
32. Burkitt MJ, Milne L, Raafat A. A simple, highly sensitive and improved method for the measurement of bleomycin-detectable iron: the "catalytic iron index" and its value in the assessment of iron status in haemochromatosis. Clin Sci. 2001; **100**: 239-247.
33. Breuer W, Ermers MJJ, Pootrakul P, Abramov A, Hershko C, Cabantchik ZI. Desferrioxamine-chelatable iron, a component of serum non-transferrin-bound iron, used for assessing chelation therapy. Blood 2001; **97**: 792-798.
34. Wheby MS, Jones LG. Role of transferrin in iron absorption. J Clin Invest 1963; **42**: 1007-1016.
35. Brissot P, Wright TL, Ma WL, Weisiger RA. Efficient clearance of non-transferrin-bound iron by rat liver. Implications for hepatic iron loading and iron overload states. J Clin Invest 1985 ; **76**: 1463-1470.
36. Zimelman AP, Zimmerman HJ, McLean R, Weintraub LR. Effect of iron saturation of transferrin on hepatic iron uptake: an in vitro study. Gastroenterology 1977; **72**: 129-131.
37. Craven CM, Alexander J, Eldridge M, Kushner JP, Bernstein S, Kaplan J. Tissue distribution and clearance kinetics of non-transferrin-bound iron in the hypotransferrinemic mouse: A rodent model for hemochromatosis. Proc Natl Acad Sci USA 1987; **84**: 3457-3461.
38. Bradbury MWB, Raja K, Ueda F. Contrasting uptakes of 59Fe into spleen, liver, kidney and some other soft tissues in normal and hypotransferrinaemic mice. Biochem Pharmacol 1994; **47**: 969-974.
39. Bernstein SE. Hereditary hypotransferrinemia with hemosiderosis, a murine disorder resembling human atransferrinemia. J Lab Clin Med 1987; **110**: 690-705.
40. Wright TL, Brissot P, Ma WL, Weisiger RA. Characterization of non-transferrin-bound iron clearance by rat liver. J Biol Chem 1986; **261**: 10909-10914.
41. Baker E, Baker SM, Morgan EH. Characterisation of non-transferrin-bound iron (ferric citrate) uptake by rat hepatocytes in culture. Biochim Biophys Acta 1998; **1380**: 21-30.
42. Barisani D, Berg CL, Wessling-Resnick M, Gollan J. Evidence for a low Km transporter for non-transferrin-bound iron in isolated rat hepatocytes. Am J Physiol 1995; **269**: G570-576.
43. Iancu TC, Shiloh H, Raja KB, Simpson RJ, Peters TJ, Perl DP, Hsu A, Good PF. The hypotransferrinaemic mouse: ultrastructural and laser microprobe analysis observations. J Pathol 1995; **177**: 83-94.
44. Kaplan J, Jordan I, Sturrock A. Regulation of the transferrin-independent iron transport system in cultured cells. J Biol Chem 1991; **266**: 2997-3004.
45. Parkes JG, Randell EW, Olivieri NF, Templeton DM. Modulation by iron loading and chelation of the uptake of non-transferrin-bound iron by human liver cells. Biochim Biophys Acta. 1995; **1243**: 373-380.
46. Randell EW, Parkes JG, Olivieri NF, Templeton DM. Uptake of non-transferrin-bound iron by both reductive and nonreductive processes is modulated by intracellular iron. J Biol Chem. 1994; **269**: 16046-16053.

47. Simpson RJ, Konijn AM, Lombard M, Raja KB, Salisbury JR, Peters TJ. Tissue iron loading and histopathological changes in hypotransferrinaemic mice. J Pathol 1993; **171**: 237-244.
48. Link G, Pinson A, Hershko C. Heart cells in culture: A model of myocardial iron overload and chelation. J Lab Clin Med. 1985; **106**: 147-153.
49. Parkes JG, Hussain RA, Olivieri NF, Templeton DM. Effects of iron loading on uptake, speciation, and chelation of iron in cultured myocardial cells. J Lab Clin Med. 1993; **122**: 36-47.
50. Ueda F, Raja KB, Simpson RJ, Trowbridge IS, Bradbury MWB. Rate of ^{59}Fe uptake into brain and cerebrospinal fluid and the influence thereon of antibodies against the transferrin receptor. J Neurochem 1993; **60**: 106-113.
51. Dickinson TK, Devenyi AG, Connor JR. Distribution of injected iron 59 and manganese 54 in hypotransferrinemic mice. J Lab Clin Med. 1996; **128**: 270-8.
52. Egyed A. Carrier mediated iron transport through erythroid cell membrane. Br J Haematol 1988; **68**: 483-486.
53. Hodgson LL, Quail EA, Morgan EH. Iron transport mechanisms in reticulocytes and mature erythrocytes. J Cell Physiol 1995; **162**: 181-190.
54. Garrick LM, Dolan KG, Romano MA, Garrick MD. Non-transferrin-bound iron uptake in Belgrade and normal rat erythroid cells. J Cell Physiol 1999; **178**: 349-358.
55. Fawwaz RA, Winchell HS, Pollycove M, Sargent T. Hepatic iron deposition in humans. First-pass hepatic deposition of intestinally absorbed iron in patients with low plasma iron-binding capacity. Blood 1967; **30**: 417-424.
56. Batey RG, Fong LC, Shamir S, Sherlock S. A non-transferrin-bound serum iron in idiopathic hemochromatosis. Digestive Dis Sci 1980 ; **25**: 340-346.
57. Gutteridge JMC, Rowley DA, Griffiths E, Halliwell B. Low-molecular-weight iron complexes and oxygen radical reactions in idiopathic haemochromatosis. Clin Sci 1985; **68**: 463-467.
58. Aruoma OI, Bomford A, Polson RJ, Halliwell B. Nontransferrin-bound iron in plasma from hemochromatosis patients: Effect of phlebotomy therapy. Blood 1988; **72**: 1416-1419.
59. de Valk B, Addicks MA, Gosriwatana I, Lu S, Hider RC, Marx JJM. Non-transferrin-bound iron is present in serum of hereditary haemochromatosis heterozygotes. Eur. J Clin Invest. 2000; **30**: 248-251.
60. Fargion S, Cappellini MD, Sampietro M, Fiorelli G. Non-specific iron in patients with beta-thalassaemia trait and chronic active hepatitis. Scand J Haematol. 1981; **26**: 161-167.
61. Anuwatanakulchai M, Pootrakul P, Thuvasethakul P, Wasi P. Non-transferrin plasma iron in ß-thalassaemia/Hb E and haemoglobin H diseases. Scand J Haematol 1984; **32**: 153-158.
62. Heilmeyer VL, Keller W, Vivell O, Keiderling W, Betke K, Wöhler F, Schultze HE Kongenitale atransferrinämie bei einem sieben jahre alten kind. Deutsch Med Wschr 1961; **86**: 1745-1751.
63. Goya N, Miyazaki S, Kodate S, Ushio B. A family of congenital atransferrinemia. Blood 1972 ; **40**: 239-245.
64. Hayashi A, Wada Y, Suzuki T, Shimizu A. Studies on familial hypotransferrinemia: unique clinical course and molecular pathology. Am J Hum Genet 1993; **53**: 201-213.
65. Anderson GJ. Non-transferrin-bound iron and cellular toxicity. J Gastroenterol Hepatol 1999; **14**: 105-108.
66. Brock J.W., Halliday J.W., Pippard J., Powell L.W.[edts], Iron Metabolism in Health and Disease. Saunders, London, 1994, pp. 311-351.
67. Halliwell B, Gutteridge JMC. Role of free radicals and catalytic metal ions in human disease: an overview. in L Packer, AN Glazer edts. Methods in Enzymology 1990; **186**: 1-85.
68. Stadtman ER. Metal ion-catalyzed oxidation of proteins: biochemical mechanism and biological consequences. Free Rad Biol Med 1990; **9**: 315-325.
69. Houglum K., Filip M., Witztum J.L., Chojkier M. Malondialdehyde and 4-hydroxynonenal protein adducts in plasma and liver of rats with iron overload. J Clin Invest 1990; **86**: 1991-1998.
70. Morel I, Lescoat G, Cillard J, Pasdeloup N, Brissot P, Cillard P. Biochem Pharmacol. 1990; **39**: 1647-1655.
71. Mak IT, Weglicki W. Characterization of iron-mediated peroxidative injury in isolated hepatic lysosomes. J Clin Invest 1985; **75**: 58-63.
72. Bacon BR, O'Neill R, Britton RS. Hepatic mitochondrial energy production in rats with chronic iron overload. Gastroenterology 1993; **105**: 1134-1140.
73. Ceccarelli D, Gallesi D, Giovannini F, Ferrali M, Masini A. Relationship between free iron level and rat liver mitochondrial dysfunction in experimental dietary iron overload. Biochem Biophys Res Comm 1995; **209**: 53-59.
74. Link G, Saada A, Pinson A, Konijn AM, Hershko C. Mitochondrial respiratory enzymes are a major target of iron toxicity in rat heart cells. J Lab Clin Med 1998; **131**: 466-474.
75. Pigeon C, Turlin B, Iancu TC, Leroyer P, Le Lan J, Deugnier Y, Brissot P, Loréal O., Carbonyl-iron supplementation induces hepatocyte nuclear changes in BALB/CJ male mice. J Hepatol. 1999; **30**: 926-934.
76. Meneghini R, Benfato MS, Bertoncini CR, Carvalho H, Gurgueira SA, Robalinho RL, Teixeira HD, Wendel CMA, Nascimento ALTO. Iron homeostasis and oxidative DNA damage. Cancer J. 1995; **8**: 109-113.
77. Brissot P, Zanninelli G, Guyader D, Zeind J, Gollan JL. Biliary excretion of non-transferrin-bound iron in rats: pathogenetic importance in iron overload disorders. Am J Physiol 1994; **267**: G135-G142.

78. Brissot P, Deugnier Y, Guyader D, Zanninelli G, Loréal O, Moirand R, Lescoat G. Iron overload and the biliary route. in Progress in Iron Research. Herschko C, ed, Plenum Press, New York 1994, 277-283.
79. Brissot P, Bolder U, Schteingart CD, Arnaud J, Hofmann AF. Intestinal absorption and enterohepatic cycling of biliary iron originating from plasma non-transferrin-bound iron in rats. Hepatology 1997; **25**: 1457-1461.
80. Peters SW, Jones BM, Jacobs A, Wagstaff M. "Free iron" and lipid peroxidation in the plasma of patients with iron overload. In Proteins of Iron Storage and Transport. G Spik, Montreuil J, Crichton RR, Mazurier J Edts. Elsevier Science Publishers B.V. 1985; pp 321-324.
81. Livrea MA, Tesoriere L, Pintaudi AM, Calabrese A, Maggio A, Freisleben HJ, D'Arpa D, D'Anna R, Bongiorno A. Oxidative stress and antioxidant status in ß-thalassemia major: iron overload and depletion of lipid-soluble antioxidants. Blood 1996; **88**: 3608-3614.
82. Tavazzi D, Duca L, Graziadei G, Comino A, Fiorelli G, Cappellini MD. Membrane-bound iron contributes to oxidative damage of ß-thalassaemia intermedia erythrocytes Br J Haematol 2001; **112**: 48-50.
83. Mahoney JR, Hallaway PE, Hedlund BE, Eaton JW. Acute iron poisoning. Rescue with macromolecular chelators. J Clin Invest. 1989; **84**: 1362-1366.
84. Rosenmund A, Haeberli A, PW Straub. Blood coagulation and acute iron toxicity. Reversible iron-induced inactivation of serine proteases in vitro. J Lab Clin Med. 1984; **103**: 524-533.
85. Tenenbein M, Kopelow ML, deSa DJ. Myocardial failure and shock in iron poisoning. Human Toxicol 1988; **7**: 281-284.

INTRACELLULAR AND EXTRACELLULAR LABILE IRON POOLS

Z. Ioav Cabantchik[*], Or Kakhlon, Silvina Epsztejn, Giulianna Zanninelli and William Breuer

1. INTRODUCTION

Labile forms of iron present in biological systems are defined as ionic Fe complexes that are redox active. They comprise a heterogeneous population of organic anions (phosphates and carboxylates), poly-functional ligands (i.e. chelates, siderophores and polypeptides) or surface components of membranes (e.g. phospholipid head groups) or extracellular matrix (e.g. glycans and sulfonates), which bind both forms of iron (II and III). Collectively, they define the respective labile iron pools (LIP), which can be of cellular (CLIP) or extracellular (ECLIP) nature. Operationally, those pools are characterized in terms of their propensity to engage in redox-cycling in an oxygenated environment and/or following pro-oxidant challenges. Methodologically, CLIP and ECLIP can be assessed in terms of iron reactivity and/ or the ability of the metal to undergo chelation by high affinity binding siderophores or chelators. Therapeutically, the LIPs are the immediate targets of chelators designed to reduce iron load in the entire organism, with emphasis on organs of accumulation such as the liver.

CLIP. In most mammalian cells, the total cell associated iron spans a wide range of concentrations (20-100 µM), mostly as iron intimately associated with proteins.[1] Only a minor fraction (<1%) appears to be loosely bound, presumably to a variety of ligands, and is readily available for numerous uses: a. physiologically, for metabolism; b. pharmacologically, for chelation by permeant chelators or metal scavengers and c. toxicologically, for promotion of free radical formation when redox-challenged. This putative iron fraction, which normally might not even exceed 1µM, is referred to as the Cellular Labile Iron Pool or CLIP [2,3]. The concept of a CLIP was postulated by earlier studies of A. Jacobs[5] and R.R. Chrichton[1] and attempts to define its levels and composition were made by them and others [17,24]. CLIP was often assumed to constitute a regulated or regulatory iron pool, in which some "representative" form of the metal is sensed (directly or indirectly) by iron responsive proteins (IRPs) and its ultimate cellular levels are controlled by IRP-induced processes[8,9].

Normally, the iron-binding capacity of transferrin (Tf) is sufficient for scavenging all inorganic iron appearing in the circulation. However, in some situations, such as iron-overload associated with thalassemia (THS),[10-12] hemochromatosis[13-15] and other conditions of iron imbalance, not all serum iron might be bound to Tf[16]. ECLIP, denotes those forms of

---

- Z. Ioav Cabantchik, O. Kakhlon, G. Zanninelli and W. Breuer. Dept. of Biological Chemistry, Alexander Silberman Institute of Life Sciences, Hebrew University, Jerusalem 91904, Israel.

labile iron in extracellular fluids that are bound to ligands other than transferrin (Tf), hence, its identification as 'non-transferrin bound iron' (NTBI). The most obvious case of NTBI appearance in serum is in heavily iron-loaded patients whose Tf iron binding capacity is surpassed. On the other hand, the occurrence of NTBI in cases of less-than-full Tf saturation has remained difficult to explain [15,17,18]. This includes forms of NTBI that are present in the serum of iron overloaded patients, including forms that are virtually inaccessible to potent chelators such as deferoxamine (DFO) [19]. In principle, whenever large amounts of iron are abruptly introduced into the circulation, such as during oral or intravenous iron supplementation, [19] NTBI forms might appear, some transiently and some more persistently. The predictable serum ligands for ECLIP are citrate, phosphates, albumin and even non-specific sites in Tf. As with CLIP, NTBI apparently does not form fully coordinated complexes, so that it may potentially catalyze redox reactions that could give rise to reactive oxygen species and lead to oxidative damage [19]. Moreover, the cellular uptake of non-Tf iron complexes is not as rigorously regulated as that of Tf-iron, and hence it could give rise to cellular iron-overload [21,22]. Therefore, the presence of ECLIP in the circulation could have pathological consequences.

A third (mixed) component of LIP is associated with the red cell membrane of iron overloaded patients with transfusional hemosiderosis (THS). That form of LIP is apparently associated with the inner aspects of the membrane, since it is differentially accessible to permeant chelators such as deferiprone (L1) but not to the poorly permeant DFO [23].

Various attempts have been made to assess LIP in cells and biological fluids using different types of methodologies. The latter comprised either isolation and chemical characterization of the putative iron complexes or complexation of the metal associated with the LIP by chelators added to the biological object (cell, tissue or biological fluid) prior to, or after its disruption and/or partial separation into compartments or components . A major caveat of any analytical method based on cell disruptive steps is the likelihood of iron species de-compartmentalization and metal re-distribution among the cell components. Moreover, disintegration of the cells usually triggers oxidative processes, which further disrupt the native distribution of the labile metal among cell components.

The approach we have found to be most versatile for tracing LIP in biological compartments is one that allows not only its quantification in the *intact biological system* but also its on-line follow-up in the native setting (*in situ*) both spatially and temporally[24]. This approach, which we refer as FDM, fluorescence detection of metals, relies on the use of fluorescent metalo-sensors, namely fluorescent probes that undergo reversible spectroscopic changes upon binding of the metal[25]. FDM is in principle applicable to solutions as well as to membrane model systems and to living cells appearing either as individual entities or as functional units in a homogeneous or heterogeneous cell suspension, tissue or organ. An important property of FDM is its capacity to provide a quantitative measure of metal ions. In biological settings, quantification of metal concentration demands appropriate calibrations and adjustments for each experimental object. Once a quantitative relationship between a change in fluorescence signal and a metal concentration has been established, it is used for setting up a look-up table, proportionality constant or an equation for converting signal changes into actual metal concentrations

2. FDM. FLUORESCENCE DETECTION OF METALS

2.1 Iron sensors

FDM relies on probes or metal sensing devices that undergo swift changes in signal properties as a function of the metal concentration[24,25]. For that purpose, the metalosensors are designed to carry a signaling moiety (the fluorophore) linked to a guest (i.e. metal) binding site (the receptor) via a spacer. The most common means of metal modulation of fluorophore properties is by an interaction that is often redox in nature[26]. The metal binding moiety determines both the affinity and specificity of the sensor as well as the speed of the response to the metal. The first probes designed to meet the metal sensor criteria were fluorescent analogs of deferioxamine (DFO)),[27] DTPA[28] and transferrin[29] (Fig. 1).

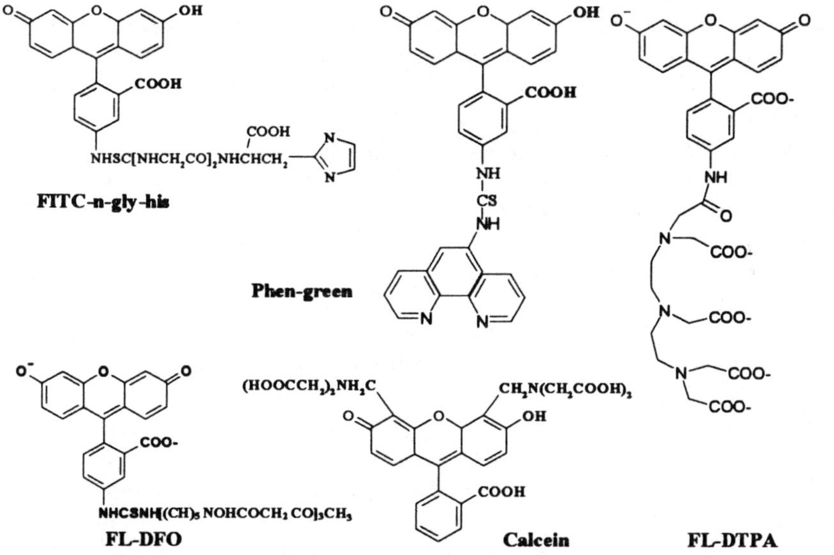

Fig. 1 Chemical structures of fluorescent metalo-sensors.
All the probes carry the green fluorescent fluorescein (FL) as the Fluorophore conjugated to a metal binding moiety or moieties: FITC-n-glyhis (or FL-glyhis) is obtained by reacting FITC (fluoresceinisothiocyanate) with N-gly-his resulting in a Cu-binding probe; phen green (or FL-phen) is obtained by reacting isothiocyanophenanthroline with Fl-amine (resulting in a Fe(II) binding probe), FL-DTPA is obtained by reacting DTPA-anhydride with amino-FL and DFO-green (or FL-DFO) is obtained by reacting FITC with deferrioxamine (DFO) yielding a Fe(III) binding probe. Calcein (CAL), which is composed of FL conjugated with an EDTA-like moiety, can bind di and trivalent metals[24].

Metalosensors such as DFO display specific and stoichiometric binding of Fe(III) and/or Fe(II) that results in marked quenching of the fluorophore[3]. The fluorophore alone, or fluorescent derivatives lacking the high-affinity metal binding moiety, are essentially unaffected by the metals even at concentrations that are higher by orders of magnitude. Fluoresceinated (FL) derivatives of the chelators DFO (DFO-green = FL-DFO) or phenanthroline (Phen-green or FL-phen)[31] were also prepared and are now offered commercially by Molecular Probes Inc. (Eugene, Oregon, USA). The different probes differ in their iron binding stoichiometry (1:1 for FL-DFO, 1:1 and 2:1 for FL-Tf and 3:1 for FL-phen), the specificity and speed of binding Fe(III) versus Fe(II) or other metals. However, in practice, in ambient conditions, the probes tightly complex both forms of the metal, as they are capable of oxidizing or reducing *in situ* one form of the metal to another - in the presence of O_2, DFO will rapidly oxidize the water soluble and low-affinity binding Fe(II) to the high-affinity binding Fe(III), while Fl-phen could slowly favor the reduction of Fe(III) to the high-affinity binding Fe(II). Thus, paradoxically, CAL binds Fe(III) with much higher affinity than Fe(II), yet the latter leads to a faster and more effective quenching of the fluorescence, due to a combination of Fe(II)'s better solubility and accessibility and the ability of the probe to promote oxidation of the bound Fe(II).

The quantitative relationship between metal binding and fluorescence quenching can be used for tracing iron quantitatively in extracellular fluids and in cell-free systems[24]. Unfortunately, the relatively high metal binding affinity of a given chelator might also limit its use as a probe for assessing CLIP and cell iron dynamics. This is because such a chelator might extract iron from otherwise non-labile sources such as ferritin, or inaccessible sources in membrane enclosed compartments or organelles, and thereby bias the measurements in addition to disrupting cell iron homeostasis. CLIP is comprised of various iron forms whose steady state levels might vary with time following intrinsic and extrinsic physiological signals. In our view, the ideal cell metalo-sensor is one that can provide a dynamic measure (quantitative or even semi-quantitative) of steady state cytosolic CLIP, which serves as the cross-roads of cell iron traffic. *In practical terms, the metal sensor should undergo spectral changes commensurate with the levels of CLIP and in a real time-scale that reflects upward and downward changes in cell iron content.*

The 1st generation of cell iron sensors is represented by the fluoresceinated analog of EDTA 4,4'-- Bis[N,N-bis(carboxymethyl)aminomethyl]fluorescein, named calcein (CAL) or fluorexon[2-4, 31-35]. CAL was originally designed as a divalent metal sensor, but was virtually abandoned due to its spectral insensitivity to Ca and Mg. On the other hand, the probe was found to be instrumental for assessing cell integrity due to its high quantum yield and relatively poor permeation across membranes[24]. As an intracellular probe, CAL is used primarily as the non-fluorescent hexa-acetoxymethyl ester analogue, CAL-AM, which readily permeates into cells and upon enzymatic hydrolysis generates the fluorescent impermeant CAL. A few cell types that express high levels of P-glycoprotein and have multidrug resistant character were found to be partially resistant to CAL-loading via CAL-AM, since the latter is apparently pumped out from those cells[34]. We have adopted CAL as a probe for monitoring

LIP dynamics, within living cells, across living cells, in solutions or in biological fluids[24]. Although CAL is relatively less iron specific than FL-Tf, FL-DFO or even FL-Phen, it is nonetheless more versatile than the others for assessing steady state CLIP levels, the mode of action of chelators in solution, in membrane model systems and in living cells.

The unique feature of CAL is the combination of properties that it displays:

a. high quantum yield; b. 1:1 stoichiometry of binding iron and commensurate quenching of fluorescence; c. rapid on/off kinetics of metal binding; d. relatively slow kinetics of Fe(III) binding in physiological solutions as compared to binding of Fe(II), despite the 10^{13} higher apparent affinity of CAL for Fe(III) than for Fe(II) (10^{24} as compared to 10^{11}); e. fast conversion of Fe(II) into Fe(III) upon binding to CAL in ambient conditions and its reversal (albeit slower) upon reduction with ascorbate; f. the possibility of loading the probe into cells as a hydrophobic precursor CAL-AM and its retention within cells as free CAL; g. relatively low toxicity of CAL loaded into mammalian cells; h. low selectivity for binding Fe as compared to Cu in solution, that is compensated by the prevalence of Fe versus Cu in cells. The basic properties of the CAL-based FDM are depicted in Fig. 2.

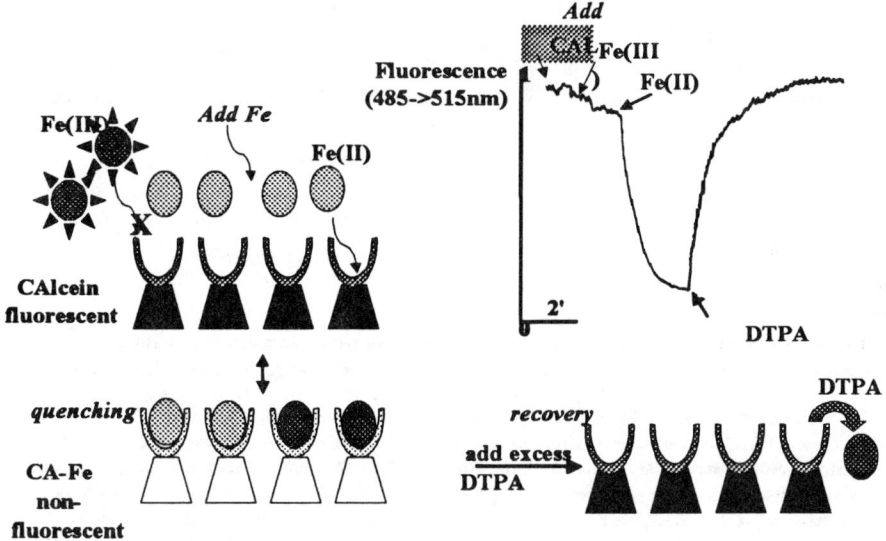

Fig. 2. The principle of FDM using CAL as the metalo-sensor.
The fluorescent CAL binds Fe(II) or Fe(III) (when complexed with agents such as NTA). Fe(III) salts prepared in physiological salt-media comprise ionic forms (filled stars) that are poorly accessible to CAL (indicated by X). In ambient conditions, Fe(II) (empty circles) binds CAL weakly but swiftly quenches fluorescence (measured at 480nm ex-515 nm em.), while becoming oxidized to Fe(III) (filled circles) and increasingly tightly bound to CAL. Thus, oxidation shifts the low-affinity CAL-Fe(II) to the high affinity CAL-Fe(III). Addition of excess DTPA (hatched rectangle) restores the original fluorescence by scavenging iron (II or III) from CAL-Fe.

2.2. Time-dependent changes of LIP in closed membrane model systems

Membrane vesicles isolated from cells or generated artificially can be used for assessing time-dependent changes in intravesicular LIP (Fig.3). For that purpose impermeant metal sensors such as CAL, Fl-DFO, FL-DTPA, Fl-Tf or FL-phen, are first encapsulated into vesicles (isolated from red cells or human placenta). The LIP is modulated by the following two major means: a. addition of permeant forms of iron (e.g. ferrous ammonium sulfate = FAS) that evoke a time-dependent rise in LIP b. addition of permeant chelators that evoke a fall in LIP by complexation of the intravesicular metal.

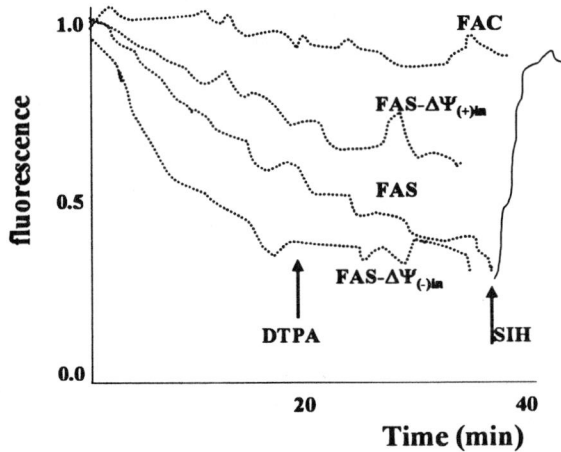

Fig. 3 Changes in LIP elicited by electrogenic ingress of iron and its accessibility to chelators.
Resealed ghosts were prepared from 20 μl packed human red cell ghosts encapsulated with CAL (50 μM) in Tris-NaCl (TRIS 10 mM, NaCl 75 mM, KCl 75, pH 7.0, 37°C) and resuspended in the same medium containing anti-CAL antibodies (to quench extra-vesicular CAL). Upon exposure to ferrous ammonium sulfate (FAS: 20 μM), but not ferric ammonium citrate (FAC: 50 μM), there is a time-dependent fall in fluorescence that represents the entry of Fe(II) into the CAL-containing compartment. Addition of the impermeant DTPA (200 μM) stops Fe(II) influx but does not affect intravesicular fluorescence, while addition of the permeant chelator SIH (50 μM) leads to fluorescence recovery. Substituting external Na for K ($K_o \gg K_i$) and adding valinomycin (1 μM) induces membrane depolarization ($\Delta\psi$ (+)) and decreases Fe(II) influx whereas replacing the external K for Na ($K_i > K_o$) hyperpolarizes the vesicle interior ($\Delta\psi$ (-)) and increases Fe(II) influx. Fluorescence is given in arbitrary units.

Validation that changes in fluorescence reflect commensurate changes in vesicular LIP requires the demonstration of vesicle impermeability to the metal sensor. That is accomplished by suspending the probe-laden vesicles in the presence of probe-quenching antibodies (anti-CAL or anti-FL, respectively) such that any fluorescence decay with time provides an indication of probe leakage. An increase in time-dependent quenching of the intra-vesicular fluorescence as a result of exposure to Fe (II) can therefore be equated to a rise in LIP due to ingress of iron. Ascorbate is usually supplemented in order to maintain the external iron in the reduced state, although that maneuver might also produce ROS and damage membranes. The Fe(II) taken up by vesicles will swiftly oxidize to Fe(III). Once the fluorescence has reached a steady state (low) level, extra-vesicular iron is removed by washing the vesicles or by sequestration by supplementation with impermeant chelators (e.g. DTPA). On the other hand, a time dependent rise in fluorescence evoked by addition of permeant chelators to CAL-iron laden vesicles, indicates chelator ingress and diminution of the LIP[24]. Determination of the intravesicular concentration of CAL allows conversion of the fluorescence signal into actual concentration of the LIP.

2.3. Time-dependent changes of CLIP in living cells

The major physiological source of cellular iron is serum TBI (Tf-bound iron), except for cells of the duodenum exposed to dietary non-TBI (NTBI). Although serum NTBI is detected in pathological conditions of iron overload, such as in HC and THS, it constitutes a heterogeneous mixture whose composition varies with the character of the disease. For convenience, but not necessarily because of physiological or pathological relevance, most uptake studies of NTBI have relied on substrates such as organic salts of iron (III) (citrate, ammonium citrate, NTA, etc.) or of iron (II) (sulfate, ammonium sulfate or even ascorbate), usually in buffered saline- medium.

For assessing LIP changes in living cells, the method based on FDM entails the initial loading of CAL into cells via the non-fluorescent membrane permeant precursor CAL-AM. Once inside cells, CAL-AM is hydrolyzed enzymatically to yield fluorescent CAL, some of which binds iron associated with LIP and undergoes quenching[2]. The target intracellular concentrations of CAL that are aimed for are in the 1-5 µM range[10]. Fluorescence-quenching anti-CAL antibodies (anti-CAL) are added to the external medium in order to eliminate signal associated with external probe that might have leaked out from the cells or remained surface-bound. Measurements on substrate-attached cells can be done by fluorescence microscopy under perfusion, while focusing the microscope-attached detection system on the cell layer. Fig. 4 depicts on-line changes in CLIP of K562 cells loaded with CAL and exposed to TBI and NTBI (in the form of FAS). The graph depicts the two major iron uptake pathways found in mammalian cells. In the first, Tf-bound iron serves as the substrate for receptor-mediated endocytosis of Tf leading to iron delivery into the cytosol via endosomes. In the second, NTBI (artificially given as FAS) is taken up (via DCT1) directly into the cytosol[3]. In K562 cells,

---

[2] This fraction of fluorescence quenched CAL (i.e. CAL-Fe), can be estimated by adding a permeant chelator and assessing the de-quenching by the rise in fluorescence.

[3] The iron transporter that mediates pathological NTBI ingress has not been identified for iron overload diseases, or for iron-deficiency conditions such as ESRD, in which EPO-treated dialysis patients have to be supplemented with parenteral iron.

iron derived from FAS or Tf enters into the LIP, as visualized by a time-dependent decrease in fluorescence[4].

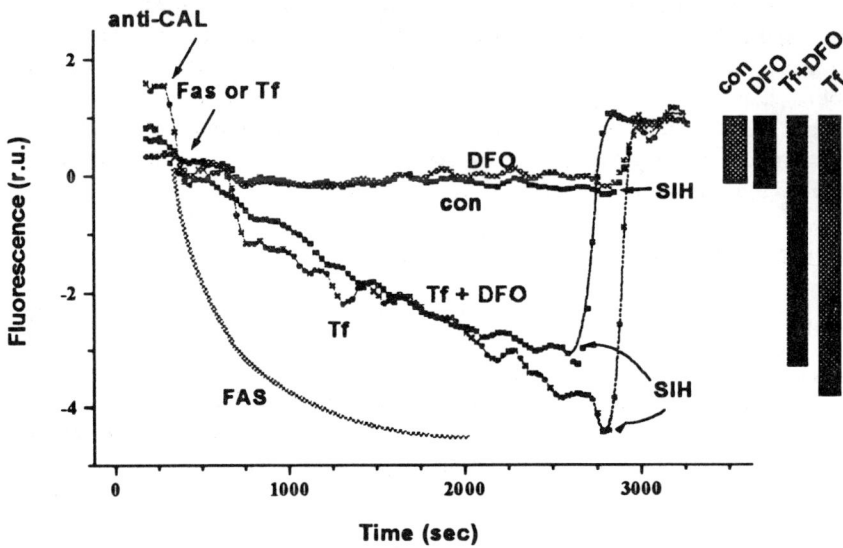

Fig. 4. Iron ingress into the CLIP of living K562 cells grown in suspension.
CAL-AM was added to cells in order to generate intracellular fluorescent CAL, some of which binds iron associated with the labile (or chelatable) iron pool LIP, whereupon quenched CAL-Fe is formed. Anti-CAL antibodies are added in order to quench extracellular probe. A stable line is established in control cells (con) or in those treated with 50 µM DFO. When supplemented with SIH (100 µM), the fluorescence associated with those cells rises, indicating the release of intracellular CAL-Fe by SIH and providing a measure for LIP. The latter can be increased in a time-dependent manner by supplying to the cells iron as Fe(II) or Tf, the 1st as FAS and the 2nd as Fe saturated Tf (20 µM each). Fe(II) enters into cells via DCT1 and Tf-Fe by receptor mediated endocytosis of Tf, both leading to a time-dependent decrease in the fluorescence signal, that can be restored by addition of the permeating SIH. The level of the latter attained at the point of addition of SIH is depicted as bars on the right. The height of each bar represents the level of fluorescence quenched by incoming iron, namely the intracellular concentration CAL-Fe, which for the control cells is 300nM.

In the short term, the iron taken up by the cells is apparently inaccessible to the poorly permeant DFO. An important property seen in Fig. 4 is that the final level of intracellular CAL-associated fluorescence attained after addition of a permeant chelator such as SIH is similar for all the experimental systems and higher than the initial level of CAL fluorescence.

---

[4] Murine erythroleukemia (MEL) cells that display a substantial uptake of Tf-^{55}Fe, showed quenching of cytosolic CAL only after prolonged incubation periods. (Epsztejn, Glickstein and Cabantchik, unpublished observations). The possibility that in erythroid cells Fe derived from Tf becomes associated with a CAL-inaccessible component is supported by other studies[35].

The increment in fluorescence signal attained (ΔFL) by addition of a permeant chelator, is equivalent to the cell iron originally complexed CAL (i.e. CAL-Fe), which is quantitatively related to the cellular labile iron pool (CLIP). Calculation of CAL-Fe concentrations and its relationship to CLIP requires knowledge of the intracellular concentration of CAL and the apparent Kd of the CAL-Fe complexes present in cells[2].

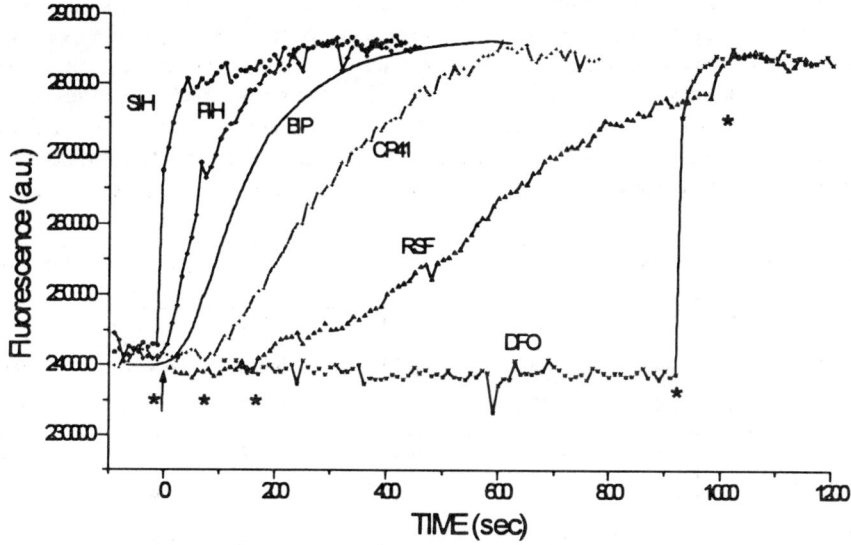

Fig. 5. Differential accessibility of iron chelators into the CLIP of living cells.
CAL is generated intracellularly by loading the cells with CAL-AM (as indicated in Fig. 4). The fluorescence-quenched CAL-Iron is revealed by addition of 100 μM of a permeant chelator CH (arrow ↑), whereupon fluorescence recovery ensues, representing the entry of the CH into the cell cytosol compartment and scavenging of the iron from CAL. The final value of fluorescence recovered is about the same for all the CH used. For systems with slowly permeating CHs such as DFO, the final fluorescence (arbitrary units, a.u.) is attained by addition of 100 μM of the permeant CH SIH, indicated by the star (*). SIH and PIH are arylaldehydeisonicotynoylhydrazones, BIP is bipyridyl, CP41 a hydrophobic hydroxypyridyn-on, and RSF a reversed siderophore. Modified from ref. 36&37.

2.4 Steady state levels of CLIP in living cells

CLIP is the first cellular compartment encountered by ionic iron entering into cells or recycled within cells and the one from which iron is distributed to all cell compartments or components. As a cross-road of iron traffic, CLIP is bound to be heterogeneous and variable with time, although, as a regulated entity, CLIP levels are likely to be maintained within certain limits.

The rationale of the approach for measuring CLIP rests on the assumptions that: a. CLIP comprises all the chelatable and rapidly exchangeable forms of cell Fe (II and III), b. all the CLIP forms are in dynamic equilibrium, c. addition of CAL shifts the equilibrium towards the

CAL-bound form of iron but does not affect cellular metabolism and d. the total CAL-Fe fraction is revealed by addition of excess amounts of the permeant chelator SIH. The actual concentration of CAL-Fe in a cell can be determined from the total CAL concentration measured in a given number of cells with the aid of an appropriate calibration curve and the determination of the number of cells and the mean cellular volume[2]. The relationship between labile iron and CAL-Fe depends on the apparent K_d of CAL-Fe in the cellular milieu. The latter parameter can be determined by fluorescence titration of internal CAL with cell Fe concentrations clamped with the divalent metal ionophore A23187 in the presence of set concentrations of external $Fe(II)$[5]. We have demonstrated that, depending on experimental conditions, measurements of cellular (CAL-Fe) and/or (Fe) provide, separately or together, representative measures or estimates of CLIP at discrete time points or on a continuous basis. A compilation of steady state CLIP values obtained with cell lines and primary cultures of hepatocytes exposed to iron overload and iron deprivation conditions is given in Table 1.

Cells & treatments	[CAL] (μM)	[CAL-Fe] (μM)	[CAL] (μM)	[Fe] (μM)	[CAL-Fe]+[Fe] (μM)	N-CLIP
K562	2.8	0.29	2.51	0.025	0.315	0.113
K562+DFO	2.8	0.15	2.75	0.012	0.162	0.058
U937	2.6	0.27	2.53	0.023	0.293	0.113
U937	9.3	1.60	8.70	0.040	1.640	0.176
U937 + DFO	15	1.05	13.95	0.017	1.067	0.071
U937 + FAC	6.1	2.40	4.70	0.112	2.512	0.412
Hepatocytes	3.7	0.70	3.00	0.051	0.751	0.203
Hepatocytes + DFO	4.6	0.20	4.40	0.010	0.210	0.046
Hepatocytes + FAC	2.9	2.10	0.80	0.578	2.678	0.923

Table 1. CLIP parameters in steady state conditions and following iron loading and chelator-mediated iron deprivation. The various parameters were calculated from data similar to those shown in Fig. 4 for K562 cells. The representative experiments were chosen on the basis of similar $[CA]_a$ attained in the various cells. Rat hepatocytes (1 day in culture conditions), U937 cells and K562 cells were incubated with either DFO (100μM) or FAC (100μM) for 18h in full growth medium. $[CA]_a$ and [CA-Fe] were experimentally determined for both treated and untreated cells. $[Fe]_a$ values were determined experimentally for untreated cells using the K_d value of CAL-FE of 0.22 μM (13). $[Fe]_a$ for treated cells was calculated using the K_d values determined for untreated cells. N-LIP represents the total LIP measured with CA i.e. [Fe]+[CA-Fe], normalized to the CA level attained in cells, i.e. ([Fe]+[CA-Fe])/$[CA_a]$ (modified from ref. 2&24.)

---

[5] The method for measuring of LIP is analogous to routinely used methods for estimating changes in intracellular concentrations of ions such as Ca(II) and H(I) (i.e. pH)[25]. A recently exploited property in metal sensor design is that of PIET (photoinduced intramolecular electron transfer).[30] This property can in principle be lowered significantly when the fluorophore is electron-deficient, and in some instances PIET can be enhanced upon metal binding, leading to increased rather than decreased fluorescence (i.e. quenching).

CLIP can be represented either by [CAL-Fe], {Fe} or the sum [CAL-Fe + Fe]. Since the CAL concentration in the cell might affect these parameters of LIP, we opted for representing CLIP by the value of [CAL-Fe+Fe] normalized to the total cell concentration of [CAL], to which we refer as N-CLIP. As seen in Table 1, in each cell type CLIP rises with iron load and diminishes following chelation treatment. In the case of U937 cells, the N-CLIP values obtained with cells loaded to different extents with CAL were similar, but not identical, indicating that CAL itself might, at relatively high concentrations, increase the CLIP by mobilizing iron. Thus, the CAL-based FDM, similarly to analogous fluorescence-based methods for assessing other metals, provides only approximate values of CLIP (Table 1) (reviewed in ref. 24).

2.5 Modulation of CLIP by ferritin expression: chemical and biological implications

The current concept of cell iron homeostasis assumes that CLIP is regulated by cellular feedback loops, which involve sensing CLIP levels by Iron Responsive Proteins (IRP's) and adjusting them by a coordinated modulation of expression of transferrin receptor (TfR) and FT[8]. The major trigger for the IRP response is generally a change in CLIP itself that eventually generates corrective measures, typical of feedback mechanisms. It is also assumed that cells, by actively manipulating the CLIP, could control biological processes in a manner independent of the original CLIP level and by means that override the IRP. Thus, for instance, repression of FT at the mRNA level by various oncogenes[38-44] has been interpreted to serve as a mechanism for increasing iron availability, which is a prerequisite for the oncogene-mediated stimulation of growth. However, in such studies a cause-effect relationship between FT and CLIP was never established. Namely, while alterations in CLIP are known to modulate FT levels, changes in FT expression have not been proved to precede changes in CLIP. Thus, demonstration of FT as an active modulator of CLIP and as an initiator of cellular responses (that depend on CLIP), is especially important since FT expression can also *respond* to many factors, besides CLIP itself[45].

Genetic manipulation of FT expression can potentially lead to FT overexpression[34, 46-48] or specific suppression of FT synthesis[49, 50]. The iron-independent overexpression of FT heavy subunit (H-FT) was accomplished by stable transfection of cells with the H-FT gene mutated in its iron responsive element (IRE). The iron-independent repression of FT heavy and/or light subunits was accomplished by antisense technology, based on oligodeoxynucleotides (ODN)[49] or on transient transfections with appropriate antisense constructs[50]. As can be seen in Fig. 6, manipulation of H-FT expression can influence CLIP levels: FT inhibition expanded the CLIP, while FT overexpression depleted it. Moreover, CLIP was inversely related to the expression level of the heavy subunit of FT (H-FT): In systems overexpressing H-FT, clones expressing higher levels of H-FT exhibited proportionally lower CLIP[46, 47] and in systems where the expression of H-FT was inhibited, CLIP expanded according to the intensity of H-FT repression[49]. The fact that in the genetically manipulated cells FT and CLIP were maintained at distinctly different levels from those of untreated cells, indicate that iron-independent modulation of FT expression can indeed override the physiological cellular mechanisms of CLIP homeostasis. Moreover, with the advent of novel means for monitoring CLIP *in vivo*, it is now possible to use FT as a tool for actively controlling CLIP and assessing the downstream cellular responses that are directly affected by it.

The consequences of CLIP changes were reflected in various cell properties. First, the production of reactive oxygen species (ROS) was repressed by FT overexpression whereas FT repression did the opposite, indicating the redox active nature of FT-modified CLIP. Moreover, antisense-evoked repression of both FT subunits (H-FT and FT's light subunit, L-FT) caused not only an increase in ROS production but also in protein oxidation[49]. The biological activity of the modified CLIP was also reflected in TfR surface expression[48, 49] that is modulated by IRP.

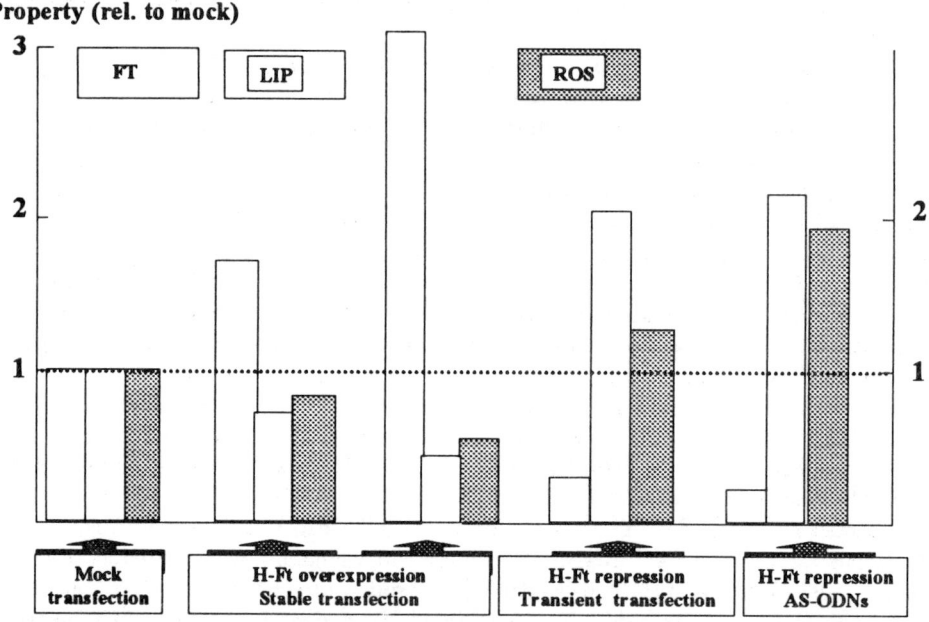

Fig. 6 Changes in H-FT expression and their effects on CLIP and on ROS production following a pro-oxidant challenge (H_2O_2, 5 µM). LIP was measured as described in the text and the rate of ROS production was measured using the fluorescent probe 2'-7'-carboxydichlorodihydrofluorescein- diacetate diacetoxymethyl ester.

Second, (i) Repression of FT synthesis facilitated the proliferation of human erythroleukemia K562 cells, pre-arrested at the G1/S phase and rendered them less dependent on external iron supply[49]. This study clearly linked the effect of FT repression on growth stimulation to the expansion of CLIP. Conversely, major H-FT overexpression in HeLa cells, accompanied by CLIP depletion, moderated overall cell growth[48]. (ii) In both HeLa[48] and mouse erythroleukemia (MEL) cells[34], overexpression of H-FT led to higher resistance to various toxic challenges such as ROS, iron overload, or cytostatic drugs. Whether the indicated resistance was dependent on H-Ft overexpression per se, or on the down modulation of CLIP, remains to be determined.

3. FDM. DETERMINATION OF LIP IN EXTRACELLULAR FLUIDS (ECLIP)

The question of the capacity of chelators to bind extracellular iron has been raised repeatedly. It has been reasoned that targeting this fraction of body iron would have a number of overwhelming advantages: (i) physically it is readily accessible to chelators, (ii) chemically, it is in equilibrium with all the intracellular pools and (iii) pharmacologically, being present in extracellular fluids, it is not only more amenable to chelation but also likely to circumvent the penetration of chelators into cells for metal scavenging, thus minimizing cytotoxicity. The obstacle to this approach is serum transferrin (Tf), which carries virtually all of serum iron (the serum transferrin iron-binding capacity is 45–82 µM, while serum iron concentration is normally 14–32 µM). Thus, the binding capacity is likely to be overwhelmed and exceeded only in cases of severe iron overload, where serum iron concentrations reach levels well above 30 µM. It is only in such cases, characterized by the appearance of NTBI, that serum iron may be expected to become chelatable.

3.1 Chelator accessibility of Transferrin-bound iron (TBI)

The capacity for removing iron from Tf has been studied for the two chelators in current clinical use, deferoxamine (DFO) and deferriprone (L1). DFO fails to remove significant amounts of iron from Tf, even though its affinity for iron exceeds that of Tf by many orders of magnitude[51]. The barrier appears to be kinetic, since chelation from Tf by DFO does occur, but requires many hours to reach appreciable levels[52,53]. Similarly to DFO, L1 does not efficiently remove iron from Tf *in vivo*. This conclusion is supported by several observations, despite early claims to the contrary, based on the use of high concentrations (4 mM) of L1[54]. The relative affinity of Tf exceeds that of L1 by an order of magnitude, as reflected by its pM value of 20–21,[55] compared to 19.4 for L1[51]. Therefore, transfer of iron from L1 to apo-Tf is favored at the estimated *in vivo* concentrations of the two ligands (10–50 µM), and was confirmed *in vitro* by direct and indirect binding determinations[19]. That such transfer can occur *in vivo* was indicated by the gradual rise of Tf-saturation (from 20 % to 80% over 6 hrs)[56] that followed L1 administration to an healthy individual. In conclusion, Tf does not constitute a likely source of readily available iron for either of the currently used chelators.

3.2 Chelator accessibility of non-Transferrin bound iron (NTBI) in serum.

In contrast to Tf-bound iron, the possibility that NTBI constitutes a chelatable serum pool of iron has been raised repeatedly and indirectly evaluated in several studies. NTBI levels in thalassemia patients were found to decrease during long-term therapy with DFO[12,57-60] and L1[60]. However, these studies did not address the question whether NTBI declined because it is directly available to the chelators, or because of a redistribution of the iron load resulting from the chelation therapy. The latter possibility is supported by the observation that lowering the iron load in hemochromatosis patients by regular venesection treatments, without the use of chelators, also lowered their NTBI levels[14,15].

3.3 Deferoxamine-chelatable iron in serum.

A direct approach to measuring "deferoxamine-chelatable iron" (DCI) in serum samples of iron-overloaded patients was made possible by the probe fluorescein-deferoxamine (FL-DFO), whose fluorescence is stoichiometrically quenched by iron[19]. The assay is schematized in Fig. 7. The advantage of FL-DFO is its high sensitivity, permitting the detection of DCI down to 0.5 µM. Unlike assays for NTBI, which require mobilization of the NTBI by high concentrations of intermediate-affinity, anionic chelators, DCI is obtained without manipulation of the serum sample, and thus represents the fraction of serum iron that is directly DFO-accessible. One drawback of virtually all fluorescence-based probes is their tendency to be affected by their environment, such as serum color or turbidity, pH, etc. This is overcome by performing the measurements in the absence and presence of an excess of unlabeled DFO, such that the difference between the two measurements represents the net iron-related signal[19].

DCI was found to be absent from controls but was found in the serum of most thalassemia major patients (range 1.5–8.6 µM), and only in a small minority of hereditary hemochromatosis patients (range 0.4–1.1 µM). In chelator-treated patients, DCI is a fluctuating entity, which (predictably) disappears during intravenous infusion of DFO, and rebounds once the chelator is lost from the circulation[19]. It is of interest that oral administration of L1 results in the appearance of DCI in many patients (Fig. 8, left panel). The capacity of L1 to scavenge iron from intracellular pools is well documented[37], so the L1-mediated increase in DCI is attributable to L1-iron complexes of intracellular origin that appear in the circulation and are detected by Fl-DFO. The latter is not surprising in view of the fact that DFO's affinity for iron in physiological solutions exceeds that of L1 by many orders of magnitude[51], resulting in efficient DFO-scavenging of L1-bound iron, as confirmed by direct binding studies *in vitro*[19]. These findings provide a mechanistic support for L1-DFO combination therapies based on the simultaneous, rather than sequential, administration of the two chelators[61-63].

3.4 Detection of NTBI.

Several assays for NTBI have been described[10,13,17,18,29,64,65] since its discovery in 1978 [10], all involving multiple steps such as mobilization of the NTBI by an intermediate-affinity chelator that does not attack TBI (Tf-bound iron), separation of the mobilized NTBI from TBI and quantification of the iron content. The development of sensitive fluorescent detectors with high affinity for iron has made it possible to introduce some technical simplifications to NTBI testing.

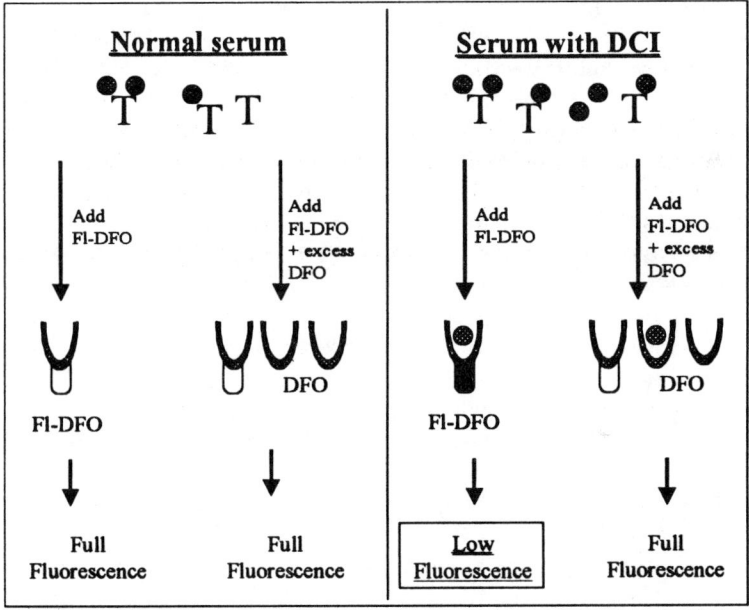

Fig. 7: Design of the DCI assay. Depiction of the steps of the assay for both normal (left) and DCI-containing sera (right). Iron is denoted by a filled circle and Tf molecules by 'T'. Step 1: Serum samples are mixed in 96-well plates with reagent A (Hepes-Buffered-Saline containing 2.5 µM Fluorescein-DFO, Fl-DFO) or reagent B (same as reagent A, but

containing excess, 100 µM, DFO). In reagent A, DFO-chelatable Fe binds to the Fl-DFO and quenches its fluorescence, whereas in reagent B the full fluorescence is obtained since the Fe is bound by the excess non-fluorescent DFO rather than by Fl-DFO. Step 2: Fluorescence is determined after 1hour incubation. In normal serum, the ratio of fluorescence of samples treated with reagent A and B approaches 1, whereas in Fe-containing serum, the fluorescence in reagent A is lower than in B, giving a ratio < 1. The ratio of the fluorescence readings (A/B) is inversely proportional to the concentration of DCI in the original sample[19].

The probe that has so far proven most useful for NTBI determinations is fluorescein labeled transferrin (Fl-Tf)[29]. Fl-Tf contains one Fluorescein per transferrin, obtained by reacting dichlorotriazinyl-fluorescein with apo-Tf at 1:1 stoichiometry. The probe undergoes quenching upon binding iron, similarly to Fl-DFO and calcein, and is used in a manner analogous to Fl-DFO (Fig. 7). It was developed specifically for the detection of NTBI mobilized with oxalate (10 mM), because, it binds oxalate-iron complexes rapidly and with high affinity, yet does not extract iron from Tf. An additional refinement involves the inclusion of Ga(III) in the mobilizing mix. Serendipitously, Ga(III) does not block the detection of iron by Fl-Tf, but significantly inhibits the iron-binding activity in the serum that might interfere with the detection. The mechanism of the resistance of Fl-Tf to Ga(III) inhibition is still unclear, however, it appears to be due to the specific localization of the Fluorescein moiety on the protein, which is determined by the reaction conditions used. The use of Ga(III) is essential, as it permits the detection of NTBI in sera with significant residual iron-binding capacity.

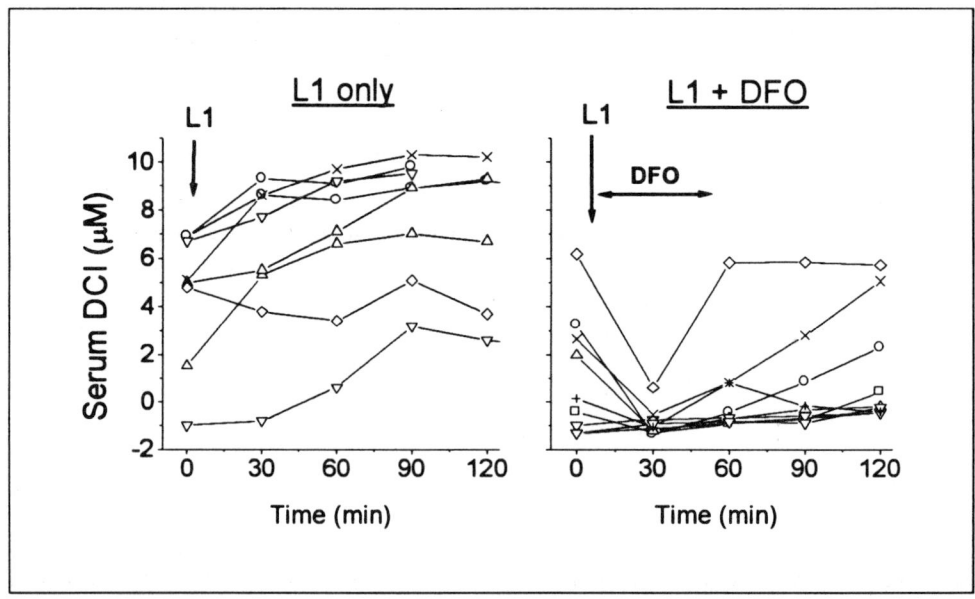

Fig. 8. Serum DCI following administration of L1 alone and in combination with DFO. The effect of L1 administration on DCI levels in the serum of thalassemia major patients was determined. Left (*L1 only*): Serum samples were taken immediately before (0 min.) and at intervals of 30 min. after administration of L1 (75 mg/ kg) *per os*, indicated by arrow. Right (*L1 + DFO*): The patients received L1 *per os* (indicated by arrow), and within 15 min were infused intravenously with DFO at a total dose 0.5 – 1.0 g given over approx. 60 min, as indicated by arrow. Serum samples were taken immediately before (0 min.) and at intervals of 30 min. after administration of L1 (75 mg/ kg). Each line represents an individual patient[19].

3.5 Can NTBI and apo-Tf exist simultaneously in serum?

The existence of NTBI in sera with Tf saturation below 100% might appear paradoxical, but it has been repeatedly observed in hemochromatosis[15,17,18] and dialysis patients[18]. Evidently, in these cases iron must be entering the circulation in a form that is inaccessible to Tf (and to DFO in hemochromatosis patients, as discussed above). Such an explanation could hold for iron introduced into the circulation from exogenous sources, such as intravenous iron supplements, which are administered as colloidal polysaccharide complexes that are poorly bound by Tf. These are apparently processed by macrophages and by the liver[21,66] and discharged into the circulation as Tf-accessible iron. Although most of the infused iron is cleared from the circulation within hours, a fraction might persist in the serum and later appear as NTBI. In the case of hemochromatosis, it is not inconceivable that NTBI is derived from hemosiderin formed within iron-loaded cells in the liver[67], pancreas and macrophages[68], which might have been released into the circulation by extrusion or as a result of cell damage. NTBI could also arise as a result of incomplete degradation of aged erythrocytes. It has been shown[69] that iron from monocyte-phagocytosed erythrocytes is released as a heterogeneous mixture composed of hemoglobin, ferritin and a low-molecular-weight fraction. This mechanism would apply particularly in situations such as the hemolytic anemias, macrophages are iron-burdened due to accelerated erythrocyte turnover.

The presence of unsaturated Tf in sera has represented a major technical obstacle to the precise measurement of NTBI, and has led to misleading results. In many cases of iron-overload (e.g. in hemochromatosis) Tf-saturation may vary between 45 - 100%, with variable degrees of residual iron-binding capacity. While the residual apo-Tf may not bind NTBI *in situ*, as discussed above, it can effectively scavenge NTBI after its mobilization by nitrilotriacetate or oxalate. Attempts to overcome this problem have involved blocking the endogenous apo-Tf with alternate metals that do not interfere with the detection, such as Co(III)[15,17] and Ga(III)[29]. These maneuvers have permitted detection of NTBI in sera with Tf-saturations in the normal range (< 45%), that would be judged NTBI-free by previous assays [15,17,29]

3.6 Relationship between DCI and NTBI

Interestingly, in a number of cases the presence of detectable NTBI is not accompanied by the presence of DCI. However, since NTBI is probably a heterogeneous mixture of iron complexes, its accessibility to chelators may be variable. Studies using high-resolution NMR analysis showed the chelation of NTBI by DFO to be a very slow process, requiring several hours even at 1 mM chelator concentration[70]. Similarly, DCI was undetectable with Fl-DFO in hemochromatotic sera, with Tf-saturation levels in the range of 80 – 85%. Yet, NTBI was present in these sera and could be detected after mobilization with oxalate[19]. In contrast, thalassemic sera, with Tf-saturation levels > 90%, were found to contain both DCI and NTBI that was mobilized with oxalate. These observations indicate that there are at least 2 forms of NTBI, one accessible to DFO and the other not. Hemochromatosis patients tend to have only the DFO-inaccessible form whereas β-thalassemic patients have both. Presumably these iron complexes differ in their ligand types and degrees of aggregation.

3.7 Potential applications of detection methods for ECLIP.

The attractive feature of the fluorescence-based assays is their ease of performance, making them potentially useful for routine screening of large numbers of serum samples. The availability of such tests would permit the undertaking of large-scale studies of the relationship between chelation treatment and DCI in thalassemic patients, and of the distribution, magnitude and kinetics of NTBI among various patient populations.

4. ACKNOWLEGMENTS

This work was supported in part by the Zalman Cohen Fund via the ARD of the Hebrew University of Jerusalem.

5. REFERENCES

1. Crichton, R.R. (1991). Inorganic Biochemistry of Iron Metabolism. New York, London, Toronto, Sydney, Tokyo, Singapore, Ellis Horwood
2. Epsztejn, S., Kakhlon, O., Breuer, W. Glickstein, H. and Cabantchik, Z.I.,(1997) A fluorescence assay for the labile iron pool (LIP) of mammalian cells. Anal. Biochem. 248: 31-40.
3. Breuer, V.W., Epstejn, S., Milgram, P. and Cabantchik, Z.I. (1995). Transport of iron and other related metals into cells as revealed by a fluorescent probe. Am. J. Physiol (Cell) 268: 1354-1361.
4. Breuer, W., Epstejn, S, and Cabantchik, Z.I. (1995). Iron acquired from transferrin by K562 cells is delivered into a cytoplasmic pool of chelatable iron(II). J. Biol. Chem. 270: 24209-24215
5. Jacobs, A. (1977) An intracellular transit iron pool. Blood 50: 4331-4336.
6. Kozlov, A.V., Yegorov, D.Y., Vladimirov, Y.A. and Azizova,O.A. (1992). Intracellular iton in liver tissue and liver homegenate-studies with electron paramagnetic resonance onthe formation of paramagnetic complexes with desferal and nitric oxide. Free Rad. Biol. Med. 13: 9-16.
7. Rothman, R. J., Serroni, A. and Farber, J. L. (1992). Cellular pool of transient ferric iron, chelatable by deferoxamine and distinct from Ft, that is involved in oxidative cell injury. Mol. Pharmacol. 42: 703-710

8. Eisenstein, R.D. (2000) Iron regulator proteins and the molecular control of mammalian iron metabolism Annu.Rev.Nutr.20:627–62
9. Aisen P, Wessling-Resnick M, Leibold EA. Iron metabolism (1999). Curr Opin Chem Biol 3:200-6
10. Hershko H, Graham G, Bates GW, Rachmilewitz E: Nonspecific serum iron in thalassaemia: an abnormal serum iron fraction of potential toxicity. Brit. J. Haematol. 1978; 40: 255-263.
11. Graham G, Bates GW, Rachmilewitz EA, Hershko C: Nonspecific serum iron in thalassemia: quantitation and chemical reactivity. Am. J. Hematol. 1979; 6:207-217.
12. Porter JB, Abeysinghe RD, Marshall L, Hider RC, Singh S: Kinetics of removal and reappearance of non-transferrin-bound plasma iron with deferoxamine therapy. Blood. 1996; 88:705-13.
13. Batey RG, Lai Chung Fong P, Shamir S, Sherlock S: A non-transferrin-bound serum iron in idiopathic hemochromatosis. Dig Dis Sci. 1980; 25:340-6.
14. Aruoma OI, Bomford A, Polson RJ, Halliwell B: Nontransferrin-bound iron in plasma from hemochromatosis patients: effect of phlebotomy therapy. Blood. 1988; 72:1416-9.
15. Loreal O, Gosriwatana I, Guyader D, Porter J, Brissot P, Hider RC: Determination of non-transferrin-bound iron in genetic hemochromatosis using a new HPLC-based method. J Hepatol. 2000; 32:727-33.
16. Breuer W, Hershko C, Cabantchik ZI. (2000) The importance of non-transferrin iron in disorders of iron metabolism. Transfusion Sci. 2000; 23: 185-92.
17. Gosriwatana I, Loreal O, Lu S, Brissot P, Porter J, Hider RC: Quantification of non-transferrin-bound iron in the presence of unsaturated transferrin. Anal Biochem. 1999; 273:212-20.
18. Breuer W, Ronson A, Slotki IN, Abramov A, Hershko C, Cabantchik ZI: The assessment of serum nontransferrin-bound iron in chelation therapy and iron supplementation. Blood. 2000; 95:2975-82.
19. Breuer W, Ermers, MJJ, Pootrakul P, Abramov A, Hershko C, Cabantchik ZI: Desferrioxamine-chelatable iron (DCI), a component of serum non-transferrin bound iron (NTBI) used for assessing chelation therapy. Blood. 97:792-8.
20. Halliwell B, Gutteridge JM: Role of free radicals and catalytic metal ions in human disease: an overview. Methods Enzymol. 1990; 186:1-85.
21. Wright TL, Brissot P, Ma WL, Weisiger RA: Characterization of non-transferrin-bound iron clearance by rat liver. J Biol Chem. 1986; 261:10909-14.
22. Kaplan J, Jordan I, Sturrock A: Regulation of the transferrin-independent iron transport system in cultured cells. J Biol Chem. 1991; 266:2997-3004
23. Shalev O, Repka T, Goldfarb A, Grinberg L, Abrahamov A, Olivieri NF, Rachmilewitz EA, Hebbel RP.1995. Deferiprone (L1) chelates pathologic iron deposits from membranes of intact thalassemic and sickle red blood cells both in vitro and in vivo. Blood. 86:2008-13
24. Cabantchik ZI, Breuer W, Slotki I, Beaumont C. (2001) Development and application of novel fluorescent assays for probing labile iron pools in biological systems. In: Badman DG, Bergeron RJ, Brittenham GM, eds. *Iron Chelators: New development strategies*. Ponte Vedra, FL: The Saratoga Group; 2000:353-383.
25. Tsien, R. Y. (1989) Fluorescent probes of cell signaling. *Ann.Rev. Neurosci*. 12: 227-53
26. Ramachandram, B. and A. Samanta, A.1998. How important is the quenching influence of the transition metal ions in the design of fluorescent PET sensors? Chemical Physics Letters 290: 9–16
27. Lytton, S.J., Mester, B., Libman, J., Shanzer, A. and Cabantchik, Z.I.. (1992). Monitoring of iron(III) removal from biological sources using a novel fluorescent siderophore. *Anal. Biochem*. 205: 326-333
28. Werts, M.H.V., Hofstraat, J.W., Geurts, F.A.J., and Verhoeven, J.W. 1997 Fluorescein and eosin as sensitizing chromophores in near- infrared luminescent ytterbium(III) , neodymium(III) and Terbium(III) chelates Chemical Physics Letters 276:196-201
29. Breuer W and Cabantchik ZI. (2001) A fluorescence-based one-step assay for serum non-transferrin bound iron (NTBI). *Anal. Biochem*. (submitted).
30. Ramachandram, B. and A. Samanta, A.1998. Transition Metal Ion Induced Fluorescence Enhancement of 4-(N, N-Dimethylethylenediamino)- 7- nitrobenz- 2- oxa- 1,3- diazole *J. Phys. Chem. 102*: 10579-10587.
31. Petrat F, Rauen U, de Groot H. 1999. Determination of the chelatable iron pool of isolated rat hepatocytes by digital fluorescence microscopy using the fluorescent probe ‚phen green SK .*Hepatology*. 29:1171-9.
32. Thomas F, Serratrice G, Beguin C, Aman ES, Pierre JL, Fontecave M, Laulhere JP . 1999 Calcein as a fluorescent probe for ferric iron. Application to iron nutrition in plant cells .J Biol Chem. 274:13375-83
33. Staubli A, Boelsterli UA 1998 .The labile iron pool in hepatocytes: prooxidant-induced increase in free iron precedes oxidative cell injury Am J Physiol. 274:G1031-7.
34. Epsztejn, S. Picard. V, Breuer, W.V., Glickstein, H. Slotki, I.N., Beaumont C. and Cabantchik, Z.I. (1999). Functional consequences of H-ferritin over-expression in transfected cells. *Blood*. 94:3593-3603..

35. Richardson, D.R. and Ponka, P. 1997 The molecular mechanisms of the metabolism and transport of iron in normal and neoplastic cells. *Biochim. Biophys Acta.* 1331: 1-40
36. Cabantchik, Z.I., Milgram, P., Glickstein, H. and Breuer, W., (1995) A method for assessing iron chelation in membrane model systems and in living mammalian cells. *Anal. Biochem.* 233: 221-227.
37. Zanninelli, G., Brissot, P., Hider, R.R., Konijn, A.P., Shanzer, A. and Cabantchik, Z.I. (1997). Chelation and mobilization of cellular iron by different classes of iron chelators. *Mol. Pharmacol.* 51: 842-852.
38. Bevilacqua, M.A., Faniello, M.C., Quaresima, B., Tiano, M.T., Giuliano, P., Feliciello, A., Avvedimento, V.E., Cimino, F. & Costanzo, F. (1997) A common mechanism underlying the E1A repression and the cAMP stimulation of the H ferritin transcription. *J Biol.Chem* 272: 20736-20741
39. Bevilacqua, M.A., Faniello, M.C., Russo, T., Cimino, F. & Costanzo, F. (1998) P/CAF/p300 complex binds the promoter for the heavy subunit of ferritin and contributes to its tissue-specific expression. *Biochem.J* 335: 521-525
40. Fuhrmann, G., Rosenberger, G., Grusch, M., Klein, N., Hofmann, J. & Krupitza, G. (1999) The MYC dualism in growth and death. *Mutat.Res.* 437: 205-217
41. Tsuji, Y., Kwak, E., Saika, T., Torti, S.V. & Torti, F.M. (1993) Preferential repression of the H subunit of ferritin by adenovirus E1A in NIH-3T3 mouse fibroblasts. *J.Biol.Chem.* 268: 7270-7275
42. Tsuji, Y., Akebi, N., Lam, T.K., Nakabeppu, Y., Torti, S.V. & Torti, F.M. (1995) FER-1, an enhancer of the ferritin H gene and a target of E1A-mediated transcriptional repression. *Mol.Cell Biol.* 15: 5152-5164
43. 43. Tsuji, Y., Moran, E., Torti, S.V. & Torti, F.M. (1999) Transcriptional regulation of the mouse ferritin H gene. Involvement of p300/CBP adaptor proteins in FER-1 enhancer activity. *J Biol.Chem* 274: 7501-7507
44. Wu, K.J., Polack, A. & Dalla Favera, R. (1999) Coordinated regulation of iron-controlling genes, H-ferritin and IRP2, by c-MYC. *Science* 283: 676-679
45. Pinero, D.J., Hu, J., Cook, B.M., Scaduto, R.C. & Connor, J.R. (2000) Interleukin-1beta increases binding of the iron regulatory protein and the synthesis of ferritin by increasing the labile iron pool. *Biochim.Biophys.Acta* 1497: 279-288
46. Picard, V., Renaudie, F., Porcher, C., Hentze, M.W., Grandchamp, B. & Beaumont, C. (1996) Overexpression of the ferritin H subunit in cultured erythroid cells changes the intracellular iron distribution. *Blood* 87: 2057-2064
47. Picard, V., Epsztejn, S., Santambrogio, P., Cabantchik, Z.I. & Beaumont, C. (1998) Role of ferritin in the control of the labile iron pool in murine erythroleukemia cells. *J.Biol.Chem.* 273: 15382-15386
48. Cozzi, A., Corsi, B., Levi S, Santambrogio, P., Albertini, A. & Arosio, P. (2000) Overexpression of wild type and mutated human ferritin H-chain in HeLa cells: in vivo role of ferritin ferroxidase activity. *J.Biol.Chem.* 275: 25122-25129
49. Kakhlon, A., Gruenbaum, Y. & Cabantchik, Z.I (2001) Repression of ferritin expression increases the labile iron pool, oxidative stress, and short term growth of human erythroleukemia cells. *Blood* 97: 2863-2871.
50. Kakhlon O, Gruenbaum Y, Cabantchik ZI. (2001) Repression of the heavy ferritin chain increases the labile iron pool of human K562 cells. *Biochem J*. 2001 Jun 1;356 (Pt 2):311-316.
51. Tilbrook, G.S. and Hider, R.C. Iron chelators for clinical use. In: Sigel, A. and Sigel, H., eds. *Metal Ions in Biological Systems: Iron Transport and Storage in Microorganisms, Plants and Animals*. 1998;Vol. 35:691-730. Marcel Dekker, New York.
52. Pollack S, Aisen P, Lasky FD, Vanderhoff G. (1976) Chelate mediated transfer of iron from transferrin to desferrioxamine. *Br J Haematol.* 34:231-5.
53. Pollack, S., Vanderhoff, G. and Lasky, F. (1977) Iron removal from transferrin- an experimental study. *Biochim. Biophys. Acta*, 497: 481-487.
54. Kontoghiorghes, G.J. (1995) New concepts of iron and aluminium chelation therapy with oral L1 (Deferiprone) and other chelators. *Analyst*, 120:845-851.
55. Aisen P, Leibman A, Zweier J. (1978) Stoichiometric and site characteristics of the binding of iron to human transferrin. *J Biol Chem*. 253:1930-7.
56. Evans, R.W., Sharma, M., Ogwang, W., Patel, K.J., Bartlett, A.N. and Kontoghiorghes. (1992) The effect of alpha-ketohydropyridine chelators on transferrin saturation in vitro and in vivo. *Drugs of Today*; 28 (Suppl. A):19-23.
57. Wang WC, Ahmed N, Hanna M. (1986) *J. Pediatr*; 108:552-7 Non-transferrin-bound iron in long-term transfusion in children with congenital anemias.
58. Ahmed NK. Hanna M. Wang W. (1986) Nontransferrin-bound serum iron in thalassemia and sickle cell patients. *Intern. J. Biochem*. 18:953-956.
59. Araujo A, Kosaryan M, MacDowell A, Wickens D, Puri S, Wonke B, Hoffbrand AV. (1996) A novel delivery system for continuous desferrioxamine infusion in transfusional iron overload. *Br J Haematol*; 93:835-7.

60. al-Refaie FN., Wickens, DG., Wonke B. Kontoghiorghes GJ. and Hoffbrand AV. (1992). Serum non-transferrin-bound iron in beta-thalassaemia major patients treated with desferrioxamine and L1. *Brit. J. Haematol.* 82:431-436.
61. Wonke, B., Wright, C. and Hoffbrand, A.V. (1998) Combined therapy with deferiprone and desferrioxamine. *Br. J. Haematol.*, 103:361-4.
62. Grady, R.W., Berdoukas, V.A., Rachmilewitz, E.A. and Giardina, P.J. Combining deferiprone and desferrioxamine to optimize chelation. 10th International Conference on Oral Chelators, Limassol, Cyprus, Mar.22-26, 2000.
63. Grady, R.W. and Giardina, P.J. Iron Chelation: Rationale for combination therapy. In: Badman, D.G., Bergeron, R.J. and Brittenham, G.M. eds. *Iron Chelators: New development strategies.* 2000; 293-310. The Saratoga Group, Ponte Vedra, FL.
64. Singh, S., Hider, R.C. and Porter, J.B. (1990) A direct method for quantification of non-transferrin-bound iron. *Anal. Biochem.* 186:320-323.
65. Evans, PJ, Halliwell, B. (1994). Measurement of iron and copper in biological systems: bleomycin and copper-phenanthroline assays. *Meth. Enzymol.* 233:82-89.
66. Scheiber B, Goldenberg H. (1998). The surface of rat hepatocytes can transfer iron from stable chelates to external acceptors. *Hepatology.* 27:1075-80.
67. Iancu TC, Deugnier Y, Halliday JW, Powell LW, Brissot P. (1997). Ultrastructural sequences during liver iron overload in genetic hemochromatosis. *J. Hepatol.* 27:628-38.
68. Simpson RJ, Deenmamode J, McKie AT, Raja KB, Salisbury JR, Iancu TC, Peters TJ. (1997): Time-course of iron overload and biochemical, histopathological and ultrastructural evidence of pancreatic damage in hypotransferrinaemic mice. *Clin. Sci. (Colch).* 93:453-62.
69. Moura E, Noordermeer MA, Verhoeven N, Verheul AF, Marx JJ. (1998). Iron release from human monocytes after erythrophagocytosis in vitro: an investigation in normal subjects and hereditary hemochromatosis patients. *Blood.* 92:2511-9.
70. Grootveld, M., Bell, J.D., Halliwell, B., Aruoma, O.I., Bomford, A. and Sadler, P.J. (1989) Non-transferrin bound iron in plasma or serum from patients with idiopathic hemochromatosis. *J. Biol. Chem.* 264:4417-22.

CARDIOPROTECTIVE EFFECT OF IRON CHELATORS

Chaim Hershko*, Gabriela Link and Abraham M Konijn

1. INTRODUCTION

Following the introduction of regular iron chelating therapy with deferoxamine (DFO), a significant improvement in the life expectancy of patients with transfusional iron overload has been witnessed. This is largely attributed to the prevention of heart disease in well-treated patients and, in a few, to the reversal of existing heart disease by aggressive intravenous DFO therapy[1,2]. Understanding the role of DFO in the protection of the heart from the toxic effects of iron is therefore a crucial issue in designing rational strategies of iron chelating treatment in thalassemic patients. Other chapters in this volume will cover in detail the mechanism of iron toxicity[3], non-transferrin-bound plasma iron[4], the chelatable pool of iron[5] and the results of iron chelating therapy[6]. In this chapter these topics will be only covered in the context of their relevance to myocardial iron toxicity and cardioprotection.

2. MECHANISM OF IRON OVERLOAD IN THALASSEMIA

In thalassemic patients, ineffective erythropoiesis results in a drastic increase in plasma iron turnover which is 10 to 15 times normal[7,8]. This wasteful production of non-viable RBC stimulates iron absorption in addition to the iron burden contributed by transfusions. The importance of inappropriately increased iron absorption is most convincingly demonstrated by the severe hemosiderosis observed in patients with thalassemia intermedia who have never been transfused. Most of the iron released to the

---

* Chaim Hershko Author, Shaare Zedek Medical Center, Jerusalem, Israel 91031, Gabriela Link and Abraham M Konijn, Department of Human Metabolism and Nutrition, Hebrew U. Hadassah Med. School, Jerusalem 91120 Supported by Grant # 197/99-2 of the Israel Science Foundation

Iron Chelation Therapy.
Edited by Chaim Hershko, Kluwer Academic/Plenum Publishers, 2002.

form of non-transferrin-bound plasma iron or NTBI. Our original description of the existence of a chelatable, low molecular weight plasma iron fraction in patients with severe iron overload has been confirmed by many subsequent studies using a variety of methods[9-15]. NTPI was shown to promote the formation of free hydroxyl radicals and to accelerate the peroxidation of membrane lipids *in vitro*[13]. Recent studies by Porter et al[15] have demonstrated that plasma NTPI is removed by intravenous DF therapy in a biphasic manner and that upon cessation of DFO infusion it reappears rapidly, lending support to the continuous, rather than intermittent, use of DFO in high risk patients. The rate of low molecular weight iron uptake by cultured rat heart cells is over 300-times greater than that of transferrin iron[16]. Moreover, unlike transferrin-iron uptake which is inhibited at high tissue iron concentrations by down-regulation of transferrin receptor production, non-transferrin iron uptake is increased by high tissue iron content[17]. Such uptake was shown to result in increased myocardial lipid peroxidation and abnormal contractility, and these effects were reversed by *in vitro* treatment with DFO[18]. Recognition of NTPI as a potentially toxic component of plasma iron in thalassemic siderosis has important practical implications for designing better strategies for the effective administration of DFO and other iron chelating drugs.

3. ORIGIN OF CHELATABLE IRON

The intracellular labile iron pool: Current models of iron acquisition, sequestration and storage by mammalian cells are based on a regulated adjustment of membrane transferrin receptor and cytosolic ferritin levels. Iron in transit between these two iron-binding proteins is believed to exist in a weakly bound low-molecular-weight complex which is also available for interaction with iron chelating drugs[19]. This chelatable labile iron pool (LIP) is assumed to be sensed by a cytosolic iron-responsive protein (IRP) which coordinately represses ferritin mRNA translation and increases transferrin receptor mRNA stability[20]. Efficient regulatory mechanisms prevent fluctuations in the size of the labile iron pool under conditions of moderate iron deprivation and iron loading. However, massive iron loading results in uncontrollable expansion of the chelatable pool, which fails to be matched by the sequestrating capacity of cellular ferritin[21]. This expanded LIP is an obvious target of intracellular iron chelation by drugs that are able to cross the barrier of the cytoplasmic membrane.

Role of RE and parenchymal iron stores : Although excess iron may be deposited in almost all tissues, most of it is found in association with two cell types: reticuloendothelial (RE) cells in the spleen, hepatic Kupffer cells and bone marrow, and parenchymal tissues such as the myocard, liver and endocrine cells. In contrast to RE cells in which iron accumulation is relatively harmless, parenchymal siderosis may result in significant organ damage. The source of iron and the proportion retained in ferritin stores or recycled into the circulation from the 2 cell types is quite different. RE cells have a limited ability to assimilate transferrin iron and they derive iron from the catabolism of hemoglobin in non-viable erythrocytes[22]. Most of this catabolic iron is recycled to plasma transferrin or NTPI within a few hours. In contrast, hepatic parenchymal cells maintain a dynamic equilibrium with plasma transferrin, with iron uptake predominating when transferrin saturation is high, and release when serum iron and transferrin saturation are low. Unlike RE cells, the turnover of parenchymal iron stores is extremely low. In general, iron overload associated with increased intestinal absorption such as hereditary hemochromatosis results in predominant parenchymal siderosis, whereas in iron overload caused by multiple blood transfusions the primary site of siderosis is the RE cells. Considerable redistribution of iron may take place subsequently.

Experimental and clinical observations indicate that the urinary excretion of iron chelated by DFO is derived mainly from RE cells. Studies in hypertransfused rats have shown that in contrast to hepatocellular radioiron excretion which is confined entirely to the bile, most of the radioiron excretion derived from RE cells is recovered in the urine . Moreover, when DTPA or IRC11, water-soluble synthetic chelators which do not enter cells easily, are employed in the same experimental model, there is no enhancement at all of hepatocellular iron excretion, but the enhancement of urinary RE radioiron excretion is similar, or higher than that observed previously with DFO[23]. Hence, DFO obtains iron for chelation by one of two alternative mechanisms; (a) in situ interaction with hepatocellular iron and subsequent biliary excretion, and; (b) chelation of iron derived from RBC catabolism in the RE system directly or following its release into the plasma in the form of NTPI with subsequent urinary excretion .

4. REGULATION OF HEART CELL IRON-PROTEINS

An essential feature of normal iron physiology is the capacity of cells to control the levels of the indispensible, but potentially hazardous iron which is associated with the (chelatable) labile iron pool, through a combination of constitutive (translational) and inducible (transcriptional) mechanisms. Detoxification of excess iron is achieved through its encasing and storage within the hollow, globular structure of the ferritin molecule. This is accomplished through a tight control of ferritin synthesis in response to cellular iron status as part of an intricate mechanism of post transcriptional coordinated regulation of iron acquisition, storage, and utilization. Ferritin subunit synthesis is regulated by translation of its mRNA. The elegant mechanism of interaction between cytosolic iron levels, activation of iron regulatory proteins IRP1 and IRP2, and their reciprocal binding to the iron responsive elements of ferritin and transferrin-receptor mRNAs have been described in several excellent recent reviews[24,25]. This mechanism controls the uptake and storage of cellular iron and, under normal conditions prevents the accumulation of cytosolic labile iron at toxic concentrations.

Heart cells are among the most vulnerable to the toxic effects of iron overload, and cardiac complications of transfusional siderosis are the most important cause of mortality in thalassemic patients[1,2]. This increased sensitivity of cardiac cells to iron poses the question whether the detoxifying mechanisms in response to increased iron influx into cardiac cells may not be as efficient in the heart as in other, less vulnerable organs because it may be unable to meet the accumulation of toxic labile iron at concentrations by an appropriate rise in ferritin synthesis. The above hypothesis may be tested by direct investigation of the iron regulatory mechanisms in cultured cardiomyocytes. Preliminary observations comparing the regulation of hepatic as against cardiac ferritin subunit synthesis in rats appear to support a fundamental difference between these organs: whereas liver-specific isoferritin profile is determined by a combination of pre- and post-translational mechanisms, in heart the post-translational regulation does not seem to be relevant, and the tissue-specific pattern is determined at the level of mRNA accumulation[26].

5. MECHANISM OF MYOCARDIAL IRON TOXICITY

The failure to reproduce hemochromatosis in animals has greatly hindered research on the pathogenesis of iron toxicity. Consequently, we have developed an experimental model of iron-loaded rat myocardial cells in culture for studying the harmful effects of iron and the protective effects of iron chelation[16,18,27-31]. A unique feature of these cultures is their ability to differentiate into spontaneously contracting cells, offering an opportunity to study simultaneously the biochemical and functional effects of iron toxicity in heart cells[16]. We have used ferric ammonium citrate at concentrations ranging from 5 to 40 micrograms/ ml to simulate the effect of NTPI on cultured heart cells. Employing iron-loading for 24 h, this method resulted in the uptake of 20 to 30 % of NTPI from the culture medium. It also resulted in marked aberrations in heart cell function expressed in decreased rates and amplitude of contractility, and severe arrhythmia resembling ventricular fibrillation[18]

The role of iron in promoting the conversion of superoxide and hydrogen peroxide into the highly toxic free hydroxyl radicals through the Haber-Weiss reaction has been extensively studied[32]. Metal-catalyzed oxidation systems involve the reduction of Fe(III) to Fe(II) and the production of active oxygen species which are capable of attacking amino-acid side chains, marking the damaged proteins for enzymatic degradation[33]. Increased lipid peroxidation is the most easily measurable effect of active oxygen species and is usually regarded as the most significant event in the pathogenesis of iron-induced cellular damage[34,35].

A number of cellular lipid membrane structures have been considered as possible targets of iron-induced peroxidative damage. Decreased latent activity of lysosomal enzymes implying increased fragility of the lysosomal membrane has been demonstrated *in vitro*[36] and in biopsies obtained from the livers of patients with primary and secondary hemochromatosis[37]. Our previous studies in rat heart cells, employing lysosomal β-hexosaminidase as an indicator of iron-induced lysosomal damage, have shown that in vitro iron-loading is associated with a marked increase in total ? -hexosaminidase activity, and that this may be attributed to increased lysosomal fragility as evidenced by the loss of lysosomal latency and increased free enzyme activity[38].

Another organelle implicated in iron toxicity is the cytoplasmic or, in the case of heart cells, <u>sarcolemmal membrane</u>. Our previous studies implicated sarcolemmal damage by the reversible decrease in beating rate, loss of transmembrane potential and severe arrhythmias produced by in vitro iron-loading of cultured, beating rat myocardial cells [18,27]. Our subsequent studies of sarcolemmal thiolic enzymes in iron-loaded heart cells[39] have shown a loss of 5'-nucleotidase and Na,K,ATPase activity attributed to the direct, or indirect (via lipid peroxidation products) effects of increased oxidative stress on sarcolemmal proteins[40,41].

However, of all heart cell organelles, the <u>mitochondria</u> play a primary role in the pathogenesis of peroxidative damage, since mitochondria consume about 90% of inhaled oxygen and since reactive oxygen species such as superoxide, hydrogen peroxide and the hydroxyl radical are formed in the mitochondria as physiological metabolites of the respiratory chain[42]. Indeed, studies by Bacon et al[43] in iron-loaded rat livers have shown decreased succinate-cytochrome c reductase (complex II+III), cytochrome c oxidase (complex IV), and ATP levels, and a decrease in oxygen consumption. More recent studies by Ceccarelli et al[44] have demonstrated a correlation between hepatic free iron levels in iron loaded rat liver, increased lipid peroxidation, and an energy-dissipating mitochondrial Ca^{2+}-cycling process which may also be responsible for the decrease in endogenous mitochondrial ATP content.

In line with previous reports by Bacon et al[43] our own studies in cultured, iron-loaded rat heart cells have shown[45] a marked loss in the activity of succinate dehydrogenase, NADH-ferricyanide reductase, succinate ubiquinone oxidoreductase (complex II), succinate cytochrome c oxidoreductase (complex II+III), and in particular rotenone sensitive NADH-cytochrome c oxidoreductase (complex I+III). Considering that complex III (ubiquinol cytochrome c oxidoreductase) activity was unaffected by iron-loading in these studies and in view of the dominant role of complex I in complex I+III activity[46,47], the sharp decrease in complex I+III activity must be attributed to complex I injury. The slightly discrepant results between previous studies in iron-loaded liver compared with our heart cell studies may be explained by a difference in the sensitivity of cardiac mitochondrial respiratory chain components to oxidative inactivation[48], a difference in cellular iron concentrations, or variations among some of the methods employed to measure respiratory complex activities. Thus, our results indicate that mitochondrial respiratory enzymes represent a major target of iron toxicity to the heart as well as the liver, lending futher support to the use of cultured heart cells as a model for studying the pathogenesis of iron-induced heart disease[16].

Iron-loading results in the generation of toxic hydroxyl radicals, as demonstrated <u>in vivo</u> in rats with chronic dietary iron overload using an electron spin resonance spin-trapping method[49]. A number of possible mechanisms may explain the observed loss of respiratory enzyme activity: (a) Free radicals may interact with lipid components of the mitochondrial membrane such as cardiolipin, which are particularly rich in polyunsaturated fatty acids and therefore very susceptible to peroxidative damage[42]. Loss of cardiolipin and other vital lipid constituents of the mitochondrial membrane may indirectly impair respiratory chain assembly and activity[50,51] ; (b) Peroxidation of polyunsaturated fatty acids results in the formation of highly reactive aldehydes such as malonyldialdehyde and 4-hydroxynonenal. These reactive aldehydes form covalent links with proteins, or protein adducts[40], involving mainly the amino group of lysine residues and sulfhydryl groups, resulting in altered enzyme activity. Lastly; (c) the direct binding of Fe(II) to a metal-binding site on the respiratory enzyme and its reaction with H_2O_2 yields active oxygen species leading to the conversion of some amino acid residues to carbonyl derivatives, loss of catalytic activity and increased susceptibility to proteolytic degradation[33]. A vicious circle may be created by the oxidative inactivation of enzymes participating in the mitochondrial respiratory chain wherein loss of enzyme activity may lead to incomplete reduction of molecular oxygen and increased formation of free radicals [52-54].

The toxicity of iron to heart cells is not a simple function of iron concentrations. It may be aggravated or inhibited by a number of coexistent variables: Reduction of ferric to ferrous iron promotes hydroxyl radical formation via the iron-driven Haber-Weiss reaction. <u>Ascorbic acid</u>, a natural reducing agent accelerates iron-induced lipid peroxidation in biological systems at low concentrations, but acts as an antioxidant at high concentrations[29,56,57]. Moreover, clinical observations indicate that ascorbate supplementation may aggravate or accelerate the development of cardiac disease in patients with iron overload[58]. Conversely, α<u>-tocopherol</u>, a natural lipid-soluble antioxidant, is able to interrupt the chain reaction of membrane lipid peroxidation initiated by free radicals and to interfere with iron-induced lipid peroxidation in liposomes [28,29,57]. Finally, as shown in our heart cell culture studies, <u>deferoxamine</u> removes iron directly from iron loaded heart cells, inhibits lipid peroxidation[27,28] and reverses the abnormalities in cellular contractility and rhythmicity induced by iron.

6. REVERSAL OF MYOCARDIAL IRON TOXICITY BY CHELATING TREATMENT IN VITRO

That the observed loss of mitochondrial inner membrane respiratory enzyme activity was a specific effect of iron toxicity is clearly demonstrated by the complete reversal of enzyme inhibition by <u>in vitro</u> iron chelating therapy (Figures 3 A and B). This beneficial effect of DFO is in line with its ability to prevent <u>in vivo</u> hydroxyl radical formation in iron-loaded rats[49]. Likewise, iron chelating therapy in iron-loaded heart cells results in the restoration of sarcolemmal thiolic enzymes such as 5'-nucleotidase and Na,K,ATPase[39]. Moreover, DFO treatment of cultured iron-loaded beating heart cells abolishes a wide range of functional abnormalities induced by iron, such as the irregular shape and duration of action potentials, abnormal excitatory depolarization and inefficient excitation-contraction couplings[18,27]. These observations are the experimental analogues of acute iron poisoning and are relevant to the ability of DFO to prevent the severe cardiac complications and mortality associated with accidental iron ingestion[55]. The design of these studies does not allow a distinction between reversible inactivation of the enzymes as against their replacement by newly produced intact protein, although the latter explanation is more reasonable. Although mitochondria and the sarcolemmal membrane are both affected by iron toxicity in these short term in vitro studies, it is more likely that in the context of chronic iron overload mitochondrial damage is the most improtant single variable associated with abnormal myocardial function.

7. CLINICAL EVIDENCE OF CARDIOPROTECTIVE EFFECT

7.1. Impact on survival: The impact of DFO treatment on the life expectancy of thalassemic patients in general and cardiac mortality in particular is eloquently demonstrated by comparison of survival in well chelated versus poorly chelated patients. In a major study of 1127 thalassemic patients at 7 Italian teaching hospitals it was shown that 70% of patients born before 1970 and hence prior to the modern era of chelation survived to the age 20 y. compared to 88% in patients born after 1970 and therefore receiving effective chelation from an early age[59]. Most of the improvement in survival was attributed to decreased cardiac mortality. This cohort-of-birth related improvement in survival is reflected in a mirror-like inverse decrease in cardiac mortality, supporting the assumption that prevention of cardiac mortality is the most important beneficial effect of DFO therapy . A subsequent update on this group of patients[60] has shown that the prevalence of heart failure and diabetes declines with subsequent birth cohorts. Conversely, hypothyroidism is becoming more frequent. Overall, diabetes was present in 5.4%, heart failure in 6.4%, thrombosis in 1.1%, HIV infection in 1.8%, and hypogonadism in 55% of those reaching pubertal age. This update on the cohort of patients described in earlier reports confirms the remarkable improvement in life expectancy comparing successive cohorts of thalassemic patients. This improvement is mainly due to decreased cardiac mortality. Conversely, mortality caused by other complications of thalassemia has not decreased. Improved survival in well chelated thalassaemic patients has also been reported in several other major studies from the U.K.[61] and North America[2,62,63]. The increasing incidence with age of hypothyroidism, and the emergence of HIV as a significant cause of mortality has also been noticed in a group of thalassemic patients living in the New York area[64].

7.2. Reversal of established heart disease. The strongest direct evidence supporting the beneficial effect of DFO on hemosiderotic heart disease is the reversal of established myocardiopathy in some far-advanced cases. Earlier experience in hereditary hemochromatosis has shown that the myocardiopathy of iron overload is potentially curable by effective iron mobilization through phlebotomy. However, in transfusional hemosiderosis, the course of established myocardial disease was uniformly fatal and, until recently, believed to be non-responsive to iron chelating therapy. Several reports indicate that such patients may still be responsive to aggressive chelating treatment.

Marcus et al[65] described first in 1984 the reversal of established symptomatic myocardial disease in 3 of 5 patients by continuous high-dose (85-200 mg/kg/d) i.v. DFO therapy at the cost of severe reversible retinal toxicity. Likewise, Hyman et al described in 1985 two patients in whom standard s.c. DFO treament was reinforced by intermittent

48 h terms of intensive (0.5 gm/h) i.v. DFO who were alive and well 7 and 8 yeas after the onset of symptomatic heart failure[66]. These early observations were followed by an increasing recognition of the extreme importance of aggressive iron chelating therapy in thalassemic patients with established myocardiopathy in view of the dismal prognosis of siderotic myocardiopathy on one hand , and its potential reversibility by intensive chelation on the other. The development of intensive chelating programs employing continuous i.v. administration of DFO has been facilitated by the experience gained in the use of implantable venous access (Port-a-Cath) catheters, as well as the introduction of convenient portable and disposable balloon pumps (Baxter Healthcare Ltd, Newbury , UK[69]). In most cases, reversal of symptomatic myocardiopathy has been achieved without significant drug toxicity[66-75].

Because of a lack of uniformity in the dose of DFO used; continuous versus intermittent i.v. therapy; duration of treatment; methods of assessing clinical response; and the limited number of patients involved in most of these reports, no clear conclusions regarding optimal methods of treatment can be derived from this cumulative experience. However, in their recent description of cumulative experience of continuous 24-hour DFO treatment in 17 patients over a period of 16 years[68], several points were emphasized by Davis and Porter:

(i) Cardiac arrhythmias, previously unresponsive to medical treatment were reversed in 6 of 6 patients. This occurred in some cases within a few days of starting treatment and can not be attributed to normalization of iron stores but to the removal of a toxic labile iron pool by continuous DFO infusion, in analogy with our observations in heart cell cultures[18].

(ii) Left ventricular ejection fraction improved in 7 of 9 patients, and survival in patients with demonstrable cardiac disease was 62% at 13 years, with deaths occurring only in patients who failed to comply with therapy. These survival data represent a major improvement compared with historical controls.

(iii) Continuous 24 DFO administration has a distinct advantage over discontinuous 8 or 12 h i.v treatment. As NTBI reaccumulates rapidly after discontinuation of i.v. DFO (15), on theoretical grounds only continuous infusion offers better protection against iron toxicity. There is no evidence to indicate that by increasing DFO dose NTBI is removed more rapidly or that treatment outcome is improved. Indeed, the excellent results of Davis and Porter were achieved using maximal doses of 50 mg/kg, employing the therapeutic index of DFO to serum ferritin ratio of 0.025 as recommended for conventional subcutaneous DFO treatment[76].

The above experience should be the guiding principle in every case in which heart transplantation is considered as an alternative treatment modality in thalassemic patients with symptomatic heart disease. A decision to transplant should only be made after careful consideration of predictors of survival[77] and only after a therapeutic trial of intravenous iron chelating treatment performed under optimal conditions and of sufficient duration[78].

8. UNRESOLVED PROBLEMS AND FUTURE OBJECTIVES

Because damage to the heart is the most important cause of mortality in thalassemic patients with secondary iron overload, its prevention should be the main objective of chelating therapy and the most important determinant in monitoring the usefulness of an iron chelating program. Despite the spectacular reversal of myocardial disease in a limited number of patients, prevention of heart disease is a more sensible goal than its reversal. Of all the variables affecting cardiac involvement, compliance is by far the most important obstacle[79]. Other significant determinants of cardiac disease are older age; late introduction of chelating treatment; high liver iron and serum ferritin and ; number of transfusions[80]. Such patients deserve special attention and active measures to decrease liver iron and serum ferritin to within a safe range where the risk of developing heart disease is minimal[2]. Unfortunately, even in countries with excellent facilities and major centers specializing in the care of thalassemic patients, these goals are still far from being achieved. In the year 2000 about 50% of UK thalassemic patients die before reaching the age of 35 years, mostly of cardiac complications in non-compliant subjects[81]. Similar rates of mortality and non-compliance are reported from specialized treatment centers in other countries.

Some of the technical developments introduced in recent years and discussed earlier, including implantable devices allowing easy access for 24 h administration of intravenous DFO, and disposable infusors with weekly home-delivery, may allow not only more efficient treatment but also improved compliance. The oral chelator L1 is less effective than DFO and is unable to prevent cardiac mortality in patients with established heart disease[82]. However, the effects of L1 and DFO are additive and their combined use has been shown to be effective in increasing iron excretion and decreasing serum ferritin in patients formerly failing to respond to single drug therapy[83]. As emphasized by Modell et al[81], patient-specific adjustment of chelating therapy may still salvage a large proportion of adolescent thalassemic patients presently labeled as "non-compliant" if they are offered a range of these new and improved methods of drug delivery.

REFERENCES

1. Gabutti V, Piga A: Results of long-term iron chelating therapy. *Acta Haemat* 1996; **95**: 26-36.
2. Olivieri NF, Nathan DG, MacMillan JH, Wayne AS, Liu PP, McGee A, Martin M, Koren G, Cohen AR. Survival in medically treated patients with homozygous β-thalassemia. *N Engl J Med* 1994; **331**: 574-578.
3. Pietrangelo A. Mechanism of iron toxicity. In Chaim Hershko Ed. Iron Chelating Therapy. Kluwer Academic/Plenum Publishers. New York.
4. Brissot P. Role of NTBI in the pathogenesis of iron toxicity and hemosiderosis. In Chaim Hershko Ed. Iron Chelating Therapy. Kluwer Academic/Plenum Publishers. New York.
5. Cabantchik ZI. Identification and quantitation of the chelatable iron pool. In Chaim Hershko Ed. Iron Chelating Therapy. Kluwer Academic/Plenum Publishers. New York.
6. Porter J. Results of long-term iron chelating therapy with deferoxamine. In Chaim Hershko Ed. Iron Chelating Therapy. Kluwer Academic/Plenum Publishers. New York.
7. Cook JD, G. Marsaglia JW, Eschbach et al. Ferrokinetics: A biologic model for plasma iron exchange in man. *J. Clin. Invest.* 1970; **49**: 197-205.

8. Hershko C, Rachmilewitz EA. . Mechanism of desferrioxamine-induced iron excretion in thalassaemia. *Brit J Haematol* 1979; **42**: 125-132.
9. Hershko C, Graham G, Bates GW, Rachmilewitz EA. Non-specific serum iron in thalassaemia: an abnormal serum iron fraction of potential toxicity. *Brit J Haematol* 1978; **40**: 255-263.
10. Singh S, Hider RC, Porter JB. A direct method for quantification of non-transferrin bound iron (NTBPI) *Anal Biochem* 1990; **186**: 320-323.
11. Anuwatanakulchai M, Pootrakul P, Thuvasethakul P, Wasi P. Non-transferrin plasma iron in ? -thalassaemia/ HbE and haemoglobin H diseases. *Scand J Haematol* 1984; **32**: 153-158 .
12. Wagstaff M., Peters SW, Jones BM, Jacobs A. Free iron and iron toxicity in iron overload. *Brit J Haematol* 1985; **61**: 566 -567.
13. Gutteridge JMC, Rowley DA, Griffiths E, Halliwell B. Low-molecular-weight iron complexes and oxygen radical reactions in idiopathic haemochromatosis. *Clin Sci* 1985; **68**: 463-467.
14. Al-Refaie, FN, Wickens DG, Wonke B, Kontoghiorghes GJ, Hoffbrand AV. Serum non-transferrin-bound iron in beta-thalassaemia major patients treated with desferrioxamine and L1. *Brit J Haemat* 1992; **82**: 431-436.
15. Porter JB, Abeysinghe RD, Marshall L, Hider RC, Singh S. Kinetics of removal and reappearance of non-transferrin-bound plasma iron with deferoxamine therapy. *Blood* 1996; **88**: 705-713.
16. Link, G, Pinson A, Hershko C. Heart cells in culture: a model of myocardial iron overload and chelation. *J Lab Clin Med* 1985; **106**: 147-153.
17. Randell, EW, Parkes JG, Olivieri NF, Templeton DM. Uptake of non-transferrin-bound iron by both reductive and nonreductive processes is. modulated by intracellular iron. *J Biol Chem* 1994; **269**: 16046-16053.
18. Link G, Athias P, Grynberg A, Pinson A, Hershko C. Effect of iron loading on transmembrane potential, contraction and automaticity of rat ventricular muscle cells in culture. *J Lab Clin Med* 1989; **113**: 103-111.
19. Rothman RJ, Serroni A, Farber JL. Cellular pool of transient ferric iron, chelatable by deferoxamine and distinct from ferritin, that is involved in oxidative cell injury. *Molec Pharm* 1992; **42**: 703-710.
20. Klausner RD, Rouault TA, Harford JB. Regulating the fate of mRNA: the control of cellular iron metabolism. *Cell* 1993; **72**: 19-28.
21. Breuer W, Epsztejn S, Cabantchik ZI. Dynamics of the cytosolic chelatable iron pool of K562 cells. *FEBS Letters* 1997; **382**: 304-308.
22. Hershko C, Weatherall JD. Iron chelating therapy. *CRC Critical Reviews in Cinical Laboratory Sciences*. CRC Press Inc. Boca Raton, Florida. Vol 26, Issue **4**, 1988: 303-345.
23. Rivkin G, Link G, Simhon E, Cyjon RL, Klein JY, Hershko C. IRC11, a new synthetic chelator with selective interaction with catabolic red blood cell iron. Evaluation in hypertransfused rats with hepatocellular and reticuloendothelial radioiron probes and in iron-loaded rat heart cells in culture. *Blood* 1997; **90**: 4180-4187.
24. Leibold EA, Guo B. Iron-dependent regulation of ferritin and transferrin receptor expression by the iron-responsive element binding protein. *Ann Rev Nutr* 1992; **12**: 345-368.
25. Klausner RD, Rouault TA, Harford JB. Regulating the fate of mRNA: The control of cellular iron metabolism. *Cell* 1993; **72**: 19-28.
26. Cairo G, Rappocciolo E, Tacchini L, Schiaffonati L. Expression of the genes for the ferritin H and L subunits in rat liver and heart. Evidence for tissue-specific regulations at pre- and post-translational levels. *Biochem J* 1991; **275**:813-816
27. Moreb J, Hershko C, Hasin Y. Effects of acute iron loading on contractility and spontaneous beating rate of cultured rat myocardial cells. *Basic Res in Cardiol* 1988; **83**: 360-368.
28. Hershko C, Link G, Pinson A. . Modification of iron uptake and lipid peroxidation by hypoxia, ascorbic acid, and ? -tocopherol in iron-loaded rat myocardial cell cultures. *J Lab Clin Med* 1987;**110**: 355-361.
29. Link G, Pinson A, Kahane I, Hershko C. . Iron loading modifies the fatty acid composition of cultured rat myocardial cells and liposomal vesicles: Effect of ascorbate and ? -tocopherol on myocardial lipid peroxidation. *J Lab Clin Med* 1989; **114**: 243-249.
30. Hershko C, Link G, Pinson A, Peter HH, Dobbin P, Hider RC. Iron mobilization from myocardial cells by 3-hydroxypyridin-4-one chelators: Studies in rat heart cells in culture. *Blood* 1991; **77**: 2049-2055.
31. Link G, Athias P, Grynberg A, Hershko C Pinson A. Iron loading modifies ? -adrenergic responsiveness of cultured ventricular myocytes. *Cardioscience* 1991; **2**: 27-30.
32. Halliwell B, Gutteridge JMC. Role of free radicals and catalytic metal ions in human disease: an overview. in L Packer, AN Glazer eds. *Methods in Enzymology* 1990; **186**: 1-85.

33. Stadtman ER. Metal ion-catalyzed oxidation of proteins: biochemical mechanism and biological consequences. *Free Radic Biol Med* 1990; **9**: 315-325.
34. Goddard JG, Sweeney GD. Ferric nitriloacetate: a potent stimulant of in vivo lipid peroxidation in mice. *Biochem Pharm* 1983; **32**: 3879-3882.
35. Dillard CJ, Downey JE, Tappel AL. Effect of antioxidants on lipid peroxidation in iron-loaded rats. *Lipids* 1984; **19**: 127-33.
36. Mak IT, Weglicki WB. Characterization of iron-mediated peroxidative injury in isolated hepatic lysosomes. *J Clin Invest* 1985; **75**: 58-63.
37. Seymour CA, Peters TJ. Organelle pathology in primary and secondary hemochromatosis with special reference to lysosomal changes. *Brit J Haemat* 1978; **40**: 239-253.
38. Link G, Pinson A, Hershko C. Iron loading of cultured cardiac myocytes modifies sarcolemmal structure and increases lysosomal fragility. *J Lab Clin Med* 1993; **121**: 127-134.
39. Link G, Pinson A, Hershko C: The ability of orally effective iron chelators dimethyl- and diethyl-hydroxypyrid-4-one and of deferoxamine to restore sarcolemmal thiolic enzyme activity in iron-loaded heart cells. *Blood* 1994; **83**: 2692-2697.
40. Houglum K, Filip M, Witztum JL, Choikier M. Malondialdehyde and 4-hydroxynonenal protein adducts in plasma and liver of rats with iron overload. *J Clin Invest* 1990; **86**: 1991-1998.
41. Schauenstein E, Esterbauer H, Zollner H. *Aldehydes in Biological Systems*. Pion Limited, London. 1977, p 102.
42. Richter C, Gogvadze V, Laffranchi R, Schlapbach R, Schweizer M, Suter M, Walter P, Yaffee M : Oxidants in mitochondria: from physiology to diseases. *Biochim Biophys Acta* 1995; **1271**: 67-74.
43. Ceccarelli D, Gallesi D, Giovannini F, Ferrali M, Masini A. Relationship between free iron level and rat liver mitochondrial dysfunction in experimental dietary iron overload. *Biochem Biophys Res Comm* 1995; **209**: 53-59.
44. Bacon BR, O'Neill R, Britton RS. Hepatic mitochondrial energy production in rats with chronic iron overload. *Gastroenterology* 1993; **105**: 1134-40.
45. Link G, Saada A, Pinson A, Konijn AM, Hershko C. Mitochondrial respiratory enzymes are a major target of iron toxicity in rat heart cells. *J Lab Clin Med* 1998; **131**: 466-474.
46. Robinson BH, Glerum DM, Chow W, Petrova-Benedict R, Lightowlers R, Capaldi R. The use of skin fibroblast cultures in the detection of respiratory chain defects in patients with lactic acidemia. *Ped Res* 1990; **28**: 549-555.
47. Hatefi Y. Preparation and properties of NADH:ubiquinone oxidoreductase (complex I) ECI6.5.3. in S Fleisher, L Packer eds. *Methods in Enzymology* 1978; **53**: 11-14.
48. Zhang Y, Marcillat O, Giulivi C, Ernster L, Davies KJ. The oxidative inactivation of mitochondrial electron transport chain components and ATPase. *J Biol Chem* 1990; **265**: 16330-16336.
49. Kadiiska MB, Burkitt MJ, Qun-Hui Xiang, Mason RP. Iron supplementation generates hydroxyl radical in vivo. An ESR spin-trapping investigation. *J Clin Invest* 1995; **96**: 1653-1657.
50. Fry M, Green DE. Cardiolipin requirement by cytochrome oxidase and the catalytic role of phospholipid. *Biochem Biophys Res Comm* 1980; **93**: 1238-1246.
51. Goormaghtigh E, Brasseur R, Ruysschaert JM. Adriamycin inactivates cytochrome C oxidase by exclusion of the enzyme from its cardiolipin essential environment. *Biochem Biophys Res Comm* 1982; **104**: 314-320.
52. Heales SJR, Davies SEC, Bates TE, Clark JB. Depletion of brain glutathione is accompanied by impaired mitochondrial function and decreased N-acetyl aspartate concentration. *Neurochem Res* 1995; **20**: 31-38.

53. Zager A. Mitochondrial free radical production induces lipid peroxidation during myohemoglobinuria. *Kidney Internat* 1996; **49**: 741-751.
54. Pitkainen S, Robinson BH. Mitochondrial complex I deficiency leads to increased production of superoxide radicals and induction of superoxide dismutase. *J Clin Invest* 1996; **98**: 345-351.
55. Reynolds LG. Diagnosis and management of acute iron poisoning. *Clin Haematol* 1989; **2**: 423-434.
56. Heys AD, Dormandy TL. Lipid peroxidation in iron loaded spleens. *Clin Sci* 1981; **60**: 295-301.
57. O'Connell MJ, R.J. Ward RJ, Baum H, Peters TJ. The role of iron in ferritin- and haemosiderin-mediated lipid peroxidation in liposome; *Biochem J* 1985; **229**: 135-139.
58. Nienhuis AW. Vitamin C and iron. *N Engl J Med* 1981; **304**: 170-171.
59. Gabutti V, Borgna-Pignatti C. Clinical manifestations and therapy of transfusional haemosiderosis. *Clin Haemat* 1994; **7**: 919-940.
60. Borgna-Pignatti C, Rugolotto S, De Stefano P, Piga A, Di Gregorio F, Gamberini MR, Sabato V, Melevendi C, Cappellini MD, Verlato G. Survival and disease complications in thalassemia major. *Ann N Y Acad Sci* 1998; **850**: 227-231.
61. Hoffbrand AV, Wonke B. Results of long-term subcutaneous desferrioxamine therapy. *Bailliere's Clin Haemat* 1989; **2**: 345-359.
62. Brittenham GM, Griffith PM, Nienhuis AW. et al. Efficacy of deferoxamine in preventing complications of iron overload in patients with thalassemia major. *New Engl J Med* 1994; **331**: 567-569.
63. Giardina PJ, Grady RW, Ehlers KH, Burstein S, Graziano JH, Markenson AL, Hilgartner MW. Current therapy of Cooley's anemia. A decade of experience with subcutaneous desferrioxamine, In: A. Bank, Ed. *Sixth Cooley's Anemia Symposium, Ann N Y Acad Sci* 1990; **612**: 275-285.
64. Calleja EM, Shen JY, Lesser M, Grady RW, New MI, Giardina PJ. . Survival and morbidity in transfusion-dependent thalassemic patients on subcutaneous desferrioxamine chelation. *Ann N Y Acad Sci* 1998; **850**: 469-470.
65. Marcus RE, Davies SC, Bantock HM, Underwood SR, Walton S, Huehns ER. Desferrioxamine to improve cardiac function in iron-overloaded patients with thalassaemia major. *Lancet* 1984; **1**: 392.
66. Hyman CB, Agness CL, Rodriguez-Funes R,. Zednikova M. Combined subcutaneous and high-dose intravenous deferoxamine therapy of thalassemia, *Ann N Y Acad Sci* 1985; **445**: 293-303.
67. Cohen AR, Martin M, Schwartz E. Current treatment of Cooley's anemia. Intravenous chelation therapy. In: A. Bank, Ed. Sixth Cooley's Anemia Symposium, *Ann N Y Acad Sci* 1990; **612**: 286-292.
68. Davis BA, Porter JB. Long-term outcome of continuous 24-hour deferoxamine infusion via indwelling intravenous catheters in high-risk beta-thalassemia. *Blood* 2000;**95**:1229-36
69. Araujo A, Kosaryan M, MacDowell A, Wickens D, Puri S, Wonke B, Hoffbrand AV. A novel delivery system for continuous desferrioxamine infusion in transfusional iron overload. *Brit J Haematol* 1996; **93**: 835-837.
70. Wacker P, Halperin DS, Balmer-Ruedin D, Oberhansli I, Wyss M..Regression of cardiac insufficiency after ambulatory intravenous deferoxamine in thalassemia major. *Chest* 1993 ;**103**:1276-1278
71. Tamary H, Goshen J, Carmi D, Yaniv I, Kaplinsky C, Cohen IJ, Zaizov R Long-term intravenous deferoxamine treatment for noncompliant transfusion-dependent beta-thalassemia patients. *Isr J Med Sci* 1994; **30**:658-664
72. Olivieri NF, Berriman AM, Tyler BJ, Davis SA, Francombe WH, Liu PP Reduction in tissue iron stores with a new regimen of continuous ambulatory intravenous deferoxamine. *Am J Hematol* 1992 ;41:61-63
73. Aldouri MA, Wonke B, Hoffbrand AV, Flynn DM, Ward SE, Agnew JE, Hilson AJ. High incidence of cardiomyopathy in beta-thalassaemia patients receiving regular transfusion and iron chelation: reversal by intensified chelation. *Acta Haematol* 1990; **84**: 113-117.
74. Cohen AR, Mizanin J, Schwartz E. Rapid removal of excessive iron with daily, high-dose intravenous chelation therapy. *J Pediatr* 1989; **115**: 151-155.
75. de Montalembert M, Jan D, Clairicia M, Hannedouche T, Sidi D, Girot R. Intensifying iron chelating therapy with desferrioxamine using implantable venous access catheters (Port-a-Cath). *Arch Fr Pediatr* 1992; **49**: 159-163.
76. Porter JB, Jaswon MS, Huehns ER, East CA, Hazell JWP. Desferrioxamine ototoxicity: evaluation of risk factors in thalasaemic patients and guidelines for safe dosage. *Br J Haematol.* 1989; **73**: 403-409.
77. Marti V, Ballester M, Marrugat J, Auge JM, Padro JM, Narula J, Caralps JM Assessment of the appropriateness of the decision of heart transplantation in idiopathic-dilated cardiomyopathy. *Am J Cardiol* 1997;**80** 746-750

78. Koerner MM, Tenderich G, Minami K, zu Knyphausen E, Mannebach H, Kleesiek K, Meyer H, Koerfer R. Heart transplantation for end-stage heart failure caused by iron overload. *Br J Haematol* 1997;**97**:293-296
79. Wolfe L, Olivieri N, Sallan D, Colan S, Rose V, Propper R, Freedman MH, Nathan DG. Prevention of cardiac disese by subcutaneous deferoxamine in patients with thalassemia major. *N Engl J Med* 1985; **312**, 1600-1603.
80. Richardson ME, Matthews RN, Alison JF, Menahem S, Mitvalsky J, Byrt E, Harper RW. Prevention of heart disese by subcutaneous desferrioxamine in patients with thalassemia major. *Aust N Z J Med* 1993; **23**: 656-661.
81. Modell B, Khan M, Darlison M. Survival in ? -thalassaemia major in the UK: data from the UK thalassaemia register. *Lancet* 2000; **355**: 2051-2052.
82. Hoffbrand AV, Al-Refaie F, Davis B, Siritanakathul N, Jackson BFA, Cochrane J et al. Long-term trial of deferiprone in 51 transfusion-dependent iron overloaded patients. *Blood* 1998; **91**: 295-300.
83. Wonke B, Wright C, Hoffbrand AV. Combined therapy with deferiprone and desferrioxamine. *Brit J Haematol* 1998; **103**: 361-364.

RESULTS OF LONG TERM IRON CHELATION TREATMENT WITH DEFEROXAMINE

Bernard A. Davis and John B. Porter[*]

1. HISTORICAL PERSPECTIVE OF DEVELOPMENT

1.1. Pre-Clinical Development

Deferoxamine (DFO), a naturally occurring, hexadentate siderophore derived from *Streptomyces pilosus*, was discovered by chance in 1960. During earlier work by the Swiss Federal Institute of Technology and Ciba Ltd on isolation of the iron-containing antibiotic ferrimycin from *Streptomyces griseoflavus*, it had become apparent that the antibiotic effect of ferrimycin was abolished by culture filtrates from other actinomycetes (Zahner, 1992). Further work revealed that one of the main ferrimycin antagonists was ferrioxamine B, the principal siderophore produced by *Streptomyces pilosus* (Zahner, 1992). Following its isolation by Bickel in May 1960, ferrioxamine B was initially investigated as a potential iron donor in iron deficient subjects. However, when preliminary studies in a human volunteer revealed that it was excreted intact in the urine without losing or exchanging its iron, the idea arose that an iron-free ferrioxamine might potentially chelate iron and eliminate it in the urine in the form of ferrioxamine. Later that year, the iron-free preparation, deferoxamine B (DFO) was synthesised by Bickel (Keberle, 1992). After *in vitro* and *in vivo* pre-clinical studies had shown it to be a powerful and specific chelator of storage iron, DFO was successfully tested in a patient suffering from severe hemochromatosis in 1961 and introduced into clinical use the following year, initially as an antidote to acute iron poisoning (Keberle, 1992).

---

[*] Bernard A. Davis, Department of Haematology, Royal Free and University College London Medical School, 98 Chenies Mews, London, WC1E 6HX, U.K. John B. Porter, Department of Haematology, Royal Free and University College London Medical School, 98 Chenies Mews, London, WC1E 6HX, U.K.

Iron Chelation Therapy.
Edited by Chaim Hershko, Kluwer Academic/Plenum Publishers, 2002.

1.2. Clinical development

The first reports of DFO treatment for iron overload were published in 1962 in patients with transfusional iron overload and hereditary haemochromatosis by intramuscular bolus (Bannerman et al, 1962; Sephton Smith, 1962). It has taken over 30 years of clinical experience to find a regimen of DFO that achieves acceptable iron excretion while having acceptable toxicity. The development of DFO for the treatment of transfusional iron overload has been essentially clinician-led and on an empirical basis, without detailed knowledge about pharmacokinetic/pharmacodynamic relationships being available. By the 1970s, it was clear that DFO given as a daily intramuscular bolus, over a period of 7 years decreased the rate of hepatic iron loading and the risk of hepatic fibrosis in thalassemia major patients (Barry et al, 1974). However, with intramuscular injections, iron balance was generally not achieved in transfusion-dependent individuals and a more efficient chelation approach was needed.

An important advance in the management of iron overload came when it was shown that continuous 24h intravenous (*iv*) and subsequently, 24h subcutaneous (*sc*) DFO infusions led to significantly improved urinary iron excretion (Hussain et al, 1976; Propper et al, 1976; Propper et al, 1977). Indeed, it became clear that with DFO doses of approximately 30 mg/kg given subcutaneously over 12h, iron balance could be achieved (Pippard et al, 1978a). Detailed metabolic iron balance studies using a low iron diet with collection of urine and faeces, showed that faecal iron excretion is approximately equal to that of urine excretion and that the proportion in the faeces increases as the dose increases and when erythropoiesis is suppressed by blood transfusion (Pippard et al, 1982a). The concomitant use of ascorbate was found to increase excretion of iron in urine (Modell & Beck, 1974; O'Brien, 1974; Wapnick et al, 1969), but less consistently so in faeces (Pippard et al, 1982a), and the overall increase in total iron excretion from ascorbate use was quite variable (Nienhuis et al, 1976). Reports of increased cardiotoxicity with the use of large doses of ascorbate in heavily iron-loaded patients (Henry, 1979), however, led to a recommendation that ascorbate supplementation should be given only at the time of administration of DFO and restricted to a dose not exceeding 3 mg/kg/day (Nienhuis, 1981).

In the 1980s, increasingly large doses were used intravenously to rescue patients in cardiac failure. While reversal of cardiac failure was demonstrated in many cases (Cohen et al, 1989; Marcus et al, 1984), retinal toxicity was also noted at very high doses (>100 mg/kg/day) (Davies et al, 1983). As iron overload fell with regular DFO use and as chelation was started at younger ages and at lower levels of iron loading, other toxicities of excess dosing emerged, most notably auditory toxicity (Olivieri et al, 1986; Porter et al, 1989a; Wonke et al, 1989) and effects on growth and skeletal development (Piga et al, 1988).

By the 1990s, increased evidence for the beneficial effects of *sc* DFO on survival emerged (Brittenham et al, 1994; Olivieri et al, 1994; Zurlo et al, 1989), as well as the critical importance of compliance with therapy on survival (Brittenham et al, 1994; Gabutti & Piga, 1996). It later became clear that reversal of heart failure and cardiac arrhythmias with continuous *iv* chelation led to sustained recovery of cardiac performance and excellent long-term survival, provided compliance with therapy was re-established (Davis & Porter, 2000).

Compliance with *sc* infusions for 8-12h on a minimum of 5 occasions a week remains a challenge for patients and clinicians. Although advances in pump technology, such as disposable balloon pumps make treatment more convenient (Araujo *et al*, 1996), the discomfort and local irritation at the site of infusions remain an impediment to regular use for a substantial minority of patients.

2. PHARMACOLOGY OF DEFEROXAMINE; RELEVANCE TO LONG-TERM OUTCOME

Knowledge of the pharmacology of DFO has accumulated over the last 30 years, much of it following its use in clinical practice. Nevertheless, an understanding of the pharmacology of DFO is important to maximising the benefits of long-term clinical use.

2.1. Chemical Structure and Coordination Chemistry

Deferoxamine (DFO) is a colourless, crystalline compound with a molecular weight of 657. It is composed of 3 distinct moieties - acetic acid, succinic acid and 1-amino-5-hydroxylaminopentane (Keberle, 1964). These units are interlinked to form an open-chain molecule with three hydroxamic acid groups inside and one free amino group at the end. It is the free amino group that confers on the drug its basic character and enables it to form salts with both organic and inorganic acids. DFO has six coordination sites for iron (III), two in each hydroxamic acid group. This hexadentate structure allows it to react efficiently at low concentrations with ferric iron in a 1:1 molar ratio at physiological pH values, its chain structure entwining completely around the central ferric ion. The resulting DFO-iron complex, ferrioxamine (FO), is very stable (stability constant = 10^{31}), since its organic shell protects it from enzymatic degradation. The completely coordinated iron moiety is incapable of catalysing the Haber-Weiss reaction and thus potentiating free radical formation. In contrast, the affinity of DFO for other metal ions is much lower, namely Cu^{2+} (10^{14}), Mg^{2+} (10^{14}), Co^{2+} (10^{11}), Zn^{2+} (10^{11}), Fe^{2+} (10^{10}), Ni^{2+} (10^{10}), and Ca^{2+} (10^{2}). Hence, treatment with DFO at standard doses should not result in the depletion of such physiologically important metal ions from the body. The only other metal ion showing a clinically significant interaction with DFO is Al^{3+} (stability constant = 10^{22-26}) and DFO has been successfully used to treat aluminium overload in renal dialysis patients (Ackrill & Day, 1985).

2.2. Plasma Pharmacokinetics

There have been surprisingly few studies on the pharmacokinetics of DFO and these were generally undertaken after the pharmacodynamic effects on iron excretion were known and DFO was in widespread clinical use. When considering the long-term outcome of DFO use, it is nevertheless helpful to consider our present understanding of the pharmacokinetics first.

DFO is poorly absorbed from the gut due to its large size (Sephton Smith, 1964) and parenteral routes of administration are the only realistic methods of efficient delivery. When given as an *iv* infusion at a dose of 50 mg/kg/day, mean steady state concentrations of 7.4 ± 2.73 µmol/L are achieved between 6 and 12 hours of the infusion (Lee *et al*, 1993). Elimination from the systemic circulation is rapid and biphasic in manner, with a clearance of 0.5 ± 0.24 litres/kg/hr, an initial half-life of 0.28 ± 0.10 hours and a terminal half-life of 3.05 ± 1.30 hours (Lee et al, 1993). DFO is also extensively distributed within the body and has an apparent volume of distribution during the terminal phase of 1.88 ± 1.0 litres/kg (Lee et al, 1993).

The major iron-binding metabolite (DFO-met B), which is formed intracellularly by oxidative deamination of the iron-free ligand soon after the commencement of the infusion (Singh et al, 1990), has generally lower steady state plasma levels than DFO (area under the curve [AUC] 191 ± 106 μmol/L versus 354 ± 131 μmol/L for DFO), and is also cleared rapidly from the systemic circulation with an initial half-life of 1.33 ± 0.61 hours and an elimination rate constant of 0.77 ± 0.74 hr^{-1} (Lee et al, 1993). When DFO is administered by sc infusion at 40mg/kg over 8 hours, peak drug concentrations are reached 8 hours after the infusion is begun (C_{max} = 10.1 μmol/L) and levels begin to fall rapidly after the infusion is stopped with a plasma half-life of 7.59 hours (initial $t_{1/2}$, 0.56 hours; terminal $t_{1/2}$, 9.8 hours) (Porter et al, 1997). However, in both studies, all DFO was converted to FO prior to sample analysis so that the pharmacokinetics examined were effectively those those of [DFO + FO]. It is likely, though not certain, that the slower terminal half-life described for DFO represented FO elimination and the initial phase, that of DFO.

Both DFO and its major metabolites are cleared by the liver and kidney. However, once iron is bound to form FO, plasma clearance is almost exclusively by the kidney. In renal disease, FO may therefore accumulate in plasma. However, FO is highly stable and does not redistribute iron significantly within the body. Uptake of DFO into hepatocytes results in chelation of cytosolic, and possibly, lysosomal iron to form FO, which is then excreted in the bile. In renal failure, DFO will be cleared from the plasma by the liver but FO which is present within the plasma compartment will not be eliminated without dialysis (Allain et al, 1987).

The relationship between the pharmacokinetics, pharmacodynamics and metabolism of DFO was recently investigated in two trials comparing 8-hour sc infusion of DFO 40mg/kg with a slow- release depot sc injection (Porter et al, 1997; Porter et al, 1998a). These trials showed a clear linear relationship between the area under the curve (AUC) (i.e. the product of the mean plasma concentration and the duration of action in μg/ml.h) for the depot formulation of DFO and cumulative urinary iron excretion. The low doses of DFO given by depot (6.2mg/kg over 3-5 days) were associated with a higher proportion of urinary drug recovery and a higher proportion excreted as FO (37-58%) than with conventional DFO (17-31%) as an 8h infusion at 40mg/kg. This was associated with a higher efficiency of chelation (amount of urinary iron divided by the iron binding equivalents of the dose given) with depot (27-66%) compared with conventional DFO at 40mg/kg (17-31%). The proportion of metabolite B, a product of intracellular metabolism and a possible marker of excess intracellular DFO (Porter et al, 1997), was almost undetectable at the low doses given as depot DFO. These findings suggest that the current standard sc regimen is relatively inefficient and that a continuous regimen (either by 24-hour infusion or slow-release depot preparation) at relatively low doses will result in higher chelation efficiency and lower drug toxicity than larger doses given over shorter infusion periods.

2.3. Kinetics of Access to Chelatable Iron Pools

While the plasma pharmacokinetics explain in part how the DFO dosing regimen can influence the efficiency of iron chelation, its toxicity and thus long-term outcome, consideration of chelatable iron pools and the kinetics of access to these are equally important. The majority of body iron is not chelatable at any given point in time, being either present as unchelatable heme iron (Keberle, 1964) or as iron which is so slowly chelatable at clinically relevant plasma DFO concentrations as to be effectively unavailable, such as ferritin and haemosiderin (Crichton *et al*, 1980) and transferrin iron (Hershko *et al*, 1973; Stefanini *et al*, 1991). However, small proportions of the iron within these molecules become transiently available as these proteins are catabolized or donate iron to transiently chelatable iron pools.

There are two major pools of chelatable iron by DFO. The first is that derived from the breakdown of senescent red cells by macrophages in spleen, liver and bone marrow (Hershko & Rachmilewitz, 1979). The exact site of chelation is not certain but the balance of evidence suggests that DFO competes with transferrin for iron released from macrophages (Saito *et al*, 1986). When transferrin becomes saturated in iron overload, DFO will compete with other plasma proteins or citrate for this iron, thereby in principle decreasing the formation of non-transferrin bound iron (NTBI) species. The quantity of chelatable iron from this turnover is 20mg/day in healthy individuals and appreciably more if ineffective erythropoiesis or extravascular haemolysis is present. Iron chelated from this pool by DFO is excreted in the urine with no excretion by the faecal route (Hershko *et al*, 1973; Hershko *et al*, 1978).

The second major pool of iron available to DFO is that derived from the breakdown of ferritin and haemosiderin, the ferritin being catabolized every 72h in hepatocytes (Drysdale & Munro, 1966), predominantly within lysosomes (Cooper *et al*, 1988). DFO can chelate iron that remains within lysosomes shortly after ferritin catabolism (Laub *et al*, 1985; Pippard *et al*, 1982b) or once this iron reaches a dynamic, transiently chelatable, cytosolic low molecular weight iron pool, (Bailey-Wood *et al*, 1975; Konijn *et al*, 1999; Pippard *et al*, 1982b; White *et al*, 1976). The magnitude of this labile iron pool is influenced by cellular iron status, the rate of ferritin catabolism, the rate of uptake of exogenous iron (eg from transferrin or plasma non-transferrin bound iron) and the rate of incorporation of this iron pool into newly synthesised iron-containing molecules (Breuer *et al*, 1995; Epsztejn *et al*, 1997). The major cell where this occurs is the hepatocyte, as the majority of body iron is present here in iron overload. Iron chelated by DFO within hepatocytes is excreted in the faeces (Hershko & Rachmilewitz, 1979) accounting for about half of total iron excretion with DFO (Pippard *et al*, 1982a). The liver is adapted to store iron in times of plenty and release it in time of scarcity. The control of ferritin catabolism is likely to be more responsive to need in the liver than in organs that are not specifically adapted for iron storage and release. Thus in myocytes, a major target for iron toxicity, it is likely though not certain, that excess ferritin and haemosiderin is turned over less often than in the hepatocyte. This may explain why the introduction of intensive chelation with DFO results in a more rapid decrease in liver iron than heart iron as monitored by T2* MRI imaging (Anderson *et al*, 2000). The more rapid chelation of the potentially toxic labile intracellular iron, may explain the rapid reversal of cardiac dysrhythmias and improvement in ventricular function with continuous DFO infusion, prior to achieving large decrements in myocardial iron (Anderson *et al*, 2000; Davis & Porter, 2000).

The rate at which DFO gains access, not only to different cell types, but also to iron pools within these cells, is critical to both its effectiveness and its toxicity. DFO, being hydrophilic and of relatively high molecular weight, tends to equilibrate relatively slowly across biomembranes (Hoyes & Porter, 1993; Porter et al, 1989b). Studies of access to cellular and subcellular iron pools show that unlike hydroxypiridinones, DFO does not cause significant decrements in lysosomal iron until 4 h after incubation with cell lines (Hoyes & Porter, 1993). The rate of formation of intracellular iron-chelate complexes (as FO) is also slow, taking over 4h for equilibration in K562 cells in marked contrast to hydroxypyridinones (Cooper et al, 1996). The toxicity resulting from excess chelator will be also influenced by the rate at which key metalloenzymes are inhibited by DFO. The dimensions and hydrophilicity of DFO mean that inhibition of some intracellular enzymes such as ribonucleotide reductase are significantly slower than with hydroxypyridinones (Cooper et al, 1996) and some non-heme iron containing enzymes such as lipoxygenase are not inhibited by DFO, unlike with the smaller hydroxypyridinone molecules (Abeysinghe et al, 1996). These same properties also slow the rate of chelation of intracellular zinc and induction of apoptosis in thymocytes relative to smaller hydroxypyridinones (MacLean et al, 2001). Thus, the properties that limit the oral bioavailability of DFO also limit access to metabolically important intracellular metal ion pools, thereby decreasing its potential toxicity. A detailed study of the rates of FO formation in different cell types *in vivo* has not yet been reported, but the presence of a facilitated mechanism in some cell types in addition to hepatocytes could lead to disproportionate DFO uptake into these cells and thus potentially result in toxicity.

3. EVIDENCE OF BENEFIT FROM LONG-TERM SC INFUSION

Subcutaneous infusion of DFO, given at a dose of 30-50 mg/kg over 8-12h, 5-7 times/week has now become the standard, recommended first-line treatment and there is now overwhelming evidence for the efficacy of this regimen in achieving long-term survival and in preventing complications of iron overload. Nearly all of this evidence has accumulated from the use of DFO in the management of β-thalassemia major.

3.1. Effect on Iron Overload

Urinary iron excretion with DFO increases in proportion to iron loading with transfusions (Modell & Beck, 1974; Sephton Smith, 1962), transferrin saturation (Modell & Beck, 1974; Sephton Smith, 1962), dose of DFO used (Bannerman et al, 1962; Graziano et al, 1978; Janka et al, 1981; Modell & Beck, 1974; Pippard et al, 1978b; Sephton Smith, 1962; Sephton Smith, 1964), ascorbate status (Modell & Beck, 1974; Hussain et al, 1977) and the phase of the transfusion cycle, being proportionally higher as the haemoglobin level falls (Pippard et al, 1982a). There is very considerable day-to-day variation in urinary iron excretion with the same dose of chelation, even in fully controlled trial conditions. In view of this variability, 24h urinary iron measurements with DFO must be interpreted with caution. In some patients, iron excretion is so large that it is possible to demonstrate negative iron balance at low doses, measuring urinary iron only. For example, (Pippard et al, 1978a) showed that iron balance can be achieved in transfusion-dependent thalassemia with only a 12 hour infusion at a dose of 30 mg/kg, measuring urine iron only.

However, because DFO also induces iron excretion by the faecal route, a true indication of balance can only be achieved using metabolic balance studies (Cumming *et al*, 1969; Pippard *et al*, 1982a; Grady *et al*, 1987; Di Palma *et al*, 1994). These studies have shown faecal iron excretion in response to DFO to be also highly variable. Thus, for example, the contribution of faecal iron excretion to total iron excretion was reported to be 32-50% after an intramuscular dose of 500mg (Cumming *et al*, 1969) and 6-84% after an 8-hour *iv* infusion at 15mg/kg/h (Grady *et al*, 1987). Pippard *et al*, (1982a) found that the contribution of faecal iron excretion to total output was inversely related to the level of erythropoiesis (13% at Hb 8.3g/dL and 37% at Hb 11.7g/dL), linearly related to DFO dose and less constantly affected by ascorbate supplementation. Although such studies are not applicable in routine clinical practice, they have given insight into the achievable effects of chelation therapy. In practice, it is generally assumed that approximately equal amounts of iron are excreted in the urine and faeces after DFO infusion.

Another way of studying the effect of DFO on iron balance is to demonstrate diminishing levels of hepatic iron, because this is closely and predictably related to total body iron stores (total body iron stores in mg/kg of body weight = 10.6 x hepatic iron concentration in mg/g of liver, dry weight) (Angelucci *et al*, 2000). Long-term *sc* therapy results in depletion of hepatic iron stores with an associated fall in serum ferritin, normalization of liver function tests and stabilization of hepatic fibrosis (Hoffbrand *et al*, 1979; Janka *et al*, 1981; Cohen *et al*, 1981; Cohen *et al*, 1984; Aldouri *et al*, 1987). Accumulated experience has shown that on a regime of 40mg/kg (range 30-50mg/kg) given as 8-12h infusions on a minimum of 5 nights a week, iron balance is achieved in hypertransfused patients with thalassemia major. This regime will maintain body iron levels below those regarded as toxic (Brittenham *et al*, 1994; Olivieri *et al*, 1994). However, the dose of DFO should be adjusted according to body iron load, age and diagnosis. 40mg/kg is usually adequate in transfusion-dependent patients who comply with treatment 5 nights a week, but a mean daily dose of up to 50mg/kg given more frequently may be necessary if an unacceptable level of iron loading has already occurred. Although doses above 50-60 mg/kg have been given in severe iron overload, these are not recommended by the manufacturers and increase the risk of adverse effects (see below).

3.2. Effect on Survival

The evidence for improvement in survival with long-term *sc* DFO therapy is now beyond dispute. Before DFO became available for the treatment of iron overload, prognosis was uniformly poor with most patients dying of cardiac complications in the second or third decade (Engle *et al*, 1964). In this study, only 7 out of 41 patients lived to the age of 20 years or beyond and over half of patients who developed heart failure were dead within 3 months of its onset. The first evidence for the beneficial effects of DFO therapy emerged in the early 1980s with the report that children who received average weekly doses of more than 4g of DFO were less likely to die from iron overload than age-matched patients who received less, or no DFO (Modell *et al*, 1982). However, there was substantial heterogeneity in the mode of administration of DFO, and patients receiving the then novel *sc* infusional therapy had not been exposed to it long enough for a proper assessment of its impact on survival.

Over the subsequent decade, further studies confirmed the beneficial effects of DFO on the survival of patients who had been treated long-term with *sc* infusions (Zurlo *et al*, 1989; Ehlers *et al*, 1991). However, it was not until the mid-1990s that the striking impact of DFO on survival was established beyond all reasonable doubt. Two studies, both with follow-up periods of more than 10 years, unequivocally established that the effective use of *sc* DFO was associated with prolonged survival free of the complications of iron overload in patients with β-thalassemia major (Brittenham *et al*, 1994; Olivieri *et al*, 1994). The body iron load was found to be the main determinant of clinical outcome in both studies. In one trial, serial measurement of the serum ferritin was used to assess the effectiveness of chelation therapy (Olivieri *et al*, 1994). Ineffective chelation therapy as evidenced by elevation of the serum ferritin beyond 2500 µg/L on more than 2/3 of occasions was associated with an estimated cardiac-free survival after 15 years of <20%, whereas effective chelation therapy (serum ferritin values <2500 µg/L for most of the time) ensured a cardiac disease free survival of 91% after 15 years. Mortality in patients who had developed heart failure as a result of poor chelation was 50%.

The influence of transfusional iron load, DFO use and hepatic iron concentration on survival in 59 patients (age range: 7-31 years) with thalassemia major, followed for 4-10 years or until death, was investigated by Brittenham and co-workers (Brittenham *et al*, 1994). Hepatic iron concentration correlated with the ratio of cumulative transfusional iron load (mmol/kg) to cumulative DFO use (g/kg). The ratio of transfusional iron load to DFO use in all patients developing cardiac disease (n=15) or dying (n=9), was above 0.6 mmol/g, equivalent to a hepatic iron concentration above 15mg/g dry weight (or 80µmol/ g wet weight). Survival was 100% in patients who either had a ratio < 0.6 or a pre-chelation iron load of < 14 mmol of iron/kg body weight. Survival in patients outside these groups was only 32% at 25 years. These findings suggest that iron load prior to starting chelation, total transfusional iron burden, and total use of DFO all impact on survival.

Other studies have confirmed the finding of Brittenham *et al* (1994) that compliance with DFO therapy is of critical importance in prolonging survival. In one study involving 257 patients with thalassemia major, patients who received more than 250 infusions/year (about 5 times/week) had 95% survival at 30 years of age, whereas in those who failed to achieve this target, survival at 30 years of age was only 12% (Gabutti & Piga, 1996). In a second larger study (1146 patients born between 1960 and 1987), a favourable trend in survival was demonstrated in patients who had the advantage of early commencement of DFO therapy; the probability of survival in patients born between 1970-74 was 82% to at least age 25 years (Borgna-Pignatti *et al*, 1998).

Recently, data from the United Kingdom Thalassemia Register showed a rather disappointing outcome, with 50% of patients in the UK dying of iron overload complications before the age of 35 years (Modell *et al*, 2000). The principal reason given for this less favourable outcome was that standard *sc* DFO therapy was too painful and too arduous to permit full compliance in up to 50% of patients nationwide. However, a substantial number of patients in that study had not been treated in specialist centres where a better outcome might be expected. A more recent report from our own institution has demonstrated that a centre effect exists; the survival of thalassemia major patients treated in our centre is strikingly better than in the UK as a whole (Davis et al, 2001a). The survival probability of 103 patients born between 1957 and 1987 and followed up in our centre was 78% at 40 years. Sub-group analysis by birth cohort showed a probability of survival of 69% at 40 years among subjects born between 1955-64 and 78% at age 35 years among those born between 1965-74. There have been no deaths among patients born after 1974. Our centre has worked consistently on maximising compliance using a multi-disciplinary approach including a clinical psychologist to identify psychological as well as practical problems associated with this treatment. Aggressive intervention with continuous intravenous therapy in high risk cases has also been a strategy for nearly two decades (Davis et al, 2001b). These findings indicate that the availability of DFO does not in itself guarantee compliance leading to improved survival in iron overloaded patients; rather it is expertise in its use, together with the provision of practical and psychological support to patients taking DFO that ensure that its beneficial effects are fully realised.

The changing patterns in survival in homozygous β-thalassemia since the advent of standard DFO therapy is summarised in Table 1.

3.3. Effect on Cardiac Disease

Cardiac disease, which manifests clinically as congestive cardiac failure and/or cardiac arrhythmias, remains the leading cause of death in transfusion-dependent thalassemia (Zurlo et al, 1989; Borgna-Pignatti et al, 1998). In the days before DFO became available, the natural history of myocardial iron deposition was one of inexorable progress to death from heart disease mainly towards the end of the second decade (Engle et al, 1964). The biggest impact of DFO on survival in thalassemia has been through the reduction of morbidity and mortality from heart disease (Zurlo et al, 1989). Therapy with sc DFO has been shown to be efficacious both in preventing heart disease and in reversing asymptomatic left ventricular dysfunction. Traditionally, intensive iv regimens have been used to prevent the onset of cardiac disease in high risk patients and to reverse existing clinical disease (see below).

Freeman et al (1983) were the first to demonstrate the beneficial effect of sc DFO on cardiac function in a small group of thalassemia patients with sub-clinical heart disease in response to intensive sc DFO therapy. Improvement in left ventricular function during exercise was observed in 4 patients after 12 months of intensive sc therapy at a mean dose of 64 mg/kg/day. These same workers later showed that with longer follow up, resting left ventricular ejection fraction values also normalized in the same group of patients and that these improvements in left ventricular function were sustained (Freeman et al, 1989). Despite an approximate doubling in the transfusional iron overload, there was a sequential fall in serum ferritin values with DFO therapy, suggesting that removal of iron was the likely mechanism for the improvement in cardiac function. Similar observations were made in another study involving 60 patients with β-thalassemia major in the United Kingdom (Aldouri et al, 1990). Twenty-two patients showed severe impairment of ventricular function and a further 19 patients had a mild abnormality. Repeat radionuclide ventriculography in 17 of these patients after intensification of subcutaneous infusion for a period of 6-28 months showed a significant improvement in the resting left ventricular ejection fraction from 45% to 52% with an associated fall in serum ferritin values. However, evidence for the role of sc DFO in reversing clinically overt cardiac disease is very limited. Rahko et al, (1986) treated an untransfused, iron-overloaded patient who had a megaloblastic anaemia of undetermined origin and gross congestive heart failure with sc DFO for a period of more than 2 years. Sequential echocardiograms and radionuclide ventriculograms showed progressive improvement in cardiac function, with normalization of the intra-cardiac pressures and cardiac output, and improvement in the left ventricular ejection fraction. Right ventricular endomyocardial biopsy 13 months after commencement of DFO revealed a marked reduction in iron content compared to the pre-treatment biopsy. The patient remained in improved condition, came off diuretics and was able to return to work part time.

Evidence for the efficacy of long-term sc DFO in preventing iron-induced cardiac disease first emerged in the mid-1980s in a study of 36 thalassemia patients who had begun sc DFO therapy (25-50 mg/kg/day over 12 hours) relatively late, after the age of 10 years (Wolfe et al, 1985). Only one patient in the compliant group (n = 17) developed cardiac failure and died of congestive cardiac failure. In contrast, 12 out of 19 non-compliant patients developed cardiac failure and 7 died. Subsequent larger studies with longer follow-up demonstrated unequivocally the efficacy of long-term sc DFO therapy in preventing cardiac complications of transfusional iron overload. In brief, these studies showed that cardiac disease is preventable if chelation is started early (Zurlo et al, 1989), hepatic iron loading kept below 15 mg/g dry weight (Brittenham et al, 1994), serum ferritin levels maintained below 2500 µg/L (Olivieri et al, 1994) and >260 infusions/year administered (Gabutti & Piga, 1996).

3.4. Effect on Endocrine Dysfunction

Endocrine abnormalities are common in subjects with transfusional iron overload. In patients with thalassemia, dysfunction of the hypothalamic-pituitary-gonadal (HPG) axis is the most common endocrinopathy observed. Damage may also occur to the thyroid, pancreas, parathyroid and adrenal glands but at a much lower frequency (Italian Working Group on Endocrine Complications in Non-endocrine Diseases, 1995).

3.4.1. Effect on HPG Dysfunction

Dysfunction of the HPG axis results in hypogonadotrophic hypogonadism (Kletzky et al, 1979; Wang et al, 1989; Chatterjee et al, 1993; Landau et al, 1993), defects of growth hormone secretion (Shehadeh et al, 1990) and reduced responsiveness of growth hormone to its release hormone (Pintor et al, 1986). Defects of the growth hormone receptor (Masala et al, 1984; Tolis et al, 1987; Leger et al, 1989) and deficiency of insulin-like growth factor (Saenger et al, 1980; Herington et al, 1981; Werther et al, 1981) have also been described. Iron toxicity plays a major role in all of these defects, inducing hypogonadism by a selective central mechanism (Costin et al, 1979; Kletzky et al, 1979a; De Sanctis et al, 1988a; Wang et al, 1989; Landau et al, 1993; Chatterjee et al, 1993) and by interfering with the production of insulin-like growth factor. Evidence for the role of iron in the aetiology of hypogonadotrophic hypogonadism has been provided histologically by its preferential deposition in pituitary gonadotrophs (Bergeron & Kovacs, 1978) and by the finding of loss of anterior pituitary volume in poorly compliant, iron overloaded thalassemia patients on MRI scanning (Chatterjee et al, 1998). Further evidence for the role of iron in the genesis of pituitary failure has also come from the finding of a significant correlation between anterior pituitary dysfunction and moderate-to-severe iron deposition in the anterior pituitary, as assessed by MRI (Berkovitch et al, 2000). Although there have been reports of iron deposition in the testes and ovaries (Canale et al, 1974), primary gonadal failure appears to be less important than defective function of the hypothalamic-pituitary axis.

Clinically, these endocrine abnormalities present as failure of growth and sexual development. Arrested or failed puberty occurs in 55% of both male and female patients (Borgna-Pignatti et al, 1998). A recent large study reported the frequency of arrested puberty in thalassemia to be 15.7% in males and 12.6% in females. The frequency of secondary amenorrhoea was 23%; approximately 13% of females had oligomenorrhoea and 2.3% had irregular menstrual cycles (Italian Working Group on Endocrine Complications in Non-endocrine Diseases, 1995)

The efficacy of DFO in the prevention of growth failure and hyponadotrophic hypogonadism was first described in a group of thalassemia patients who had been optimally chelated with DFO since mid-childhood (Bronspiegel-Weintrob et al, 1990). Normal puberty was attained in 90% of these subjects compared with only 38% of a cohort who had started DFO treatment in their early teens and had received a lower total dose of DFO. However, there was no significant difference in final and mid-parental heights in both groups. Further evidence for the beneficial effects of DFO on sexual development is provided by the marked increase in fertility in women with thalassemia since the late 1980s (Jensen et al, 1995a). Although conflicting observations have been made during this same period (Chatterjee et al, 1993), the overall weight of evidence is in favour of improved sexual maturation in response to properly administered DFO therapy among thalassemia patients (Italian Working Group on Endocrine Complications in Non-endocrine Diseases, 1995; Jensen et al, 1997; Borgna-Pignatti et al, 1998). It is disappointing to note, however, that secondary amenorrhoea may occur in up to 25% of thalassemic women with previously normal pituitary function, despite adequate chelation therapy (Chatterjee et al, 1993).

Although normalization of pituitary growth hormone reserve has been reported (Schafer et al, 1985), there are as yet no reports of reversal of established anterior pituitary failure.

3.4.2. Effect on Diabetes Mellitus

Diabetes mellitus is a relatively common complication in poorly chelated patients with thalassemia and occurs with a frequency of 4.9% in transfused and chelated individuals. (Italian Working Group on Endocrine Complications in Non-endocrine Diseases, 1995). Although a number of other mechanisms have been proposed as a cause for the development of diabetes in iron overloaded individuals, it is clear that iron-induced damage to the pancreas is a major factor, as evidenced by the association between severity and duration of iron overload and the development of diabetes in patients with thalassemia major (De Sanctis et al, 1988b). Standard sc DFO therapy has been proven to reduce the risk of diabetes mellitus and glucose intolerance in thalassemia major in compliant patients (Brittenham et al, 1994; Gabutti & Piga, 1996; Gamberini et al, 1998). Improvements in glucose tolerance and a decreased need for oral hypoglycaemic agents has been reported following intensive DFO therapy in patients with symptomatic iron overload (Fosburg & Nathan, 1990).

3.4.3. Effect on Hypothyroidism, Hypoparathyroidism and Adrenal Insufficiency

Mild abnormalities of thyroid function as well as clinical hypothyroidism have been described in iron overloaded thalassemia patients, including individuals who have been managed with transfusion and chelation therapy (Lassman *et al*, 1974; Flynn *et al*, 1976; Sabato *et al*, 1983; Magro *et al*, 1990; Grundy *et al*, 1994). A strong correlation exists between serum ferritin levels and the presence of thyroid dysfunction (Jensen *et al*, 1997). Clinical and sub-clinical hypoparathyroidism are well documented in thalassemia patients (Flynn *et al*, 1976; McIntosh, 1976; Costin *et al*, 1979; Gertner *et al*, 1979; De Sanctis *et al*, 1992). Adrenal dysfunction affecting androgen secretion, but sparing corticosteroid synthesis has also been described (Sklar *et al*, 1987). These complications are all preventable in patients managing at least 260 infusions/year of *sc* DFO (Gabutti & Piga, 1996). Although there are reports of an apparent increase in incidence of hypothyroidism in younger cohorts of transfused and chelated patients, this is probably due to underdiagnosis in the past (Borgna-Pignatti *et al*, 1998). Improvement in thyroid function after intensive DFO therapy has been reported (Flynn *et al*, 1982; Landau *et al*, 1993), but not when frank hypothyroidism has supervened (Gabutti & Piga, 1996). However, there have been no reports of reversal of parathyroid and adrenal dysfunction with DFO.

3.5. Effect on Haemopoiesis

An unexpected beneficial effect of long-term DFO therapy on haemopoiesis has been reported in 11 patients with myelodysplastic syndromes treated for up to 60 months (Jensen *et al*, 1996). Reduction in red cell requirements was observed in 7/11 patients with 5/11 becoming transfusion-independent. Platelet counts increased in 7/11 patients and neutrophil counts in 7/9 evaluable patients. An increase in serum transferrin receptor concentration was demonstrated in all patients with improvements in blood counts, consistent with increased erythropoietic activity, but the mechanism underlying the improved counts is unknown.

4. BENEFIT OF CONTINUOUS IV INFUSION IN HIGH-RISK PATIENTS

Long-term *iv* therapy has been used as a preparative measure for pregnancy and bone marrow transplantation (Gabutti & Piga, 1996). However, its principal and most important use is for poorly chelated patients (approximately 10%), who by reason of late commencement of, or poor compliance with chelation therapy are at increased risk of premature death from cardiac disease. The indications for *iv* therapy in this subgroup are documented cardiac dysfunction with or without clinical evidence of heart failure, significant cardiac arrhythmias, massive iron overload and intolerance of *sc* DFO on account of severe local reactions (Davis & Porter, 2000; Piga *et al*, 2001).

Intravenous DFO therapy for the treatment of symptomatic cardiac disease was first reported in the 1980s (Marcus *et al*, 1984). Considerable symptomatic benefit and objective evidence of improvement in cardiac function were observed in 3 out of 5 patients treated with large doses of DFO (up to 200 mg/kg) given as 24-hour infusions for 6-12 months. The remaining 2 patients died of cardiac failure despite continuance of DFO, but the 3 patients who showed improvement were still alive at 18 months follow up. Later Cohen et al (1989), using a discontinuous, high dose regimen of 6-12g of DFO over 12 hours daily in a cohort of 16 iron overloaded patients, demonstrated rapid depletion of hepatic iron stores and drastic reductions in serum ferritin values over a 12-25 month period. There was also improvement in symptoms and echocardiographic parameters in two patients, one with congestive cardiac failure and the other with a severe ventricular arrhythmia, who no longer required cardiac medication after 12-24 months of therapy. Aldouri et al (1990) treated 4 patients with continuous 24-hour therapy and observed a marked increase in urinary iron excretion with a concomitant improvement in the left ventricular ejection fraction in 2 patients and reversal in atrial fibrillation in a third. Olivieri et al (1992a) showed that not only did continuous 24-hour infusions produce superior urinary iron excretion compared with an equivalent sc 12-hour regimen, but also that the iv regimen encouraged excellent patient compliance. These observations were supported by the finding of a dramatic decline in mean serum ferritin values from 7552 µg/L to 2186 µg/L over a 15 month period on mean daily doses of 54 mg/kg/day initially, increasing to 63 mg/kg/day. In another study, discontinuous high dose DFO (100 mg/kg over 8-12 hours) resulted in a significant fall in serum ferritin values in 13 patients with thalassemia major, but there was no improvement in the cardiac status of 2 patients with proven cardiac disease after 41-43 months of therapy (Tamary et al, 1994).

Two recent longer-term studies have established that iv therapy is a safe, life-saving therapeutic option in the management of high risk thalassemia patients. In the first study, 25 intravenous devices were inserted over a 16 year period into 17 patients and DFO (50mg/kg/day) infused continuously over 24 hours, 6-7 days a week (Davis & Porter, 2000). Resting LVEF improved significantly from 36% to 49% in 7 out of 9 patients with previously documented deterioration in LVEF and stabilized in the remaining 2 patients (p=0.002). Improvements in LVEF were documented by MUGA within 6-12 months of commencing treatment, and were often clinically apparent within days to weeks of starting infusion. Improvement was sustained on long-term follow up. Atrial fibrillation was reversed in 6 out of 6 patients within 12 months; in one patient, cardioversion occurred within 5 days without the need for conventional anti-arrhythmic drugs. Serum ferritin values fell in a biphasic fashion from a pre-therapy mean of 6281 µg/L to 3736 µg/L during the lifetime of the catheters (median 623 days), with an initial faster phase (rate of fall, 1082 µg/L/month for ferritin values >3000 µg/L) and a slower second phase (rate of fall, 133 µg/L/month). There was a linear relationship between the pre-treatment serum ferritin concentration and the initial rate of decline. Survival probability was 61% at 13 years in the group as a whole and 62% at 13 years in those with demonstrable cardiac disease. All patients who had presented with cardiac arrhythmias and 3 of the 4 who were in gross cardiac failure made a full recovery and no longer required cardiotropic drugs. Two of the 3 patients who died had initially improved on iv therapy but had been unable to persevere with sc treatment when that was re-introduced following the removal of their catheters. Drug toxicity was limited to a single early case of reversible retinopathy in a patient with pre-existing diabetes mellitus who had been on an initial dose of 80 mg/kg/day.

A further report involved the administration of continuous DFO infusions via 40 Port-a-Caths to 30 high-risk patients over a 20 year period (Berdoukas, 2001). During the lifetime of the Port-a-Caths (mean 29 months), 19 patients showed improvement in cardiac function, with a return to normal in 10. Six patients showed a stabilisation in cardiac function. There was also a significant reduction in hepatic iron levels in 25 surviving patients during the period of treatment. Four patients died from cardiac failure and the fifth from massive air embolism after insertion of the catheter.

Intravenous DFO therapy is labour intensive and clinical practice varies slightly from centre to centre. Currently, however, it is usual to give DFO as a continuous 24-hour infusion through a Port-a-Cath, 5-7 days a week, at doses not exceeding 50mg/kg/day. Disposable balloon pumps are commonly used (Davis & Porter, 2000) and needle changes are undertaken preferably by trained hospital personnel, or if this is not practical by the patients themselves (Davis & Porter, 2000; Piga *et al*, 2001).

The complications associated with long-term *iv* therapy are in order of frequency: infection, thrombo-embolism, mechanical problems with the catheter itself and complications directly related to the surgical procedure. Infective complications occur at a frequency of between 0.51 per 1000 catheter days and 1.7 per 1000 catheter days (Cohen *et al*, 1989; Tamary *et al*, 1994; Davis & Porter, 2000; Piga *et al*, 2001) and are comparable with infection rates in other patient groups (Davis & Porter, 2000). Two types of infection occur: bacteraemic infections and localised infections of the subcutaneous pocket/port. The relative frequencies of these differ from study to study; Davis & Porter (2000) found no difference in frequencies of these two types of infection, while in a much larger, multi-centre study by Piga *et al* (2001), there was a preponderance of port infections. Most catheter infections are due to Staphylococci; coagulase-negative staphylococci were predominant in the study by Davis & Porter, while *Staphylococcus aureus* was responsible for the majority of injections in the study by Piga *et al*. Gram negative infections occurred at a much lower frequency in both studies and there have been no reports of fungal infections such as mucormycosis which has been documented in immunosuppressed thalassemia patients and in renal dialysis patients receiving DFO (Sands *et al*, 1985; Boelaert *et al*, 1991; Gaziev *et al*, 1996). Infective complications may be minimized by adopting strict, aseptic policies with regard to care of the catheters. These include restricting the accessing of catheters to trained personnel only and the rotation of needle insertion sites (Davis & Porter, 2000; Piga *et al*, 2001).

Thrombo-embolic complications are reported with a frequency of between 0.13 per 1000 catheter days (Piga *et al*, 2001) and 0.48 per 1000 catheter days (Davis & Porter, 2000), rates comparable to those in other patient groups. Most commonly, these complications are line thromboses, but more extensive venous thromboses, including superior vena cava thrombosis may occur (Davis & Porter, 2000). Pulmonary embolism has also been reported (Davis & Porter, 2000; Berdoukas, 2001). Long-term thrombo-prophylaxis with warfarin to a target international normalized ratio of 2.0-3.0 has been used in an attempt to reduce the incidence of thromboses further, but insufficient follow up at this stage is available to comment on its effectiveness (Davis & Porter, 2000).

Other rare but troublesome complications include catheter occlusion, perforation of the superior vena cava, catheter dislocations and fracture, and subjective pain and discomfort. The only reported case of catheter-related death followed massive air embolism after catheter insertion (Berdoukas, 2001).

The incidence of DFO toxicity with long-term *iv* therapy is extremely low provided dosage is kept below 60 mg/kg/day and the therapeutic index below 0.025 (Davis & Porter, 2000).

5. MONITORING OF IRON LOADING TO ACHIEVE OPTIMAL LONG-TERM OUTCOME

Safe and effective chelation requires a balance to be struck between abrogating the toxic effects of excess iron and minimizing the unwanted side effects of excess chelation treatment. Monitoring of patients with iron overload involves the regular assessment of the cardiological, endocrinological, growth and developmental effects of iron overload, but is beyond the scope of this chapter. This section focuses on the value of monitoring of iron loading, using serum ferritin, liver iron and heart iron to achieve the optimal outcome. Direct monitoring for toxic effects of excess DFO is also required alongside these measures and is addressed in section 7.

5.1. Serum Ferritin

Serial measurement of serum ferritin, preferably monthly, is the simplest way of monitoring the efficacy of DFO chelation therapy but has its limitations. The target ferritin value should be approximately 1000-2000 µg/L as the risk of cardiac complications increases with values above 2500 µg/L (Olivieri *et al*, 1994). The rate of fall in serum ferritin in response to chelation therapy is also a useful tool in informing patients of their progress. Serum ferritin values below 3000 µg/L fall at a reasonably constant rate with *iv* DFO treatment (approximately 130µg/L/month on 50 mg/kg/day) and this can be used to encourage both clinicians and patients about realistic target values which can be achieved with such a regimen (Davis & Porter, 2000). Although the relationship between serum ferritin and liver iron concentrations is well established (Worwood, 1986), the correlation is not sufficiently precise to be of strong prognostic value. A variety of conditions are known to complicate the use of serum ferritin levels as an estimate of body iron stores. Serum ferritin is an acute phase protein and levels may be falsely elevated in acute infections, liver inflammation or damage, chronic inflammation, haemolysis and ineffective erythropoiesis (Solomon *et al*, 1981; Baynes *et al*, 1986; Brabec *et al*, 1990). On the other hand, spuriously low values are found in ascorbate deficiency which is common in iron overload due to increased catabolism of the vitamin to oxalate (Lynch *et al*, 1967; Roeser *et al*, 1980; Chapman *et al*, 1982). For these reasons, serum ferritin measurements must not be used as the sole or final arbiter of effective chelation therapy. The use of serum ferritin to limit DFO toxicity is discussed below (section 7).

5.2. Liver Iron Concentration

The liver is the major iron storage site in iron loading conditions, containing 70% or more of body iron stores (Modell & Berdoukas, 1984). There is a close correlation between liver iron and total body iron in transfusional iron overload, total body iron (in mg/kg of body weight) being approximately equivalent to 10.6 times the liver iron concentration (in milligrams per gram of liver, dry weight) (Angelucci *et al*, 2000). Thus, in patients with transfusional iron overload, measurement of liver iron is the most reliable and most sensitive way of determining body iron status and monitoring efficacy of chelation therapy (Olivieri & Brittenham, 1997).

A high ratio of transfusional iron burden in mmol/kg body weight) relative to cumulative DFO use (in g/kg body weight) was found to have significant prognostic value in 59 thalassaemia major patients (Brittenham *et al*, 1994) and this ratio correlated with the observed hepatic iron concentration. Because all patients dying or developing heart disease had a ratio >0.6, equivalent to 15mg iron/g dry weight of liver, this value was suggested as a threshold of "high risk" for thalassemia major patients (Olivier & Britannia, 1997). It was further suggested that target liver iron levels between 3-7 mg/g dry weight should be aimed for in iron overloaded, transfusion-dependent patients receiving chelating therapy, as heterozygotes with hereditary hemochromatosis can attain such concentrations without suffering significant pathology (Olivieri & Brittenham, 1997). However, this assumes a similar pattern of iron distribution between the heart and liver in the two groups of subjects and does not take into account the marked differences in the rate of iron loading between these groups.

As heart failure was the cause of death in all patients in the study by Brittenham *et al* (1994), liver iron concentration can only at best represent an association of risk with survival. We, and others have observed patients dying or developing de-novo heart failure, with liver iron concentrations significantly below 15mg/g dry weight. Furthermore, recent evidence suggests a discordance between the levels of liver and heart iron in individuals on DFO chelation treatment (Anderson *et al*, 2000), possibly due to more rapid emptying of liver iron than heart iron by DFO. This argues against the use of simple "thresholds" for liver iron to predict prognosis. It is likely that a formula which takes into account the duration of exposure to high body iron burden, chelation history, as well as heart iron will be necessary for optimal prognostication. Despite these caveats, this liver iron target range is the best currently available and evaluated single indicator of effective chelation (Olivieri & Brittenham, 1997).

Measurement of the iron concentration by chemical determination on an adequate liver biopsy sample (at least 1.0mg in dry weight) is an established method for assessing body iron stores (Olivieri & Brittenham, 1997). Biopsies can be performed safely under ultrasound guidance and are also useful for the histological assessment of the distribution of iron between hepatocytes and Kupffer cells, as well for investigating the presence of fibrosis, cirrhosis and inflammation. However, it must be borne in mind that inadequate sample size (<1.0 mg/g dry weight) or uneven distribution of iron in the presence of cirrhotic change (Villeneuve *et al*, 1996) may give misleading results. Liver iron stores may also be quantitated accurately and non-invasively by magnetic susceptometry using a *s*uperconducting *q*uantum *i*nterference *d*evice (SQUID) (Brittenham *et al*, 1982; Nielsen *et al*, 1995), but this is not generally available. Only 3 centres, one in the United States and two in Europe have the necessary equipment and infrastructure.

Recently, magnetic resonance imaging using a novel acquisition technique (T2*), has been evaluated for quantitation of liver iron (Anderson *et al*, 2001). Liver iron concentrations measured by this technique show good and consistent correlation to those obtained by chemical analysis of liver biopsy samples.

5.3. 24-Hour Urinary Iron

Measurement of 24-hour urinary iron excretion following a single *im* bolus or subcutaneous infusion has been used to monitor the efficacy of chelation therapy with DFO (Modell & Berdoukas, 1984). However, its clinical usefulness is limited by the wide day-to-day variability in DFO-induced urinary iron excretion.

5.4. Heart Function and Heart Iron

Regular assessment of cardiac function by non-invasive methods such as echocardiography (Henry et al, 1978; Giardina et al, 1985; Cohen et al, 1990; Richardson et al, 1993) and radionuclide ventriculography, using the technique of multi-gated acquisition (MUGA) scanning (Leon et al, 1979; Freeman et al, 1983; Marcus et al, 1984; Aldouri et al, 1990; Davis & Porter, 2000) are widely used to monitor the effectiveness of chelation treatment. Data published prior to the adoption of current standard and intensive DFO regimens had shown that death from heart failure was predicted by an abnormal left ventricular ejection fraction on echocardiography in the preceding 6-12 months (Henry et al, 1978; Giardina et al, 1985). The rationale then for regular cardiac monitoring is the early detection of cardiac dysfunction and the institution of appropriate, potentially life-saving intervention, in the form of aggressive chelation therapy. As objective evidence of left ventricular dysfunction is usually synonymous with advanced disease and as the onset of clinical decompensation is difficult to predict from single non-invasive tests, the clinical effectiveness of this practice has in the past been called into question (Borow et al, 1982). However, with the benefit of 20 years of prospective follow up of a group of 85 transfusion-dependent thalassemics at our unit who underwent annual MUGA scans, we have found that sequential monitoring of cardiac function is a valuable tool for assessing cardiac risk and can be used for modulating chelation therapy (Davis et al, 2001b). In this study, a fall in the left ventricular ejection function was significantly associated with the subsequent development of symptomatic cardiac disease and cardiac death, unless treatment intensification was successfully effected, either by improved dosing and adherence to standard *sc* DFO or by instituting a 24-hour continuous *iv* regimen through a Port-a-Cath. All patients who complied with intensified treatment lived, while all who failed to comply died.

More recently, the paramagnetic properties of iron have been exploited in the use of magnetic resonance imaging (MRI) for the quantitation of body iron stores and assessment of cardiac function (Jensen et al, 1995; Mavrogeni et al, 1998; Anderson et al, 2001). Previous MRI methods were not sufficiently reproducible to be of much clinical use (Angelucci et al, 1997), but methodology has recently been developed at our institution for the quantitation of body iron using MR T2* imaging (Anderson et al, 2001). T2* is an inherent value of any tissue and is inversely related to iron concentration. The measurement gives a high degree of reproducibility in both the liver and the heart (coefficient of variations of 3.3% and 5.0% respectively). A recent study of 151 regularly transfused patients with homozygous β-thalassemia showed a marked discordance between myocardial and liver iron deposition in many patients (Anderson et al, 2000), possibly reflecting chelating history as well as different rates of iron uptake by yachts and hepatocytes. Excess myocardial iron was present in 58% of patients despite long-term chelation treatment and there was a linear decline in left and right ventricular ejection associated with increasing myocardial iron deposition (evidenced by a falling myocardial T2*). This technique could in principle be used in any hospital with suitable MRI facilities and its main advantage over existing imaging techniques is that it would enable earlier detection of cardiac involvement by iron overload. However, its application to chelation monitoring will have to await careful prospective study.

6. UNWANTED EFFECTS OF DEFEROXAMINE AND THEIR PREVENTION

6.1. General Considerations

For a drug which is taken in gram quantities daily throughout life and whose action is to chelate iron, an essential element for normal physiological functions, DFO has a remarkably low toxicity. This can be attributed to its slow penetration of cells other than hepatocytes, due to its high molecular weight and low lipid solubility as well as the high stability of the iron-chelate complex and its high selectivity for iron. Toxicity is still observed however. After over 30 years of clinical use, much is now understood about DFO induced toxicity and how this can be minimised. The first significant toxicities to be widely recognised were retinal and ototoxicity in the 1980s, as increasingly high doses were used on increasingly less overloaded patients. Later the effects of excess dosing on growth and skeletal changes were recognised. In general, these risks are greatest with high doses, in younger patients and in patients who are relatively less iron overloaded. Other unwanted effects such as localised skin reactions and *Yersinia* infection are not directly related to iron chelation but can be minimized by appropriate intervention (see below).

6.2. Ocular Toxicity

DFO-induced retinopathy was first described in patients receiving very high doses (>120mg/kg/day) with intravenous therapy (Davies *et al*, 1983) and were subsequently confirmed by others (Borgna-Pignatti *et al*, 1984; Blake *et al*, 1985; Olivieri *et al*, 1986). Complications include night blindness, annular field loss with a geographically similar defect of dark adaptation, as well as electro-oculographic (EOG) and electroretinographic (ERG) abnormalities characteristic of degenerative disorders of retinal photoreceptors involving the rods. Together with retinal pigmentation, these features are indistinguishable from those of retinitis pigmentosa. The loss of central acuity and colour vision in association with central scotomata and EOG, ERG and visual evoked response abnormalities is characteristic of degeneration affecting the cones. Variable degrees of optic atrophy may also occur, and rarely, blindness. Apart from high doses, diabetes is another possible risk factor (Arden *et al*, 1984), possibly because this is associated with increased permeability of the blood-retinal barrier.

Some features are reversible to varying degrees on discontinuation of DFO treatment while others are not. Thus, in the study by Davies *et al* (1993), defects in EOG, ERG and dark adaptation improved in one patient but recovery of visual field loss was only partial, while in the another patient, rapid subjective improvement in visual symptoms occurred but some loss of visual acuity, dark adaptation and visual field constriction remained. In the patient series reported by Olivieri *et al* (1986), complete resolution of symptoms occurred within 4 weeks of stopping DFO in two patients who had presented with a marked decrease in central vision, eccentric fixation and severely impaired visual acuity, but optic atrophy with thinning of the nerve-fibre layer persisted. In a third patient, there was partial improvement in visual acuity but asymmetric optic disk atrophy persisted; the fourth patient had partial recovery of visual acuity and colour vision.

Other minor abnormalities have been described in iron overload patients with no visual symptoms (Arden *et al*, 1984; Olivieri *et al*, 1986; De Virgiliis *et al*, 1988; Gelmi *et al*, 1988). These include visual field constriction, hyperpigmentary and hypopigmentary changes, abnormalities of dark adaptation, and abnormalities in the electrophysiological tests.

Non-iron overloaded patients are particularly vulnerable to ocular toxicity from DFO. The occurrence of visual symptoms has been reported in renal dialysis patients after the administration of single doses of DFO (50 or 100 mg/kg *iv*) for the treatment of aluminium overload (Rubinstein *et al*, 1985; Bournerias *et al*, 1987; Pengloan *et al*, 1987). Visual disturbances have also been reported in patients with rheumatoid arthritis and related disorders after only a few doses of 50 mg/kg/day (Blake *et al*, 1985; Polson *et al*, 1985).

The pathogenesis of retinal toxicity is unclear and a variety of mechanisms have been proposed including a direct effect of DFO on the retina, or an indirect effect through chelation of trace metals in the retinal cells (Davies *et al*, 1983; Blake *et al*, 1985; Pall *et al*, 1986). The histological and ultrastructural changes in DFO-induced retinopathy have been described (Davies *et al*, 1983; Rahi *et al*, 1986). They include loss of microvilli from the apical surface, patchy depigmentation and vacuolation of the cytoplasm, mitochondrial swelling and calcification and disorganisation of the plasma membrane and abnormal thickening of Bruch's membrane.

Cataracts have been also reported in patients taking DFO (Waxman & Brown, 1969; Modell, 1979; Davies *et al*, 1983) and may be dose-related. This is seen from the occurrence of this complication in patients receiving very high doses of the drug (>200 mg/kg/day) (Davies *et al*, 1983; Modell & Berdoukas, 1984) and the paucity of reports among patients taking more conventional doses of the drug. Complete resolution of the cataracts usually occurs on stopping DFO, which may then be safely resumed at a lower dose. There have been no reports of any visual loss in patients who have had cataracts. It is noteworthy that sub-capsular punctiform lens opacities may result from the iron overload *per se* (Duke-Elder, 1969).

6.3. Ototoxicity

The first report linking DFO with ototoxicity was by (Marsh *et al*, 1981) who described a case of tinnitus in a moderately iron-overloaded patient with β-thalassemia intermedia receiving DFO. Tinnitus developed during therapy but disappeared over a period of 3 weeks after discontinuation of DFO. It reappeared on re-introduction of DFO, but this time persisted for several years. Later reports confirmed this link (Guerin *et al*, 1985; Olivieri *et al*, 1986; Porter *et al*, 1989a; Wonke *et al*, 1989). From these studies, it became clear that the hallmark of DFO-induced ototoxicity is bilateral, symmetrical, high frequency, sensorineural hearing loss which may or may not be symptomatic and which may or may not be preceded by tinnitus. The onset is usually insidious, although it may occasionally be acute (Olivieri *et al*, 1986). Audiometric studies are more sensitive than auditory brain-stem responses in its diagnosis (Olivieri *et al*, 1986). Rarely, a vestibular component may be present (Porter *et al*, 1989a). Complete or partial recovery of hearing may occur on discontinuation or reduction of dosage of DFO (Olivieri *et al*, 1986; Porter *et al*, 1989a), but stabilization of the hearing loss on reduced doses of DFO is the more usual finding. A significant minority of patients may have a deficit that is severe enough to require the use of hearing aids. A higher than expected frequency of conductive hearing loss is not indicative of DFO toxicity but may reflect a higher incidence of the abnormality in this group of patients compared with the general population (Porter et al, 1989a).

A possible mechanism for DFO ototoxicity is the presence of a greater proportion of the iron-free drug available to chelate trace metals in the cochlea, thereby inhibiting important metalloenzymes such as tyrosinase or lipoxygenase. Alternatively, DFO may exert a direct toxic effect on the cochlea (Simon et al, 1983). However, the only reported study to have examined the effect of DFO on the animal cochlea (Shirane & Harrison, 1987) failed to demonstrate changes in the sensory epithelium following a dose of 100 mg/kg/day for 64 days in non-iron overloaded animals. Furthermore, in a preliminary study in our institution, no evidence of cochlear damage in normal or iron overloaded rats receiving DFO at 200 mg/kg/day for 2 months was observed using electron microscopy (Porter, unpublished observations).

The risk of ototoxicity can be reduced by the use of the therapeutic index (the ratio of mean daily dose in mg/kg divided by the current serum ferritin in µg/L). Though it has been applied to prevent other potentially toxic DFO effects, it was first elaborated for ototoxicity (Porter et al, 1989a) which was not seen in patients in whom this ratio remained below 0.025 at all times.

6.4. Other Neurotoxicity

Reversible coma following the co-administration of DFO and prochlorperazine was reported in two patients with rheumatoid arthritis without iron overload (Blake et al, 1985). Prochlorperazine perturbs the blood-brain-barrier and may facilitate the penetration of DFO or metal ions in the cerebrospinal fluid. A case of a milder central nervous system effect (transient aphasia) after accidental high dosage has been reported (Dickerhoff, 1987), and it has been suggested that oxidative damage mediated by decompartmentalised copper ions may be responsible (Pall et al, 1986). Recently, there have been reports of a sensorimotor neuropathy, not previously described, occurring in 2 patients with β-thalassemia who were on high dose (120mg/kg/day) iv DFO therapy (Levine et al, 1997). The development of this syndrome did not correlate with serum ferritin level or the DFO dose/serum ferritin ratio, but resolved on discontinuation of DFO treatment. The syndrome did not recur on recommencement of DFO at a lower dose of 40 mg/kg/day in 1 patient, but did so in the second, on a reduced dosage of 55 mg/kg/day.

6.5. Growth and Skeletal Complications

The toxic effects of DFO on linear growth were first described in an Italian study on pre-pubertal thalassemia patients who had commenced intensive chelation with DFO (20-100 mg/kg/day) in 1979 (Piga et al, 1988). Prior to starting DFO therapy, 34% of patients were below the 3rd centile for height, presumably due to the endocrine effects of iron overload. By 1983, only 13% of patients were still below the 3rd centile, but two years later, 66% were again below the 3rd centile. On inspection of the data, it became clear that patients with serum ferritin concentrations below 1000 µg/L and on higher DFO doses (>70mg/kg/day) were significantly over-represented in the group with growth failure (i.e. below the 3rd centime for height). When the dose of DFO was halved in these patients, 16 out of 19 subjects experienced a sharp increase in growth.

Characteristically, the retardant effect of DFO on growth results in disproportionate truncal shortening and loss of sitting height (Piga et al, 1988; Rodda et al, 1995). It has become recently apparent that this is probably due to the effect of DFO on spinal cartilage (Hartkamp et al, 1993; Hatori et al, 1995; Olivieri et al, 1995).

DFO-related growth failure is usually accompanied by other radiological skeletal abnormalities including rickets and/or scurvy-like lesions of the long bones (Orzincolo et al, 1990), circumferential metaphyseal osseous defects, sharp zones of provisional calcification, and widened growth plates (Brill et al, 1991). Marked abnormalities of the metaphyseal growth plate were readily observed in the distal ulnar, radial, and tibial metaphyses (Olivieri et al, 1992b) and platyspondyly of the vertebral bodies (Orzincolo et al, 1994). Subsequently, it became clear that children commenced on DFO therapy before the age of two years were especially vulnerable (Olivieri et al, 1992b).

The principal risk factors for skeletal toxicity from DFO are commencement of chelation therapy before the age of 3 years (Brill et al, 1991; Olivieri et al, 1992b) and doses >35 mg/kg in very young children (Olivieri et al, 1995). In order to balance the risks of DFO-related skeletal toxicity against those of iron overload, it is usual to commence chelation treatment at about 3 years of age (not before) and to restrict the dose of DFO to a maximum of 35 mg/kg/day in children below 5 years and 40 mg/kg thereafter until growth is complete.

6.6. Other Toxicities from Excess Dosing

Nausea and vomiting, with hypotension and acute collapse (Modell, 1979) and transient aphasia (Dickerhoff, 1987) have been seen when sudden high dose *iv* boluses of DFO have been inadvertently administered. A non-infectious pulmonary syndrome of life-threatening severity, characterized by tachypnoea, hypoxemia, diffuse interstitial infiltrates on chest X-ray and a restrictive pattern on pulmonary function studies has been reported in patients receiving *iv* doses of DFO at rates of ≥ 10mg/kg/h (Freedman et al, 1990). Renal toxicity has been documented in patients treated with *iv* DFO at a dose of about 180 mg/kg/day (Koren et al, 1989) and thrombocytopenia has been observed in two renal dialysis patients receiving DFO (Walker et al, 1985; Dickerhoff, 1987).

6.7. Local Skin Reactions

Local reactions, with skin reddening and soreness, at the site of subcutaneous infusions may occur. These are often caused by DFO being reconstituted above the recommended concentration of 10% and may be avoided by increasing the volume of water used to dilute the DFO. On occasions when such reactions remain a problem, the instillation of small doses of hydrocortisone (5-10 mg) mixed in with the dissolved DFO may be effective. Reactions are also likely to be seen if absorption of DFO from the infusion site is impaired. Painful lumps may then occur and last for several days. The usual reason for this is the use of sites where sub-dermal fibrosis has occurred as a result of repeated long term use. Rotating sites of infusion is usually effective in preventing this type of reaction. Severe local reactions, which fail to respond to these treatment strategies, may respond to a desensitization regimen that was originally used for the treatment of anaphylactic reactions (see below). If these approaches fail, *iv* DFO therapy or an alternative chelator should be considered.

6.8. Hypersensitivity

Allergy to DFO may rarely occur and is independent of dose. This can present with true anaphylaxis or occasionally with non-specific symptoms such as fever, myalgia or arthralgia during DFO infusion. It can be effectively treated by careful desensitization carried out under close medical supervision (Miller et al, 1981; Bousquet et al, 1983). Desensitization is usually successful, but may need to be repeated. If it is unsuccessful, an alternative chelator such as deferiprone should be considered. Occasionally, severe local skin reactions at the site of subcutaneous DFO infusion can respond to desensitization (Davis & Porter, unpublished observations).

6.9. Yersinia and Other Infections

Microorganisms require iron to grow, many synthesizing their own iron chelators (siderophores) in order to extract iron for this purpose. *Yersinia enterocolitica* is an organism that cannot make its own siderophores but possesses receptors for ferrioxamine (Perry & Brubaker, 1979) which allows it acquire and use iron as a growth factor (Robins-Browne & Prpic, 1983; Gallant et al, 1986). Thus the pathogenicity of *Yersinia* (low under normal conditions) increases markedly in iron loading conditions (Robins-Browne et al, 1979; Robins-Browne et al, 1987) and this risk increases with DFO treatment. Severe, occasionally fatal infections, such as enterocolitis, peritonitis and septicaemia have been reported in patients receiving DFO (Butzler et al, 1978; Melby et al, 1982; Robins-Browne & Prpic, 1983; Boyce et al, 1985; Chiu et al, 1986; Gallant et al, 1986; Kelly et al, 1987). For this reason, iron overloaded patients on DFO who present with fever, diarrhoea or abdominal pain should be considered to have a *Yersinia* infection until otherwise proven. *Yersinia* infection can be diagnosed by appropriate stool samples, blood cultures and serological testing and, if proven or seriously suspected, should be treated promptly with antibiotics. Ciprofloxacin is the current antibiotic of choice, but co-trimoxazole is also effective. DFO should be withheld until the infection has been eliminated. Prolonged antibiotic treatment is occasionally necessary to prevent recurrence, but it is rarely necessary to withhold DFO after the initial infection has been treated.

It is important to recognise that the pathogenicity of a number of other bacteria such as *Klebsiella* may be enhanced by ferrioxamine and if there is clinical suspicion of bacterial infection, it may be wise to withhold DFO until this has been diagnosed and treated.

Mucormycosis has been reported in immunosuppressed thalassemia patients and in renal dialysis patients receiving DFO (Sands et al, 1985; Boelaert et al, 1991; Gaziev et al, 1996). Mortality from mucormycosis, contracted after bone marrow transplantation for thalassemia is high, with 3 out of 4 patients dying from this infection in one report (Gaziev et al, 1996).

There has been a single report of *Pneumocystis carinii* pneumonia in association with DFO use (Kouides et al, 1988). However, this report is at variance with later studies which demonstrated that continuous DFO infusion has significant activity against animal models of acute *Pneumocystis carinii* pneumonia at steady-state plasma concentrations that are well below plasma concentrations achieved with standard doses in human subjects (Merali et al, 1995; Merali et al, 1996).

7. MINIMIZING DEFEROXAMINE TOXICITY WITH LONG-TERM USE

7.1. Dose Adjustment for Age and Iron Loading

In general, toxicity from DFO therapy is unlikely at the doses currently employed (30-50 mg/kg, 5-7 days a week. As discussed above, it is now clear that the toxic effects of excess DFO dosing can be minimized by limiting the maximum dose before growth has ceased to < 40mg/kg (Gabutti & Piga, 1996; Olivieri & Brittenham, 1997; Thalassemia International Federation, 2000), and giving even smaller doses in very young children, particularly if DFO is started below the age of 3 years (Olivieri & Brittenham, 1997). A mean daily dose above 50mg/kg should not generally be necessary in adults, provided the patients has not developed complications, in which case continuous exposure may be as effective as dose escalation (Davis & Porter, 2000). Downward dosing adjustment is necessary as iron loading falls because of the link between risk of toxicity and low iron stores. This is particularly important when serum ferritin levels are below 1000 µg/L and in patients with diabetes. The use of measurements of ferritin every 1-2 months, with a reduction of the mean daily dose to achieve a therapeutic index < 0.025, reduces the risk of DFO toxicity as iron stores fall (Porter *et al*, 1989a).

The limitations of using the serum ferritin for dose adjustment follow from its false elevation in the presence of inflammation (Baynes *et al*, 1986) and its suppression in the presence of ascorbate deficiency (Lynch *et al*, 1967; Roeser *et al*, 1980; Chapman *et al*, 1982). The therapeutic index was initially used in thalassemia major patients with respect to audiometric toxicity (Porter *et al*, 1989a). Although this has subsequently been used to limit retinal (Davis & Porter, 2000) and recommended for skeletal (Olivieri & Brittenham, 1997) toxicity, its validity has not been formally scrutinized for these uses. Another limitation of relying solely on the serum ferritin to adjust the dose is that the relationship between serum ferritin and body iron differs in different iron loading states. Thus, in thalassemia intermedia serum ferritin appears to be disproportionately low relative to iron stores compared with thalassemia major. In sickle cell disorders, serum ferritin may remain elevated as a consequence of tissue damage for several weeks after an infarctive event and use of ferritin a guide to dosing is particularly unreliable.

As liver iron is a more reliable measurement of body iron than serum ferritin, in principle it should be possible to use a ratio of DFO dose to liver iron to adjust doses as levels fall. Such a scheme has been suggested (Olivieri & Brittenham, 1997), but the studies linking a given ratio of DFO to liver iron and DFO toxicity risk are thus far unavailable. Fischer, (2001) has shown that the therapeutic index correlates with liver iron and with an index of DFO dose divided by liver iron, but the correlation is insufficiently tight to be clinically applied directly as a substitute for the therapeutic index. The ratio of dose to liver iron that is associated with DFO toxicity needs prospective evaluation before clear recommendations can be made. Another problem with using a DFO dose/liver iron ratio as a means for dose adjustment, is that liver iron measurement cannot generally be performed as frequently as the serum ferritin. With intensive chelation therapy however the body iron can fall sufficiently rapidly to require downward dose adjustment within a few months. It is therefore likely that even when the level of liver iron at which dose reduction should occur has been established, that serum ferritin will still be used between liver iron determinations.

7.2. Monitoring for Audiometric and Retinal Toxicity

The frequency of monitoring for audiometric and retinal toxicity will depend on the doses being used, the age of the patients and the feasibility of the tests being performed in the relevant clinical setting. In general it is advisable to check audiometry yearly in all children on regular DFO, in all patients where iron loading and serum ferritin values are low or in patients given high intensity treatment. In patients with previously identified audiometric disturbance audiometry may advisable more frequently.

7.3. Monitoring for Growth and Skeletal Development

In addition to limiting the maximum dose given to young patients, as discussed above, regular monitoring of growth and truncal height relative to full height is important (standing height, sitting height, sub-ischial leg length, target height). In a recent multi-center study in thalassemic patients aged 3-36 years, about 18% exhibited short stature and disproportion between the upper and lower body segments was present in 14% due to a spinal growth impairment which started in infancy (Caruso-Nicoletti *et al*, 1998). The authors concluded that this was a function of the disease itself rather than its treatment. However, other factors such as hypogonadism, siderosis, or DFO-induced bone dysplasia could not be excluded. Clinically evident bony lesions typical of DFO over-treatment are seen in only about 3% of thalassemia major patients (De Sanctis *et al*, 2000) and bone histological findings (including abnormal chondrocytes, alteration of cartilage staining pattern, irregular columnar cartilage, and lacunae in the cartilaginous tissue) do not differ between those thalassemia patients with and without such lesions (De Sanctis *et al*, 2000). Deficiency of growth hormone was observed in one report to be more common in patients with classical DFO induced lesions than those without (De Sanctis *et al*, 1998). Thus, with present knowledge it is difficult to know if and when truncal shortening is due to DFO overdosing.

A possible link between DFO metabolism and osteoporosis was suggested by pilot studies by our group (Porter *et al*, 1992) but a subsequent study showed no intrinsic difference in DFO metabolism between patients with and without bony problems (Porter *et al*, 1998b). It must be remembered that the major cause of osteoporosis in thalassemia major is hypogonadism. It must also be appreciated that osteoporosis is not the cause of backache in many patients with thalassemia major, intervertebral disc degeneration being a more common cause. Until the relative roles of the disease itself and DFO are better understood, careful monitoring of height with downward dose adjustment in those with changes in height velocity or truncal height is the most practical approach.

7.4. Other Approaches to Minimizing Toxicity

There appears to be a positive relationship between plasma levels of the major metabolites of DFO and the risk of toxicity (Lee *et al*, 1993). Since these metabolites are formed only intracellularly, their concentrations depend on the proportion of uncomplexed drug taken up into cells, which would in turn depend on the amounts of intracellular and extracellular iron. Gross iron overload would therefore be expected to spare the toxic effects of DFO compared with minimal degrees of iron loading. A recent study has shown that this is indeed the case (Porter *et al*, 1998b); the proportion of metabolite B was proportionately greater in the blood and urine of thalassemic patients who were given a higher dose of drug in relation to their serum ferritin (i.e. a high therapeutic ratio). These differences in metabolism were quantitative rather than qualitative however, reflecting differences in iron loading (Porter *et al*, 1998b) and measurement of metabolites is impractical in routine clinical practice.

The recognition of the presence of factors other than high DFO dose or low degrees of iron loading can be important to minimising toxicity. Thus, factors which increase access to DFO to the CNS, such as diabetes mellitus (Arden *et al*, 1984) and the action of certain psychotrophic drugs (Blake *et al*, 1985) should be borne in mind. The concomitant use of other chelators such as deferiprone with DFO requires careful evaluation to ensure that this does not facilitate shuttling of metals other than excess iron.

Table 2 shows a scheme for monitoring for DFO toxicity. It is worth remembering that inspection of infusion sites not only gives valuable information about the nature of any local reactions, but also gives an indirect indication of compliance.

Table 2. Monitoring for adverse effects of deferoxamine

Effect	Monitoring	Frequency
Local skin reactions	History & examination	Clinic visits
Ototoxicity	Clinical History	Clinic visits
	Audiometry	Yearly unless symptomatic
	Therapeutic ratio	3-monthly with ferritin results
Retinal toxicity	Clinical history	Clinic visits
	Fundoscopy	Yearly unless symptomatic
	Electroretinography	Yearly unless symptomatic
Growth abnormalities	Height (sitting and standing)	6-monthly
	Dose/therapeutic ratio	3-monthly
Bony abnormalities	X-rays of knees, wrists, spine	Every 1-2 years unless abnormal
Yersinia infection	Stool/blood culture, serology	Clinical suspicion

7.5. DFO and Pregnancy

DFO is not recommended for use in pregnancy by the manufacturers. However, there have been over 40 pregnancies reported in iron-overloaded subjects given DFO at various stages of the pregnancy without apparent teratogenic effects (Hershko & Hoffbrand, 2000). A balanced judgment has to be made about the possible risks to the fetus from DFO *vis-à-vis* the maternal risks from non-chelation for the duration of pregnancy. It is our practice to give low dose DFO 20-30 mg/kg in the final trimester to mothers who have a perceived high cardiac risk.

8. CONCLUSIONS

The accidental discovery of DFO some 40 years ago heralded a departure from the use of iron chelators of low specificity and high toxicity for the treatment of transfusional iron overload. However, it was the subsequent development of the nightly subcutaneous regimen two decades ago that ushered in the era of modern and effective iron chelation therapy and radically altered the prognosis of this once fatal disease. Patients who comply fully with current treatment regimens from an early age can expect to live a full and active life, free from the complications of iron overload. Long-term outcome has steadily improved in patients who can comply with treatment and in treatment centers that are able to provide practical and psychological support for patients who are finding it difficult to comply. However, long-term outcome of DFO use with other methods of DFO delivery such as *sc* boluses, depot DFO and combined DFO with deferiprone are largely unknown and beyond the scope of this article.

The increased understanding of the mechanisms underlying iron toxicity, and the accumulating body of literature on the pharmacokinetic/pharmacodynamic relationship of DFO have resulted in the design and use of appropriate aggressive chelation regimens for poorly chelated patients with impressive results. Despite the drawbacks of high cost and cumbersome administration, DFO infusion remains the reference chelation regimen and the drug of choice for the management of transfusional iron overload.

9. REFERENCES

Abeysinghe,R.D., Roberts,P.J., Cooper,C.E., MacLean,K.H., Hider,R.C., & Porter,J.B. (1996) The environment of the lipoxygenase iron binding site explored with novel hydroxypyridinone iron chelators. *J.Biol.Chem.*, **271**, 7965-7972.

Ackrill,P. & Day,J.P. (1985) Desferrioxamine in the treatment of aluminum overload. *Clin.Nephrol.*, **24 Suppl 1**, S94-S97.

Aldouri,M.A., Wonke,B., Hoffbrand,A.V., Flynn,D.M., Laulicht,M., Fenton,L.A., Scheuer,P.J., Kibbler,C.C., Allwood,C.A., Brown,D., & . (1987) Iron state and hepatic disease in patients with thalassaemia major, treated with long term subcutaneous desferrioxamine. *J.Clin.Pathol.*, **40**, 1353-1359.

Aldouri,M.A., Wonke,B., Hoffbrand,A.V., Flynn,D.M., Ward,S.E., Agnew,J.E., & Hilson,A.J. (1990) High incidence of cardiomyopathy in beta-thalassaemia patients receiving regular transfusion and iron chelation: reversal by intensified chelation. *Acta Haematol.*, **84**, 113-117.

Allain,P., Chaleil,D., Mauras,Y., Beaudeau,G., Varin,M.C., Poignet,J.L., Ciancioni,C., Ang,K.S., Cam,G., & Simon,P. (1987) Pharmacokinetics of desferrioxamine and of its iron and aluminium chelates in patients on haemodialysis. *Clin.Chim.Acta*, **170**, 331-338.

Anderson,L., Holden,S., Davis,B., Prescott,E., Charrier,C., Bunce,N., Firmin,D., Wonke,B., Porter,J., Walker,J.M., & Pennell,D. (2001). Cardiovascular T2-star (T2*) magnetic resonance for the early diagnosis of myocardial iron overload. *European Heart Journal* (in press) .

Anderson,L., Porter,J.B., Wonke,B., Walker,J.M., Holden,S., Davis,B.A., Prescott,E., Charrier,C., Firmin,D.N., & Pennell,D.J. (2000) Cardiac iron deposition is not predicted by conventional markers of iron overload in homozygous β-thalassaemia. *Blood* **96**, 606a.

Angelucci,E., Brittenham,G.M., McLaren,C.E., Ripalti,M., Baronciani,D., Giardini,C., Galimberti,M., Polchi,P., & Lucarelli,G. (2000) Hepatic iron concentration and total body iron stores in thalassemia major. *N.Engl.J.Med.*, **343**, 327-331.

Angelucci,E., Giovagnoni,A., Valeri,G., Paci,E., Ripalti,M., Muretto,P., McLaren,C., Brittenham,G.M., & Lucarelli,G. (1997) Limitations of magnetic resonance imaging in measurement of hepatic iron. *Blood*, **90**, 4736-4742.

Araujo,A., Kosaryan,M., MacDowell,A., Wickens,D., Puri,S., Wonke,B., & Hoffbrand,A.V. (1996) A novel delivery system for continuous desferrioxamine infusion in transfusional iron overload. *Br.J.Haematol.*, **93**, 835-837.

Arden,G.B., Wonke,B., Kennedy,C., & Huehns,E.R. (1984) Ocular changes in patients undergoing long-term desferrioxamine treatment. *Br.J.Ophthalmol.*, **68**, 873-877.

Bailey-Wood,R., White,G.P., & Jacobs,A. (1975) The use of Chang cells cultured in vitro for the investigation of cellular iron metabolism. *Br.J.Exp.Pathol.*, **56**, 358-362.

Bannerman,R.M., Callender,S.T., & Williams,D.L. Effect of desferrioxamine and D.T.P.A. in iron overload. *Br.Med.J.* 2, 1573-1577. 1962.

Barry,M., Flynn,D.M., Letsky,E.A., & Risdon,R.A. (1974) Long-term chelation therapy in thalassaemia major: effect on liver iron concentration, liver histology, and clinical progress. *Br.Med.J.*, **2**, 16-20.

Baynes,R., Bezwoda,W., Bothwell,T., Khan,Q., & Mansoor,N. (1986) The non-immune inflammatory response: serial changes in plasma iron, iron-binding capacity, lactoferrin, ferritin and C-reactive protein. *Scand.J.Clin.Lab Invest*, **46**, 695-704.

Berdoukas,V. (2001) The use of portacaths in patients with transfusion dependent anaemia. The Sydney Children's Hospital experience. *S.O.S.T.E.Notiziario* **2,** 139.

Bergeron,C. & Kovacs,K. (1978) Pituitary siderosis. A histologic, immunocytologic, and ultrastructural study. *Am.J.Pathol.*, **93**, 295-309.

Berkovitch,M., Bistritzer,T., Milone,S.D., Perlman,K., Kucharczyk,W., & Olivieri,N.F. (2000) Iron deposition in the anterior pituitary in homozygous beta-thalassemia: MRI evaluation and correlation with gonadal function. *J.Pediatr.Endocrinol.Metab*, **13**, 179-184.

Blake,D.R., Winyard,P., Lunec,J., Williams,A., Good,P.A., Crewes,S.J., Gutteridge,J.M., Rowley,D., Halliwell,B., Cornish,A., & . (1985) Cerebral and ocular toxicity induced by desferrioxamine. *Q.J.Med.*, **56**, 345-355.

Boelaert,J.R., Fenves,A.Z., & Coburn,J.W. (1991) Deferoxamine therapy and mucormycosis in dialysis patients: report of an international registry. *Am.J.Kidney Dis.*, **18**, 660-667.

Borgna-Pignatti,C., De Stefano,P., & Broglia,A.M. (1984) Visual loss in patient on high-dose subcutaneous desferrioxamine. *Lancet*, **1**, 681.

Borgna-Pignatti,C., Rugolotto,S., De Stefano,P., Piga,A., Di Gregorio,F., Gamberini,M.R., Sabato,V., Melevendi,C., Cappellini,M.D., & Verlato,G. (1998) Survival and disease complications in thalassemia major. *Ann.N.Y.Acad.Sci.*, **850**, 227-231.

Borow,K.M., Propper,R., Bierman,F.Z., Grady,S., & Inati,A. (1982) The left ventricular end-systolic pressure-dimension relation in patients with thalassemia major. A new noninvasive method for assessing contractile state. *Circulation*, **66**, 980-985.

Bournerias,F., Monnier,N., Dufier,J.L., & Reveillaud,R.J. (1987) [Severe ocular toxicity of desferrioxamine in the hemodialyzed patient]. *Nephrologie*, **8**, 27-29.

Bousquet,J., Navarro,M., Robert,G., Aye,P., & Michel,F.B. (1983) Rapid desensitisation for desferrioxamine anaphylactoid reaction. *Lancet*, **2**, 859-860.

Boyce,N., Wood,C., Holdsworth,S., Thomson,N.M., & Atkins,R.C. (1985) Life-threatening sepsis complicating heavy metal chelation therapy with desferrioxamine. *Aust.N.Z.J.Med.*, **15**, 654-655.

Brabec,V.,Cermak,J., & Sebestik,V. (1990) Serum ferritin in patients with various haemolytic disorders. *Folia Haematol.Int.Mag.Klin.Morphol.Blutforsch.* , **117**, 219-227.

Breuer,W., Epsztejn,S., Millgram,P., & Cabantchik,I.Z. (1995) Transport of iron and other transition metals into cells as revealed by a fluorescent probe. *Am.J.Physiol*, **268**, C1354-C1361.

Brill,P.W., Winchester,P., Giardina,P.J., & Cunningham-Rundles,S. (1991) Deferoxamine-induced bone dysplasia in patients with thalassemia major. *AJR Am.J.Roentgenol.*, **156**, 561-565.

Brittenham,G.M., Farrell,D.E., Harris,J.W., Feldman,E.S., Danish,E.H., Muir,W.A., Tripp,J.H., & Bellon,E.M. (1982) Magnetic-susceptibility measurement of human iron stores. *N.Engl.J.Med.*, **307**, 1671-1675.

Brittenham,G.M., Griffith,P.M., Nienhuis,A.W., McLaren,C.E., Young,N.S., Tucker,E.E., Allen,C.J., Farrell,D.E., & Harris,J.W. (1994) Efficacy of deferoxamine in preventing complications of iron overload in patients with thalassemia major. *N.Engl.J.Med.*, **331**, 567-573.

Bronspiegel-Weintrob,N., Olivieri,N.F., Tyler,B., Andrews,D.F., Freedman,M.H., & Holland,F.J. (1990) Effect of age at the start of iron chelation therapy on gonadal function in beta-thalassemia major. *N.Engl.J.Med.*, **323**, 713-719.

Butzler,J.P., Alexander,M., Segers,A., Cremer,N., & Blum,D. (1978) Enteritis, abscess, and septicemia due to Yersinia enterocolitica in a child with thalassemia. *J.Pediatr.*, **93**, 619-621.

Canale,V.C., Steinherz,P., New,M., & Erlandson,M. (1974) Endocrine function in thalassemia major. *Ann.N.Y.Acad.Sci.*, **232**, 333-345.
Caruso-Nicoletti,M., De,S., V, Capra,M., Cardinale,G., Cuccia,L., Di Gregorio,F., Filosa,A., Galati,M.C., Lauriola,A., Malizia,R., Mangiagli,A., Massolo,F., Mastrangelo,C., Meo,A., Messina,M.F., Ponzi,G., Raiola,G., Ruggiero,L., Tamborino,G., & Saviano,A. (1998) Short stature and body proportion in thalassaemia. *J.Pediatr.Endocrinol.Metab*, **11 Suppl 3**, 811-816.
Chapman,R.W., Hussain,M.A., Gorman,A., Laulicht,M., Politis,D., Flynn,D.M., Sherlock,S., & Hoffbrand,A.V. (1982) Effect of ascorbic acid deficiency on serum ferritin concentration in patients with beta-thalassemia major and iron overload. *J.Clin.Pathol.*, **35**, 487-491.
Chatterjee,R., Katz,M., Cox,T.F., & Porter,J.B. (1993) Prospective study of the hypothalamic-pituitary axis in thalassaemic patients who developed secondary amenorrhoea. *Clin.Endocrinol.(Oxf)*, **39**, 287-296.
Chatterjee,R., Katz,M., Oatridge,A., Bydder,G.M., & Porter,J.B. (1998) Selective loss of anterior pituitary volume with severe pituitary-gonadal insufficiency in poorly compliant male thalassemic patients with pubertal arrest. *Ann.N.Y.Acad.Sci.*, **850**, 479-482.
Chiu,H.Y., Flynn,D.M., Hoffbrand,A.V., & Politis,D. (1986) Infection with Yersinia enterocolitica in patients with iron overload. *Br.Med.J.(Clin.Res.Ed)*, **292**, 97.
Cohen,A., Martin,M., & Schwartz,E. (1981) Response to long-term deferoxamine therapy in thalassemia. *J.Pediatr.*, **99**, 689-694.
Cohen,A., Martin,M., & Schwartz,E. (1984) Depletion of excessive liver iron stores with desferrioxamine. *Br.J.Haematol.*, **58**, 369-373.
Cohen,A.R., Martin,M., & Schwartz,E. (1990) Current treatment of Cooley's anemia. Intravenous chelation therapy. *Ann.N.Y.Acad.Sci.*, **612**, 286-292.
Cohen,A.R., Mizanin,J., & Schwartz,E. (1989) Rapid removal of excessive iron with daily, high-dose intravenous chelation therapy. *J.Pediatr.*, **115**, 151-155.
Cooper,C.E., Lynagh,G.R., Hoyes,K.P., Hider,R.C., Cammack,R., & Porter,J.B. (1996) The relationship of intracellular iron chelation to the inhibition and regeneration of human ribonucleotide reductase. *J.Biol.Chem.*, **271**, 20291-20299.
Cooper,P.J., Iancu,T.C., Ward,R.J., Guttridge,K.M., & Peters,T.J. (1988) Quantitative analysis of immunogold labelling for ferritin in liver from control and iron-overloaded rats. *Histochem.J.*, **20**, 499-509.
Costin,G., Kogut,M.D., Hyman,C.B., & Ortega,J.A. (1979) Endocrine abnormalities in thalassemia major. *Am.J.Dis.Child*, **133**, 497-502.
Crichton,R.R., Roman,F., & Roland,F. (1980) Iron mobilization from ferritin by chelating agents. *J.Inorg.Biochem.*, **13**, 305-316.
Cumming,R.L., Millar,J.A., Smith,J.A., & Goldberg,A. (1969) Clinical and laboratory studies on the action of desferrioxamine. *Br.J.Haematol.*, **17**, 257-263.
Davies,S.C., Marcus,R.E., Hungerford,J.L., Miller,M.H., Arden,G.B., & Huehns,E.R. (1983) Ocular toxicity of high-dose intravenous desferrioxamine. *Lancet*, **2**, 181-184.
Davis,B.A., O'Sullivan,C., & Porter,J.B. (2001a) Survival in β-thalassaemia major is enhanced by specialist management. Submitted.
Davis,B.A., O'Sullivan,C., & Porter,J.B. (2001b) Value of LVEF monitoring in the long-term management of β-thalassaemia. Submitted.
Davis,B.A. & Porter,J.B. (2000) Long-term outcome of continuous 24-hour deferoxamine infusion via indwelling intravenous catheters in high-risk beta-thalassemia. *Blood*, **95**, 1229-1236.
De Virgiliis,S., Congia,M., Turco,M.P., Frau,F., Dessi,C., Argiolu,F., Sorcinelli,R., Sitzia,A., & Cao,A. (1988) Depletion of trace elements and acute ocular toxicity induced by desferrioxamine in patients with thalassaemia. *Arch.Dis.Child*, **63**, 250-255.
De Sanctis.,V, Stea,S., Savarino,L., Granchi,D., Visentin,M., Sprocati,M., Govoni,R., & Pizzoferrato,A. (2000) Osteochondrodystrophic lesions in chelated thalassemic patients: an histological analysis. *Calcif.Tissue Int.*, **67**, 134-140.
De Sanctis.,V, Stea,S., Savarino,L., Scialpi,V., Traina,G.C., Chiarelli,G.M., Sprocati,M., Govoni,R., Pezzoli,D., Gamberini,R., & Rigolin,F. (1998) Growth hormone secretion and bone histomorphometric study in thalassaemic patients with acquired skeletal dysplasia secondary to desferrioxamine. *J.Pediatr.Endocrinol.Metab*, **11 Suppl 3**, 827-833.
De Sanctis.,V, Vullo,C., Bagni,B., & Chiccoli,L. (1992) Hypoparathyroidism in beta-thalassemia major. Clinical and laboratory observations in 24 patients. *Acta Haematol.*, **88**, 105-108.
De Sanctis.,V, Vullo,C., Katz,M., Wonke,B., Tanas,R., & Bagni,B. (1988a) Gonadal function in patients with beta thalassaemia major. *J.Clin.Pathol.*, **41**, 133-137.
De Sanctis.,V, Zurlo,M.G., Senesi,E., Boffa,C., Cavallo,L., & Di Gregorio,F. (1988b) Insulin dependent diabetes in thalassaemia. *Arch.Dis.Child*, **63**, 58-62.

Di Palma,A., Moratelli,S., Tolomelli,P., Giuberti,M., Tenan,R., Fagioli,F., Landi,L., Toffoli,C., Atti,G., & Vullo,C. (1994) Pattern of iron excretion in relation to haemoglobin level and iron load in 8 haematological patients following the administration of subcutaneous deferrioxamine. *Eur.J.Haematol.*, **53**, 197-200.

Dickerhoff,R. (1987) Acute aphasia and loss of vision with desferrioxamine overdose. *Am.J.Pediatr.Hematol.Oncol.*, **9**, 287-288.

Drysdale,J.W. & Munro,H.N. (1966) Regulation of synthesis and turnover of ferritin in rat liver. *J.Biol.Chem.*, **241**, 3630-3637.

Duke-Elder,S. (1969) *Diseases of the Lens and Vitreous*. St. Louis, CV Mosby.

Ehlers,K.H., Giardina,P.J., Lesser,M.L., Engle,M.A., & Hilgartner,M.W. (1991) Prolonged survival in patients with beta-thalassemia major treated with deferoxamine. *J.Pediatr.*, **118**, 540-545.

Engle,M.A., Erlandson,M., & Smith,C.H. (1964) Late cardiac complications of chronic, severe, refractory anemia with hemochromatosis. *Circulation* **30**, 698-705.

Epsztejn,S., Kakhlon,O., Glickstein,H., Breuer,W., & Cabantchik,I. (1997) Fluorescence analysis of the labile iron pool of mammalian cells. *Anal.Biochem.*, **248**, 31-40.

Fischer,R. (2001) SQUID biomagnetic liver susceptometry in the adjustment of chelation treatment. *S.O.S.T.E.Notiziario* **2**, 50-54.

Flynn,D.M., Fairney,A., Jackson,D., & Clayton,B.E. (1976) Hormonal changes in thalassaemia major. *Arch.Dis.Child*, **51**, 828-836.

Flynn,D.M., Hoffbrand,A.V., & Politis,D. (1982) Subcutaneous desferrioxamine: the effect of three years' treatment on liver, iron, serum ferritin, and comments on echocardiography. *Birth Defects Orig.Artic.Ser.*, **18**, 347-353.

Fosburg,M.T. & Nathan,D.G. (1990) Treatment of Cooley's anemia. *Blood*, **76**, 435-444.

Freedman,M.H., Grisaru,D., Olivieri,N., MacLusky,I., & Thorner,P.S. (1990) Pulmonary syndrome in patients with thalassemia major receiving intravenous deferoxamine infusions. *Am.J.Dis.Child*, **144**, 565-569.

Freeman,A.P., Giles,R.W., Berdoukas,V.A., Talley,P.A., & Murray,I.P. (1989) Sustained normalization of cardiac function by chelation therapy in thalassaemia major. *Clin.Lab Haematol.*, **11**, 299-307.

Freeman,A.P., Giles,R.W., Berdoukas,V.A., Walsh,W.F., Choy,D., & Murray,P.C. (1983) Early left ventricular dysfunction and chelation therapy in thalassemia major. *Ann.Intern.Med.*, **99**, 450-454.

Gabutti,V. & Piga,A. (1996) Results of long-term iron-chelating therapy. *Acta Haematol.*, **95**, 26-36.

Gallant,T., Freedman,M.H., Vellend,H., & Francombe,W.H. (1986) Yersinia sepsis in patients with iron overload treated with deferoxamine. *N.Engl.J.Med.*, **314**, 1643.

Gamberini,M.R., Fortini,M., Gilli,G., Testa,M.R., & De,S., V (1998) Epidemiology and chelation therapy effects on glucose homeostasis in thalassaemic patients. *J.Pediatr.Endocrinol.Metab*, **11 Suppl 3**, 867-869.

Gaziev,D., Baronciani,D., Galimberti,M., Polchi,P., Angelucci,E., Giardini,C., Muretto,P., Perugini,S., Riggio,S., Ghirlanda,S., Erer,B., Maiello,A., & Lucarelli,G. (1996) Mucormycosis after bone marrow transplantation: report of four cases in thalassemia and review of the literature. *Bone Marrow Transplant.*, **17**, 409-414.

Gelmi,C., Borgna-Pignatti,C., Franchin,S., Tacchini,M., & Trimarchi,F. (1988) Electroretinographic and visual-evoked potential abnormalities in patients with beta-thalassemia major. *Ophthalmologica*, **196**, 29-34.

Gertner,J.M., Broadus,A.E., Anast,C.S., Grey,M., Pearson,H., & Genel,M. (1979) Impaired parathyroid response to induced hypocalcemia in thalassemia major. *J.Pediatr.*, **95**, 210-213.

Giardina,P.J., Ehlers,K.H., Engle,M.A., Grady,R.W., & Hilgartner,M.W. (1985) The effect of subcutaneous deferoxamine on the cardiac profile of thalassemia major: a five-year study. *Ann.N.Y.Acad.Sci.*, **445**, 282-292.

Giardina, P.J., Ehlers, K.H., Grady, R.W., Lesser, M.L., New, M.I., Hilgartner, M.W. (1996) Progress in the management of thalassemia: over a decade and a half of experience with subcutaneous desferrioxamine. *Bone Marrow Transplant.* **19 Suppl 2**, 9-10.

Grady,R.W., Giardina,P.J., & Hilgartner,M.W. (1987) Intravenous desferrioxamine and total iron balance. *Blood* **70**, 47a.

Graziano,J.H., Markenson,A., Miller,D.R., Chang,H., Bestak,M., Meyers,P., Pisciotto,P., & Rifkind,A. (1978) Chelation therapy in beta-thalassemia major. I. Intravenous and subcutaneous deferoxamine. *J.Pediatr.*, **92**, 648-652.

Grundy,R.G., Woods,K.A., Savage,M.O., & Evans,J.P. (1994) Relationship of endocrinopathy to iron chelation status in young patients with thalassaemia major. *Arch.Dis.Child*, **71**, 128-132.

Guerin,A., London,G., Marchais,S., Metivier,F., & Pelisse,J.M. (1985) Acute deafness and desferrioxamine. *Lancet*, **2**, 39-40.

Hartkamp,M.J., Babyn,P.S., & Olivieri,F. (1993) Spinal deformities in deferoxamine-treated homozygous beta-thalassemia major patients. *Pediatr.Radiol.*, **23**, 525-528.
Hatori,M., Sparkman,J., Teixeira,C.C., Grynpas,M., Nervina,J., Olivieri,N., & Shapiro,I.M. (1995) Effects of deferoximine on chondrocyte alkaline phosphatase activity: proxidant role of deferoximine in thalassemia. *Calcif.Tissue Int.*, **57**, 229-236.
Henry,W. Echocardiographic evaluation of the heart in thalassemia major. (1979) In Nienhuis A.W.; moderator. Thalassemia major: molecular and clinical aspects. *Ann.Intern.Med.* **91**,892-894.
Henry,W.L., Nienhuis,A.W., Wiener,M., Miller,D.R., Canale,V.C., & Piomelli,S. (1978) Echocardiographic abnormalities in patients with transfusion-dependent anemia and secondary myocardial iron deposition. *Am.J.Med.*, **64**, 547-555.
Herington,A.C., Werther,G.A., Matthews,R.N., & Burger,H.G. (1981) Studies on the possible mechanism for deficiency of nonsuppressible insulin-like activity in thalassemia major. *J.Clin.Endocrinol.Metab*, **52**, 393-398.
Hershko,C., Cook,J.D., & Finch,D.A. (1973) Storage iron kinetics. 3. Study of desferrioxamine action by selective radioiron labels of RE and parenchymal cells. *J.Lab Clin.Med.*, **81**, 876-886.
Hershko,C., Grady,R.W., & Cerami,A. (1978) Mechanism of iron chelation in the hypertransfused rat: definition of two alternative pathways of iron mobilization. *J.Lab Clin.Med.*, **92**, 144-151.
Hershko,C. & Hoffbrand,A.V. (2000) Iron chelation therapy. *Rev.Clin.Exp.Hematol.* **4**, 337-361.
Hershko,C. & Rachmilewitz,E.A. (1979) Mechanism of desferrioxamine-induced iron excretion in thalassaemia. *Br.J.Haematol.*, **42**, 125-132.
Hoffbrand,A.V., Gorman,A., Laulicht,M., Garidi,M., Economidou,J., Georgipoulou,P., Hussain,M.A., & Flynn,D.M. (1979) Improvement in iron status and liver function in patients with transfusional iron overload with long-term subcutaneous desferrioxamine. *Lancet*, **1**, 947-949.
Hoyes,K.P. & Porter,J.B. (1993) Subcellular distribution of desferrioxamine and hydroxypyridin-4-one chelators in K562 cells affects chelation of intracellular iron pools. *Br.J.Haematol.*, **85**, 393-400.
Hussain,M.A., Green,N., Flynn,D.M., & Hoffbrand,A.V. (1977) Effect of dose, time, and ascorbate on iron excretion after subcutaneous desferrioxamine. *Lancet*, **1**, 977-979.
Hussain,M.A., Green,N., Flynn,D.M., Hussein,S., & Hoffbrand,A.V. (1976) Subcutaneous infusion and intramuscular injection of desferrioxamine in patients with transfusional iron overload. *Lancet*, **2**, 1278-1280.
Italian Working Group on Endocrine Complications in Non-endocrine Diseases. (1995) Multicentre study on prevalence of endocrine complications in thalassaemia major. *Clin.Endocrinol.(Oxf)*, **42**, 581-586.
Janka,G.E., Mohring,P., Helmig,M., Haas,R.J., & Betke,K. (1981) Intravenous and subcutaneous desferrioxamine therapy in children with severe iron overload. *Eur.J.Pediatr.*, **137**, 285-290.
Jensen,C.E., Tuck,S.M., Old,J., Morris,R.W., Yardumian,A., De Sanctis, V, Hoffbrand,A.V., & Wonke,B. (1997) Incidence of endocrine complications and clinical disease severity related to genotype analysis and iron overload in patients with beta-thalassaemia. *Eur.J.Haematol.*, **59**, 76-81.
Jensen,C.E., Tuck,S.M., & Wonke,B. (1995) Fertility in beta thalassaemia major: a report of 16 pregnancies, preconceptual evaluation and a review of the literature. *Br.J.Obstet.Gynaecol.*, **102**, 625-629.
Jensen,P.D., Heickendorff,L., Pedersen,B., Bendix-Hansen,K., Jensen,F.T., Christensen,T., Boesen,A.M., & Ellegaard,J. (1996) The effect of iron chelation on haemopoiesis in MDS patients with transfusional iron overload. *Br.J.Haematol.*, **94**, 288-299.
Jensen,P.D., Jensen,F.T., Christensen,T., & Ellegaard,J. (1995) Evaluation of transfusional iron overload before and during iron chelation by magnetic resonance imaging of the liver and determination of serum ferritin in adult non-thalassaemic patients. *Br.J.Haematol.*, **89**, 880-889.
Keberle,H. (1964) The biochemistry of desferrioxamine and its relation to iron metabolism. *Ann.N.Y.Acad.Sci.* **119**, 758-768.
Keberle,H. (1992) From antibiotic to chelating agent. In Gross, K., Aumiller, J., Gelzer, J. (eds) Desferrioxamine ®Desferal: History, Clinical Value, Perspectives. Munich, MMV Medizin Verlag, 26-36.
Kelly,D.A., Price,E., Jani,B., Wright,V., Rossiter,M., & Walker-Smith,J.A. (1987) Yersinia enterocolitis in iron overload. *J.Pediatr.Gastroenterol.Nutr.*, **6**, 643-645.
Kletzky,O.A., Costin,G., Marrs,R.P., Bernstein,G., March,C.M., & Mishell,D.R., Jr. (1979b) Gonadotropin insufficiency in patients with thalassemia major. *J.Clin.Endocrinol.Metab*, **48**, 901-905.
Konijn,A.M., Glickstein,H., Vaisman,B., Meyron-Holtz,E.G., Slotki,I.N., & Cabantchik,Z.I. (1999) The cellular labile iron pool and intracellular ferritin in K562 cells. *Blood*, **94**, 2128-2134.
Koren,G., Bentur,Y., Strong,D., Harvey,E., Klein,J., Baumal,R., Spielberg,S.P., & Freedman,M.H. (1989) Acute changes in renal function associated with deferoxamine therapy. *Am.J.Dis.Child*, **143**, 1077-1080.
Kouides,P.A., Slapak,C.A., Rosenwasser,L.J., & Miller,K.B. (1988) Pneumocystis carinii pneumonia as a complication of desferrioxamine therapy. *Br.J.Haematol.*, **70**, 383-384.

Landau,H., Matoth,I., Landau-Cordova,Z., Goldfarb,A., Rachmilewitz,E.A., & Glaser,B. (1993) Cross-sectional and longitudinal study of the pituitary-thyroid axis in patients with thalassaemia major. *Clin.Endocrinol.(Oxf)*, **38**, 55-61.
Lassman,M.N., O'Brien,R.T., Pearson,H.A., Wise,J.K., Donabedian,R.K., Felig,P., & Genel,M. (1974) Endocrine evaluation in thalassemia major. *Ann.N.Y.Acad.Sci.*, **232**, 226-237.
Laub,R., Schneider,Y.J., Octave,J.N., Trouet,A., & Crichton,R.R. (1985) Cellular pharmacology of deferrioxamine B and derivatives in cultured rat hepatocytes in relation to iron mobilization. *Biochem.Pharmacol.*, **34**, 1175-1183.
Lee,P., Mohammed,N., Marshall,L., Abeysinghe,R.D., Hider,R.C., Porter,J.B., & Singh,S. (1993) Intravenous infusion pharmacokinetics of desferrioxamine in thalassaemic patients. *Drug Metab Dispos.*, **21**, 640-644.
Leger,J., Girot,R., Crosnier,H., Postel-Vinay,M.C., & Rappaport,R. (1989) Normal growth hormone (GH) response to GH-releasing hormone in children with thalassemia major before puberty: a possible age-related effect. *J.Clin.Endocrinol.Metab*, **69**, 453-456.
Leon,M.B., Borer,J.S., Bacharach,S.L., Green,M.V., Benz,E.J., Jr., Griffith,P., & Nienhuis,A.W. (1979) Detection of early cardiac dysfunction in patients with severe beta-thalassemia and chronic iron overload. *N.Engl.J.Med.*, **301**, 1143-1148.
Levine,J.E., Cohen,A., MacQueen,M., Martin,M., & Giardina,P.J. (1997) Sensorimotor neurotoxicity associated with high-dose deferoxamine treatment. *J.Pediatr.Hematol.Oncol.*, **19**, 139-141.
Lynch,S.R., Seftel,H.C., Torrance,J.D., Charlton,R.W., & Bothwell,T.H. (1967) Accelerated oxidative catabolism of ascorbic acid in siderotic Bantu. *Am.J.Clin.Nutr.*, **20**, 641-647.
MacLean,K.H., Cleveland,J.L., & Porter,J.B. (2001) Cellular zinc content is a major determinant of iron-chelator induced apoptosis of thymocytes. *Blood* (in press).
Magro,S., Puzzonia,P., Consarino,C., Galati,M.C., Morgione,S., Porcelli,D., Grimaldi,S., Tancre,D., Arcuri,V., De,S., V, & . (1990) Hypothyroidism in patients with thalassemia syndromes. *Acta Haematol.*, **84**, 72-76.
Marcus,R.E., Davies,S.C., Bantock,H.M., Underwood,S.R., Walton,S., & Huehns,E.R. (1984) Desferrioxamine to improve cardiac function in iron-overloaded patients with thalassemia major. *Lancet*, **1**, 392-393.
Marsh,M.N., Holbrook,I.B., Clark,C., & Shaffer,J.L. (1981) Tinnitus in a patient with beta-thalassaemia intermedia on long-term treatment with desferrioxamine. *Postgrad.Med.J.*, **57**, 582-584.
Masala,A., Meloni,T., Gallisai,D., Alagna,S., Rovasio,P.P., Rassu,S., & Milia,A.F. (1984) Endocrine functioning in multitransfused prepubertal patients with homozygous beta-thalassemia. *J.Clin.Endocrinol.Metab*, **58**, 667-670.
Mavrogeni,S.I., Maris,T., Gouliamos,A., Vlahos,L., & Kremastinos,D.T. (1998) Myocardial iron deposition in beta-thalassemia studied by magnetic resonance imaging. *Int.J.Card Imaging*, **14**, 117-122.
McIntosh,N. (1976) Endocrinopathy in thalassaemia major. *Arch.Dis.Child*, **51**, 195-201.
Melby,K., Slordahl,S., Gutteberg,T.J., & Nordbo,S.A. (1982) Septicaemia due to Yersinia enterocolitica after oral overdoses of iron. *Br.Med.J.(Clin.Res.Ed)*, **285**, 467-468.
Merali,S., Chin,K., Del Angel,L., Grady,R.W., Armstrong,M., & Clarkson,A.B., Jr. (1995) Clinically achievable plasma deferoxamine concentrations are therapeutic in a rat model of Pneumocystis carinii pneumonia. *Antimicrob.Agents Chemother.*, **39**, 2023-2026.
Merali,S., Chin,K., Grady,R.W., & Clarkson,A.B., Jr. (1996) Trophozoite elimination in a rat model of Pneumocystis carinii pneumonia by clinically achievable plasma deferoxamine concentrations. *Antimicrob.Agents Chemother.*, **40**, 1298-1300.
Miller,K.B., Rosenwasser,L.J., Bessette,J.A., Beer,D.J., & Rocklin,R.E. (1981) Rapid desensitisation for desferrioxamine anaphylactic reaction. *Lancet*, **1**, 1059.
Modell,B. (1979) Advances in the use of iron-chelating agents for the treatment of iron overload. *Prog.Hematol.*, **11**, 267-312.
Modell,B. & Berdoukas,V. (1984) Clinical Approach to Thalassaemia. London, Grune and Stratton.
Modell,B., Khan,M., & Darlison,M. (2000) Survival in beta-thalassaemia major in the UK: data from the UK Thalassaemia Register. *Lancet*, **355**, 2051-2052.
Modell,B., Letsky,E.A., Flynn,D.M., Peto,R., & Weatherall,D.J. (1982) Survival and desferrioxamine in thalassaemia major. *Br.Med.J.(Clin.Res.Ed)*, **284**, 1081-1084.
Modell,C.B. & Beck,J. (1974) Long-term desferrioxamine therapy in thalassemia. *Ann.N.Y.Acad.Sci.*, **232**, 201-210.
Nielsen,P., Fischer,R., Engelhardt,R., Tondury,P., Gabbe,E.E., & Janka,G.E. (1995) Liver iron stores in patients with secondary haemosiderosis under iron chelation therapy with deferoxamine or deferiprone. *Br.J.Haematol.*, **91**, 827-833.
Nienhuis,A.W. (1981) Vitamin C and iron. *N.Engl.J.Med.*, **304**, 170-171.

Nienhuis,A.W., Delea,C., Aamodt,R., Bartter,F., & Anderson,W.F. (1976) Evaluation of desferrioxamine and ascorbic acid for the treatment of chronic iron overload. *Birth Defects Orig.Artic.Ser.*, **12**, 177-185.

O'Brien,R.T. (1974) Ascorbic acid enhancement of desferrioxamine-induced urinary iron excretion in thalassemia major. *Ann.N.Y.Acad.Sci.*, **232**, 221-225.

Olivieri,N.F., Basran,R.K., Talbot,A.L., Babyn,P., & Bailey,J.D. (1995) Abnormal growth in thalassemia major associated with deferoxamine-induced destruction of spinal cartilage and compromise of sitting height. *Blood* **86,** 482a.

Olivieri,N.F., Berriman,A.M., Tyler,B.J., Davis,S.A., Francombe,W.H., & Liu,P.P. (1992a) Reduction in tissue iron stores with a new regimen of continuous ambulatory intravenous deferoxamine. *Am.J.Hematol.*, **41**, 61-63.

Olivieri,N.F. & Brittenham,G.M. (1997) Iron-chelating therapy and the treatment of thalassemia. *Blood*, **89**, 739-761.

Olivieri,N.F., Buncic,J.R., Chew,E., Gallant,T., Harrison,R.V., Keenan,N., Logan,W., Mitchell,D., Ricci,G., Skarf,B., & . (1986) Visual and auditory neurotoxicity in patients receiving subcutaneous deferoxamine infusions. *N.Engl.J.Med.*, **314**, 869-873.

Olivieri,N.F., Koren,G., Harris,J., Khattak,S., Freedman,M.H., Templeton,D.M., Bailey,J.D., & Reilly,B.J. (1992b) Growth failure and bony changes induced by deferoxamine. *Am.J.Pediatr.Hematol.Oncol.*, **14**, 48-56.

Olivieri,N.F., Nathan,D.G., MacMillan,J.H., Wayne,A.S., Liu,P.P., McGee,A., Martin,M., Koren,G., & Cohen,A.R. (1994) Survival in medically treated patients with homozygous beta-thalassemia. *N.Engl.J.Med.*, **331**, 574-578.

Orzincolo,C., Castaldi,G., De,S., V, Scutellari,P.N., Ciaccio,C., & Vullo,C. (1990) [Rickets- and/or scurvy-like bone lesions in beta-thalassemia major]. *Radiol.Med.(Torino)*, **80**, 823-829.

Orzincolo,C., Castaldi,G., & Scutellari,P.N. (1994) Platyspondyly in treated beta-thalassemia. *Eur.J.Radiol.*, **18**, 129-133.

Pall,H., Blake,D.R., Good,P.A., Winyard,P., & Williams,A.C. (1986) Copper chelation and the neuro-ophthalmic toxicity of desferrioxamine. *Lancet*, **2**, 1279.

Pengloan,J., Dantal,J., Rossazza,C., Abazza,M., & Nivet,H. (1987) Ocular toxicity after a single intravenous dose of desferrioxamine in 2 hemodialyzed patients. *Nephron*, **46**, 211-212.

Perry,R.D. & Brubaker,R.R. (1979) Accumulation of iron by yersiniae. *J.Bacteriol.*, **137**, 1290-1298.

Piga, A., Longo, F., Consolati, A., Sachetti, L., De Leo, A., Caramellino, L. (1996) Mortality and morbidity in thalassemia with conventional treatment. *Bone Marrow Transplant.* **19 Suppl 2,** 11-13.

Piga,A., Luzzatto,L., Capalbo,P., Gambotto,S., Tricta,F., & Gabutti,V. (1988) High-dose desferrioxamine as a cause of growth failure in thalassemic patients. *Eur.J.Haematol.*, **40**, 380-381.

Piga,A., Muroni,P., Buffardi,S., Magnano,C., Durken,M., Nangeroni,M., Forni,G.L., Pizzarelli,G., Galanello,R., & Longo,F. (2001) Long-term central venous catheter complications during intensive chelation therapy in β-thalassemia patients: a multicenter study. *S.O.S.T.E.Notiziario* **2,** 141.

Pintor,C., Cella,S.G., Manso,P., Corda,R., Dessi,C., Locatelli,V., & Muller,E.E. (1986) Impaired growth hormone (GH) response to GH-releasing hormone in thalassemia major. *J.Clin.Endocrinol.Metab*, **62**, 263-267.

Pippard,M.J., Callender,S.T., & Finch,C.A. (1982a) Ferrioxamine excretion in iron-loaded man. *Blood*, **60**, 288-294.

Pippard,M.J., Callender,S.T., & Weatherall,D.J. (1978b) Intensive iron-chelation therapy with desferrioxamine in iron-loading anaemias. *Clin.Sci.Mol.Med.*, **54**, 99-106.

Pippard,M.J., Johnson,D.K., & Finch,C.A. (1982b) Hepatocyte iron kinetics in the rat explored with an iron chelator. *Br.J.Haematol.*, **52**, 211-224.

Pippard,M.J., Letsky,E.A., Callender,S.T., & Weatherall,D.J. (1978a) Prevention of iron loading in transfusion-dependent thalassaemia. *Lancet*, **1**, 1178-1181.

Polson,R.J., Jawed,A., Bomford,A., Berry,H., & Williams,R. (1985) Treatment of rheumatoid arthritis with desferrioxamine: relation between stores of iron before treatment and side effects. *Br.Med.J.(Clin.Res.Ed)*, **291**, 448.

Porter,J.B., Alberti,D., Hassan,I., Howes,C., Stallibrass,L., Racine,A., Alexander,E., Davis,B., Voi,V., Brookman,L., & Piga,A. (1997) Subcutaneous depot desferrioxamine (CGH 749B); Relationship of pharmacokinetics to efficacy and drug metabolism. *Blood* **90**, 265a.

Porter,J.B., Faherty,A., Stallibrass,L., Brookman,L., Hassan,I., & Howes,C. (1998b) A trial to investigate the relationship between DFO pharmacokinetics and metabolism and DFO-related toxicity. *Ann.N.Y.Acad.Sci.*, **850**, 483-487.

Porter,J.B., Huehns,E.R., & Hider,R.C. (1989b) The development of iron chelating drugs. *Baillieres Clin.Haematol.*, **2**, 257-292.

Porter,J.B., Jaswon,M.S., Huehns,E.R., East,C.A., & Hazell,J.W. (1989a) Desferrioxamine ototoxicity: evaluation of risk factors in thalassaemic patients and guidelines for safe dosage. *Br.J.Haematol.*, **73**, 403-409.
Porter,J., Singh,S., Mohammed,N., Abeysinghe,R., & Hider,R. (1992) Osteopaenia in thalassaemia major: The relevance of desferrioxamine (DFO) metabolism. *International Society of Haematology*, **24th Congress**, London, 47.
Porter,J.B., Weir,D.T., Davis,B., Alexander,E., McCombie,R.R., Walker,S.M., Osborne,S., & Lowe,P.J. (1998a) Iron balance and chelation efficiency following single dose depot desferrioxamine. *Blood* **92**, 325a.
Propper,R.D., Cooper,B., Rufo,R.R., Nienhuis,A.W., Anderson,W.F., Bunn,H.F., Rosenthal,A., & Nathan,D.G. (1977) Continuous subcutaenous administration of deferoxamine in patients with iron overload. *N.Engl.J.Med.*, **297**, 418-423.
Propper,R.D., Shurin,S.B., & Nathan,D.G. (1976) Reassessment of the use of desferrioxamine B in iron overload. *N.Engl.J.Med.*, **294**, 1421-1423.
Rahi,A.H., Hungerford,J.L., & Ahmed,A.I. (1986) Ocular toxicity of desferrioxamine: light microscopic histochemical and ultrastructural findings. *Br.J.Ophthalmol.*, **70**, 373-381.
Rahko,P.S., Salerni,R., & Uretsky,B.F. (1986) Successful reversal by chelation therapy of congestive cardiomyopathy due to iron overload. *J.Am.Coll.Cardiol.*, **8**, 436-440.
Richardson,M.E., Matthews,R.N., Alison,J.F., Menahem,S., Mitvalsky,J., Byrt,E., & Harper,R.W. (1993) Prevention of heart disease by subcutaneous desferrioxamine in patients with thalassaemia major. *Aust.N.Z.J.Med.*, **23**, 656-661.
Robins-Browne,R.M. & Prpic,J.K. (1983) Desferrioxamine and systemic yersiniosis. *Lancet*, **2**, 1372.
Robins-Browne,R.M., Prpic,J.K., & Stuart,S.J. (1987) Yersiniae and iron. A study in host-parasite relationships. *Contrib.Microbiol.Immunol.*, **9**, 254-258.
Robins-Browne,R.M., Rabson,A.R., & Koornhof,H.J. (1979) Generalized infection with Yersinia enterocolitica and the role of iron. *Contrib.Microbiol.Immunol.*, **5**, 277-282.
Rodda,C.P., Reid,E.D., Johnson,S., Doery,J., Matthews,R., & Bowden,D.K. (1995) Short stature in homozygous beta-thalassaemia is due to disproportionate truncal shortening. *Clin.Endocrinol.(Oxf)*, **42**, 587-592.
Roeser,H.P., Halliday,J.W., Sizemore,D.J., Nikles,A., & Willgoss,D. (1980) Serum ferritin in ascorbic acid deficiency. *Br.J.Haematol.*, **45**, 459-466.
Rubinstein,M., Dupont,P., Doppee,J.P., Dehon,C., Ducobu,J., & Hainaut,J. (1985) Ocular toxicity of desferrioxamine. *Lancet*, **1**, 817-818.
Sabato,A.R., De,S., V, Atti,G., Capra,L., Bagni,B., & Vullo,C. (1983) Primary hypothyroidism and the low T3 syndrome in thalassaemia major. *Arch.Dis.Child*, **58**, 120-127.
Saenger,P., Schwartz,E., Markenson,A.L., Graziano,J.H., Levine,L.S., New,M.I., & Hilgartner,M.W. (1980) Depressed serum somatomedin activity in beta-thalassemia. *J.Pediatr.*, **96**, 214-218.
Saito,K., Nishisato,T., Grasso,J.A., & Aisen,P. (1986) Interaction of transferrin with iron-loaded rat peritoneal macrophages. *Br.J.Haematol.*, **62**, 275-286.
Sands,J.M., Macher,A.M., Ley,T.J., & Nienhuis,A.W. (1985) Disseminated infection caused by Cunninghamella bertholletiae in a patient with beta-thalassemia. Case report and review of the literature. *Ann.Intern.Med.*, **102**, 59-63.
Schafer,A.I., Rabinowe,S., Le Boff,M.S., Bridges,K., Cheron,R.G., & Dluhy,R. (1985) Long-term efficacy of deferoxamine iron chelation therapy in adults with acquired transfusional iron overload. *Arch.Intern.Med.*, **145**, 1217-1221.
Sephton Smith,R. (1962) Iron excretion in thalassaemia after administration of chelating agents. *Br.Med.J.* **2**, 1577-1580.
Sephton Smith,R. (1964) Chelating agents in the diagnosis and treatment of iron overload in thalassemia. *Ann.N.Y.Acad.Sci.* **119**, 776-788.
Shehadeh,N., Hazani,A., Rudolf,M.C., Peleg,I., Benderly,A., & Hochberg,Z. (1990) Neurosecretory dysfunction of growth hormone secretion in thalassemia major. *Acta Paediatr.Scand.*, **79**, 790-795.
Shirane,M. & Harrison,R.V. (1987) The effects of deferoxamine mesylate and hypoxia on the cochlea. *Acta Otolaryngol.*, **104**, 99-107.
Simon,P., Ang,K.S., Meyrier,A., Allain,P., & Mauras,Y. (1983) Desferrioxamine, ocular toxicity, and trace metals. *Lancet*, **2**, 512-513.
Singh,S., Hider,R.C., & Porter,J.B. (1990) Separation and identification of desferrioxamine and its iron chelating metabolites by high-performance liquid chromatography and fast atom bombardment mass spectrometry: choice of complexing agent and application to biological fluids. *Anal.Biochem.*, **187**, 212-219.

Sklar,C.A., Lew,L.Q., Yoon,D.J., & David,R. (1987) Adrenal function in thalassemia major following long-term treatment with multiple transfusions and chelation therapy. Evidence for dissociation of cortisol and adrenal androgen secretion. *Am.J.Dis.Child*, **141**, 327-330.

Solomon,L.R., Hillman,R.S., & Finch,C.A. (1981) Serum ferritin in refractory anemias. *Acta Haematol.*, **66**, 1-5.

Stefanini,S., Chiancone,E., Cavallo,S., Saez,V., Hall,A.D., & Hider,R.C. (1991) The interaction of hydroxypyridinones with human serum transferrin and ovotransferrin. *J.Inorg.Biochem.*, **44**, 27-37.

Tamary,H., Goshen,J., Carmi,D., Yaniv,I., Kaplinsky,C., Cohen,I.J., & Zaizov,R. (1994) Long-term intravenous deferoxamine treatment for noncompliant transfusion-dependent beta-thalassemia patients. *Isr.J.Med.Sci.*, **30**, 658-664.

Thalassemia International Federation. (2000) Guidelines for the clinical management of thalassemia.

Tolis,G., Politis,C., Kontopoulou,I., Poulatzas,N., Rigas,G., Saridakis,C., Athanasiou,V., Mortoglou,A., Malachtari,S., & Ling,N. (1987) Pituitary somatotropic and corticotropic function in patients with beta-thalassemia on iron chelation therapy. *Birth Defects Orig.Artic.Ser.*, **23**, 449-452.

Villeneuve,J.P., Bilodeau,M., Lepage,R., Cote,J., & Lefebvre,M. (1996) Variability in hepatic iron concentration measurement from needle-biopsy specimens. *J.Hepatol.*, **25**, 172-177.

Walker,J.A., Sherman,R.A., & Eisinger,R.P. (1985) Thrombocytopenia associated with intravenous desferrioxamine. *Am.J.Kidney Dis.*, **6**, 254-256.

Wang,C., Tso,S.C., & Todd,D. (1989) Hypogonadotropic hypogonadism in severe beta-thalassemia: effect of chelation and pulsatile gonadotropin-releasing hormone therapy. *J.Clin.Endocrinol.Metab*, **68**, 511-516.

Wapnick,A.A., Lynch,S.R., Charlton,R.W., Seftel,H.C., & Bothwell,T.H. (1969) The effect of ascorbic acid deficiency on desferrioxamine-induced urinary iron excretion. *Br.J.Haematol.*, **17**, 563-568.

Waxman,H.S. & Brown,E.B. (1969) Clinical usefulness of iron chelating agents. *Prog.Hematol.*, **6**, 338-373.

Werther,G.A., Matthews,R.N., Burger,H.G., & Herington,A.C. (1981) Lack of response of nonsuppressible insulin-like activity to short term administration of human growth hormone in thalassemia major. *J.Clin.Endocrinol.Metab*, **53**, 806-809.

White,G.P., Bailey-Wood,R., & Jacobs,A. (1976) The effect of chelating agents on cellular iron metabolism. *Clin.Sci.Mol.Med.*, **50**, 145-152.

Wolfe,L., Olivieri,N., Sallan,D., Colan,S., Rose,V., Propper,R., Freedman,M.H., & Nathan,D.G. (1985) Prevention of cardiac disease by subcutaneous deferoxamine in patients with thalassemia major. *N.Engl.J.Med.*, **312**, 1600-1603.

Wonke,B., Hoffbrand,A.V., Aldouri,M., Wickens,D., Flynn,D., Stearns,M., & Warner,P. (1989) Reversal of desferrioxamine induced auditory neurotoxicity during treatment with Ca-DTPA. *Arch.Dis.Child*, **64**, 77-82.

Worwood,M. (1986) Serum ferritin. *Clin.Sci.(Colch.)*, **70**, 215-220.

Zahner,H. (1992) Microorganisms as producers of Fe(III)-transport compounds. In Gross, K., Aumiller, J., Gelzer, J. (eds) Desferrioxamine ®Desferal: History, Clinical Value, Perspectives. Munich, MMV Medizin Verlag, 11-25.

Zurlo,M.G., De Stefano,P., Borgna-Pignatti,C., Di Palma,A., Piga,A., Melevendi,C., Di Gregorio,F., Burattini,M.G., & Terzoli,S. (1989) Survival and causes of death in thalassaemia major. *Lancet*, **2**, 27-30.

Table 1. Survival in β-thalassemia major—changing patterns

Author	Year	Patients	Survival in hypertransfused and DFO-chelated patients
Modell et al	1982	92	25% at 25 years among U.K. patients born after 1963 and treated at both specialist and nonspecialist units; better survival among those who received mean DFO dose >4 g/week
Zurlo et al	1989	1087	84.2% at 15 years in 1965-69 birth cohort v 96.9% at 15 years in 1970-74 cohort; patients treated at 7 specialist centers
Brittenham et al	1994	59	100% at 25 years in well-chelated patients v 32% at 25 years in poorly chelated patients born between 1963 and 1987; periodic follow up by specialists in a single center
Olivieri et al	1994	97	91% in best chelated group v 18% in worst chelated group after 15 years of therapy; patients treated at 3 specialist units
Giardina et al	1996	88	Median survival of 29 years in patients born after 1963 and treated at a single specialist center
Piga et al	1996	257	66% at 25 years in patients born between 1958 and 1993 and treated at a single specialist center
Borgna-Pignatti et al	1998	1146	82% at 25 years in patients born between 1970 and 1974 and treated in 7 specialist centers
Modell et al	2000	796	50% mortality before the age of 35 years among U.K. patients treated at both specialist and nonspecialist units
Davis et al	2001a	103	78% at 40 years for full cohort born 1957-1997 and 100% at 25 years in patients born after 1974; patients treated at a single specialist unit in U.K.

LONG TERM DEFERIPRONE CHELATION THERAPY

Victor A Hoffbrand* and Beatrix Wonke

1. INTRODUCTION

A wide variety of compounds have been tested in animals and humans in order to find drugs which are capable of removing iron from patients with refractory anaemias receiving regular blood transfusions. Parenteral preparations which have been evaluated clinically include desferrioxamine (DFO), diethylene triamine penta-acetic acid (DTPA), desferrithiocin and HBED (N,NL bis (2-hydroxy-benzyl ethylenediamine - NNI diacetic acid). Orally active chelators include 2,3 dihydroxybenzoic acid, pyridoxal isonicotinyl hydrazone, 1,2,dimethyl-pyrid-4-one (L1, deferiprone) and 1,2 diethylpyrid-4-one). The only two compounds in wide clinical use are DFO and deferiprone. DFO discussed in detail elsewhere in this volume remains standard therapy but because of cost, failure of compliance and, less frequently, toxicity or sensitivity, it is not satisfactory for many patients worldwide, especially in poorer countries where thalassaemia major and other transfusion dependent genetic disorders of haemoglobin are most frequent and antenatal diagnosis and prevention is inadequate or non-existent. In India, deferiprone, known as Kelfer (Cipla, Bombay, India) has been licensed since 1990. In the European Community Countries deferiprone, known as Ferriprox, has been licensed for those patients for whom DFO is contra-indicated or who exhibit serious toxicity to DFO therapy. In this review we shall describe the long term results of therapy with deferiprone either alone or in combination with DFO. Other recent reviews of this drug include those of Addis et al[1], Basman Balfour et al[2], Diav-Citrin et al[3], Hershko & Hoffbrand[4], Olivieri[5].

After in vitro biochemical and cell culture experiments, in vivo, studies were carried out in mice, rats, rabbits and monkeys (reviewed[6,7]). The first short term trials of deferiprone in humans began in 1987[8,9]. Initially, multiply transfused patients with

---

* A.V. Hoffbrand, Department of Haematology, Royal Free Hospital, Pond Street, Hampstead, NW3 2QG.London, Beatrix Wonke, Consultant Haematologist, Whittington Hospital, London

Iron Chelation Therapy.
Edited by Chaim Hershko, Kluwer Academic/Plenum Publishers, 2002.

myelodysplasia were administered the drug. In the same year patients with thalassaemia major (TM) were also tested. These early studies showed that the drug was, in most cases, well tolerated at doses of 75-100mg/kg per day divided into two, three or four sub-doses, that it increased urine iron excretion in relation to the dose given and to body iron stores assessed by units of blood transfused in previously poorly chelated patients. In some cases, excretion exceeded 0.5mg/kg/day, the amount of iron usually accumulated by patients with TM and other refractory anaemias given regular blood transfusion therapy. Subsequent to these early promising results, detailed pharmacokinetic studies were performed by several groups[10-12] and long term clinical trials have been carried out in many centres throughout the world (Table I)[13-28].

2. PHARMACOKINETICS

Pharmocokinetic studies have helped to establish protocols for drug administration and to avoid toxic side effects. They have also helped to explain the variation in response in patients with apparently similar degrees of iron burden.

Deferiprone (M.W.139) appears in the blood within 5 to 10 minutes of a single oral dose and the peak plasma concentration occurs within 45-60 minutes suggesting absorption occurs from the stomach and/or duodenum. The plasma half life ($T1/2$) is about 1.5 hours and iron excretion is related to the area under the curve of the free drug in plasma. The mean elimination half life is short, about 2-3 hours in iron overloaded patients and 1-3 hours in normal volunteers. The drug is rapidly glucuronidated. This inactivates its iron chelating site and the variability of speed of glucuronidisation partly explains the variability in the efficiency of chelation by the drug between different patients. The $T1/2$ of the glucuronide (about 2.5 hours) is longer than that of the free drug and correlates with creatinine clearance. About 80% of a single oral dose appears in the urine and it is excreted in the urine as the free compound, its glucuronide derivative and bound to iron. It forms 3:1 deferiprone:iron complexes with a binding constant of 37. Deferiprone also binds aluminium, gallium, copper and zinc but magnesium, calcium and manganese do not compete with iron for binding to deferiprone in vitro. About 4-6% of the drug is excreted in urine bound to iron after a single oral dose in iron loaded patients. Although some studies have reported faecal excretion of up to 20% of a single administered dose, others have not detected the drug in faeces. There is also no definite evidence that the drug induces its own metabolism although this was suggested in one study.

Administration of the drug with food reduces the peak plasma level but not the total amount of drug absorbed. Neither vitamin C nor food has a predictable effect on iron excretion in response to a single dose of the drug.

3. LONG TERM CLINICAL TRIALS

These have been carried out with deferiprone alone for periods of 6 months to 7 years in regularly transfused patients mainly with TM in London, Toronto, several centres in Italy, Bombay, Amsterdam and Beirut (Table I). The efficacy has been monitored by urine iron excretion, serum ferritin, in some cases non-transferrin bound iron in serum[29], liver iron

Table 1. Effect of long-term deferiprone treatment on urinary iron excretion (UiE) and serum ferritin.

Authors	Reference	No. of patients	Dose (mg/kg)	duration (months)	Mean UiE mg/24 hours	Mean s. ferritin Before	After
Agarwal et al (1992)	13	52	75-100mg	(3-21)	42.3		
Al-Refaie et al (1992)	14	11	85-119	(6-12)	0.44/kg	5499	4126
Al-Refaie et al (1995)	15	84	50-100	(<1-36)	0.6/kg	4026	3069
(1998)	17	7					
Hoffbrand et al (1998)	18	51	50-79	(4-49)	32.1	2937	3223
Kersten et al (1996)	19	36	50-100	(1-36)	21.0	3563	2560
Longo et al (1998)	20	52	75	(24)	15.9	1897	2116
Mazza et al (1998)	21	29	70	(>12)	-	3748	2550
Olivieri et al (1993)	22	15	75-100	(9-30)	-		
Olivieri et al (1995)	23	21 (12)	75	(12-56)	-	5759	3273
		9	75	(12-56)	-	1596	1768
Olivieri et al (1998)	24	18	75	(24-84)	-	4455	2831
Cohen et al (2000)	25	162	75	(12)		2579	2452
Taher et al (1999)	26	17	50-75	(6)	9.9	3863	3012
Taher et al (2001)	27	16	75	(24)	20.7	3663	2599
Del Vecchio (2000)	28	9	75	(18)	22.4	3178	3269

concentration (Table II) and by changes in organ, e.g. liver and heart function. Most of the patients treated have had TM but patients with myelodysplasia and other transfusion dependent anaemias have been included in some studies.

Table 2. Effect of long-term deferiprone treatment on liver iron concentrations.

Authors	Reference	No. of patients	Dose mg/kg/d	duration mo.	Method	Liver iron Mean or median (g/kg dry weight) Before	After
Longo et al (1998)	20	52	75	24	Squid	1.43	2.03
Mazza et al (1998)	21	20	70	>12	Biopsy	16.1	21.0
Olivieri et al (1995)	23	21	75	12-50	Biopsy or Squid	80.7	46.8 (μmol)
Olivieri et al (1998)	24	15	75	24		7.5	10.5 (mg/g)
DeVecchio et al (2000)	28	9	75	18	Squid	1.69	2.30 (mg/g)

4. EFFICACY

In all studies the drug has been found to reduce serum ferritin and liver iron concentrations in a majority of patients but increases in these measures of iron status have also been found in a minority of cases in each study suggesting that in these cases the drug, as administered, was not effective at reducing the body iron burden. A recent meta-analysis of nine clinical trials including data on 129 iron-overloaded patients showed that after a mean of 16 months, 75% of patients with severe iron overload showed a fall in serum ferritin and 52% were in negative iron balance.

In general, the results differ according to the initial status of the patients; in those previously poorly chelated with serum ferritin levels greater than 2500µg/l there has usually been an overall fall in serum ferritin whereas in those previously more adequately chelated the serum ferritin has usually been reported to remain unchanged. This pattern of response was best seen in the single centre Bombay study where among previously poorly chelated patients the greatest falls in serum ferritin were in the patients with initially highest levels[13]. In this study 46 (88.5%) of 52 of the previously largely unchelated patients showed a drop in serum ferritin with a mean decrease of 2292 µg/l. In one of the Toronto studies, 12 patients starting with a serum ferritin >2500 µg/l showed a significant fall with deferiprone therapy at a dose of 75mg/kg/day whereas in those starting with a serum ferritin <2500µg/l there was no overall significant change in serum ferritin over nearly 3 years of treatment[23].

It is, however, difficult from all the studies carried out to define the proportion of patients, chelated solely with deferiprone, who are maintained at a "safe" body iron burden. This is because it is unclear how to define a safe level. Cardiac disease due to iron overload is the dominant cause of death in multiply transfused TM patients and failure of compliance with DFO is associated with a cardiomyopathy and death[30-32]. On the basis of a retrospective analysis of TM patients receiving DFO as chelation therapy Olivieri et al[33] considered a consistent serum ferritin <2500µg/l safe in terms of cardiac damage, whereas levels above this were considered to be associated with significant risk of cardiadamage. Given the variability of serum ferritin according to iron load but also hepatitis status[34], vitamin C status[35] and its poor relation to body iron burden measured by other tests it is, however, clear that serum ferritin alone cannot be used as a predictor of organ damage.

Liver iron has been regarded by some as the best measure of the success or failure of chelation therapy[36]. This is based on studies in genetic haemochromatosis showing a liver iron content of >15mg/g dry weight (>80µmol/gram wet weight) is associated with poor survival. However, in genetic haemochromatosis, iron is largely confined to parenchymal cells of the liver and other organs wherever, in transfusional iron overload, the iron is accumulated initially in the reticulo-endothelial (RE) system and then redistributes to other organs, e.g. the heart, liver parenchymal cells and endocrine organs. There are no firm data correlating liver iron (a measure of both RE and parenchymal iron) with survival in TM. The recent studies of Anderson et al[37] estimating cardiac iron by T2-star (T2*) magnetic resonance have shown no correlation between cardiac iron, serum ferritin and liver iron measured by liver biopsy or T2*

magnetic resonance. These studies show that cardiac iron rather than liver iron correlates closely with cardiac function in 106 patients with TM. Furthermore, despite long term iron chelation therapy, 58% of 151 TM patients were found to have excess myocardial iron deposition, with progressive decline in both left and right ventricular ejection fraction[38]. The authors concluded that myocardial iron content cannot be predicted from serum ferritin or liver iron and conventional assessments of cardiac function can only detect those patients with advanced cardiac disease. They suggest iron chelation therapy should be intensified when the T2* falls below the normal range even in the presence of a normal resting left ventricular ejection fraction. They also found that iron cleared more slowly from the heart than liver when overall body stores are lowered by intravenous DFO therapy. Whether deferiprone, by entering cardiac cells used alone or with DFO will lower cardiac iron more quickly than DFO, is a current focus of active research. Preliminary data by Anderson et al[39] also suggest that deferiprone is more effective in preventing iron-induced cardiomyopathy than standard therapy with DFO. Deferiprone has a lower M.W. than DFO and consequent greater ability to penetrate cells. This aspect is discussed further below (see "Shuttle Effect").

The reasons for failure of deferiprone in some patients to prevent an increase in body iron burden may relate to transfusion need. The study of Longo et al[20] showed a rise in serum ferritin of 11% during 2 years of treatment with deferiprone. There was, however, an 18% increase in transfusion need and therefore of iron input during the same period. A second major factor may be the speed of glucuronidisation of deferiprone (see earlier). A third factor is the dose of the drug. Initial short term studies by Kontoghiorghes et al[8,9] showed that iron excretion was closely related to dose given. In some patients considerable differences in urine iron occurred between doses of 75 and 100mg/kg/day. Wonke et al[40] showed that increasing the daily dose from a mean of 75mg/kg to 83-100mg/kg resulted in a significant fall in mean serum ferritin in 9 patients after 6-17 months. In one patient tested there was also a fall in liver iron at the increased dose from 17.1 to 6.5mg/g dry weight after 11 months. The higher dose also resulted in an overall increase in urine iron excretion among the 9 patients.

Diav-Citrin et al[41] analysed 19 patients with TM treated for 2-6.5 years with deferiprone 75mg/kg/day. In seven hepatic iron concentrations were reduced or maintained at less than 7mg/g dry weight of liver. In the other 12 patients the levels rose to or stabilised above that level. The 12 patients showed initially lower plasma vitamin C levels and higher hepatic iron levels. However, lower vitamin C levels may have been the result of the higher body iron burden. The authors also noted an increase in apparent volume of distribution of deferiprone in time implying a shift of compartment of iron stores to one less accessible to deferiprone. They suggested that vitamin C therapy may result in increased iron excretion but in our own studies and those of other workers, vitamin C, however, has had no significant effect on iron excretion in response to deferiprone therapy[7]. The failure of deferiprone to maintain or reduce iron stores in some patients at a safe level may also relate to the initial iron burden of the patient and the length of time of follow-up. Our own study[18] has been quoted as showing that deferiprone is ineffective[42] because liver iron in 17 patients treated with deferiprone 75mg/kg/day for 24-48 months the liver iron ranged from 5.9 to 42.1mg/g dry weight, 10 having liver iron levels above 15.0mg/g dry weight. However, patients were entered in this study because of previous inadequate chelation and gross iron overload. Without initial liver biopsy samples it is impossible to determine whether or not liver iron was increasing or decreasing during deferiprone therapy. Moreover, the patient with the highest liver iron content (41.0mg/g) in our study was a poor complier with deferiprone as well as DFO therapy. Olivieri and colleagues[23] have reported equal efficacy of DFO 50mg/kg/day and deferiprone 75mg/kg/day but subsequently have suggested deferiprone loses efficacy, a suggestion not supported by our data showing that in patients receiving deferiprone for 2-4 years, in whom serum ferritin had not significantly changed, urine iron excretion in response to deferiprone had also not significantly changed[18].

Deferiprone is a potentially important drug for the treatment of less severe forms of iron overload than thalassaemia major, e.g. in patients with thalassaemia intermedia and other anaemias in which reduction of iron stores by venesection is impracticable but who do not have the need for continuing blood transfusions at the rate of 50 units per year. Olivieri et al[43] first showed reduction of serum ferritin, liver iron and urine iron excretion to normal with deferiprone therapy in a case of thalassaemia intermedia. We have found excellent response in terms of serum ferritin and liver iron in cases of beta thalassaemia/haemoglobin E disease treated with deferiprone at the low dose of 50mg/kg daily for 12 months (Pootrakul & Hoffbrand, unpublished).

5. SIDE EFFECTS

Studies with animals at daily doses of 200-300mg/kg/day showed marrow aplasia, rise in MCV, inhibition of lymphatic tissues and adrenal steatosis. It is possible that the effects on the immune system were mediated by zinc deficiency and occurred largely in non-iron loaded animals. It has also been suggested on the basis of in vitro studies that deferiprone may form 2:1 or 1:1 complexes with iron which could promote iron-induced peroxidative damage to DNA and other cell components[44]. This has not been substantiated in studies comparing DFO and deferiprone at mobilising heart cell iron and preserving mitochondrial respiratory function in normal and iron loaded heart cells in culture, even at suboptimal deferiprone concentrations[45].

All the established side effects of deferiprone have occurred in heavily iron loaded patients and were found in early studies by the London group. They reported nausea and other gastrointestinal symptoms[46], arthropathy[46], agranulocytosis[47], transient elevation of liver enzymes[46] and zinc deficiency of varying severity[48] in different patients. Other side effects, particularly immunological, suggested by other authors have not been substantiated in large long term trials. It is of note that whereas the main toxic effects of deferiprone have occurred in heavily iron loaded patients the drug appears remarkably safe in patients with lesser degrees of iron iverload. The reverse is true for DFO with which toxicity to the eyes, hearing and bones occurs in well chelated patients with relatively low degrees of iron overload. Improvement of genu valgum and growth velocity has been reported in TM patients treated with deferiprone, the bone and cartilage abnormalities having developed with standard dose DFO therapy[49].

5.1. Gastrointestinal disturbance. Nausea, anorexia and vomiting have been the most frequent side effects in some studies. In many cases the symptoms are mild and disappear despite continuation of therapy. Nausea occurred in 41% (7 of 17) in a Lebanese study (26), 45 (24%) of 187 patients in the Italian multicentre study[25] and in 7 (8%) of 84 patients in the International Collaborative Study[15]. In some cases the drug had to be discontinued. This occurred in 5 (10%) of 51 patients in our own study but in only 4 (2%) of 187 patients in the Italian study. Nausea was associated with the accumulation of the glucuronide in plasma in one of our patients with poor renal function.

5.2. Arthropathy. This consists of pain in the joints, especially the knees and other large joints. In some cases there are effusions. Pain in the muscles around joints also may occur. The incidence of arthropathy appears to be greater in the most iron loaded patients, being as high as 35% (20 of 52) patients in the Indian study[13] but only 6% of 187 of the Italian study of less iron loaded patients[25]. It also appears to be more common at larger drug doses[13]. Usually the arthropathy resolves spontaneously or with reduction of drug dose or temporary or permanent withdrawal of the drug but in a few cases the arthropathy has persisted for six months or more despite stopping the drug. The cause for the arthropathy is uncertain. Although immune abnormalities have been described in some patients, large studies have not shown any consistent change in incidence of positive antinuclear factor anti-histone or rheumatoid factor tests of other autoantibodies or in T or B lymphocyte populations in patients receiving the drug. The suggestion that free radicals are generated in the joints due to the formation of 2:1 or 1:1 chelator to iron complexes remains theoretical. It would be consistent with joint problems occurring in the most iron loaded patients but not with the incidence being greater in patients with similar degrees of iron overload receiving larger daily doses of the drug.

5.3. Neutropenia and Agranulocytosis. The first case of agranulocytosis was in a patient with Blackfan Diamond anaemia[47]. The agranulocytosis persisted for 3 weeks. Subsequent cases with myelodysplasia and thalassaemia major have been reported. All have been reversible and in most cases with neutropenia or agranulocytosis rechallenge has led to recurrence. Females appear to be more susceptible than males, as with agranulocytosis due to sensitivity to other drugs. In vitro and in vivo studies have not shown any evidence of an immune mechanism and an idiosynchratic toxic reaction appears most likely[48,49]. The incidence of agranulocytosis (neutrophils $< 0.5 \times 10^9/l$) was 0.6/100 patient years and of milder degrees of neutropenia (0.5-$1.5 \times 10^9/l$) 5.4/100 patient years in the LA.02 trial[25]. In some cases the neutropenia appeared to be due to an intercurrent viral infection. The neutropenia was more frequent in patients with an intact spleen and with some degree of neutropenia before starting deferiprone therapy.

5.4. Zinc Deficiency. This was first reported in 7 of 35 patients receiving deferiprone between 50 and 100mg/kg/day being more frequent in patients with diabetes mellitus in whom urinary zinc excretion is particularly increased in response to the drug[48]. One patient in this study developed skin lesions attributed to zinc deficiency which responded to zinc therapy. Subnormal serum zinc levels occurred in 8 of 51 patients subsequently treated in London[18]. Mean serum zinc levels also fell during the Italian multicentre studies but subnormal levels did not occur in any of the patients studied in Toronto[22]. Zinc deficiency has been suggested to play a role in immune deficiency in patients receiving deferiprone but there is no evidence for such an immune deficiency (see later).

5.5. Liver Abnormalities. Transient changes in liver enzymes without any overall significant change over the period of study were first reported to occur in a minority of patients receiving deferiprone[46]. In the multicentre study, increased alanine-transaminase (ALT) levels were generally transient and occurred more commonly in patients with hepatitis C[25]. For patients who completed the study ALT was significantly higher than baseline mean level of 59 U/l at 3 months and 6 months and for all patients (intention-to-treat analysis) at 3,6 and 9 months. The proportion of patients with ALT levels >2 x the upper limit of normal did not differ at each quarterly assessment. They were significantly more frequent in patients with positive hepatitis C serology. Four patients all seropositive for hepatitis C discontinued deferiprone therapy because of rise in ALT. The ALT returned to baseline levels. In three patients rechallenge resulted in a second rise in ALT.

5.6. Hepatic Fibrosis. Concerns related to the accelerated development of hepatic fibrosis have been expressed based on observations made in patients on long-term LI therapy at the Toronto Hospital for Sick Children[24]. These authors also quoted previous studies in iron loaded gerbils given a compound closely related to deferiprone, 1,2 diethyl-3-hydroxypyridin-4-one, reported to show worsening hepatic and cardiac fibrosis[52] as evidence in favour of their hypothesis that deferiprone may cause liver fibrosis. These gerbil studies, however, have not been confirmed in a similar animal model using deferiprone[53]. It is likely that in Carthew's study[52] the gerbils were subject to coincidental laboratory infections since liver fibrosis did not occur under pathogen free conditions[53]. Moreover there was no evidence that deferiprone caused hepatic necrosis or portal fibrosis[53]. Because, in thalassaemia, viral hepatitis is an important cause of cirrhosis, it is more informative to focus on hepatitis C virus (HCV)-negative patients to evaluate the potential hepatotoxicity of L1. Of the 14 evaluable patients from the Toronto study, five developed progression of hepatic fibrosis on L1 treatment[24]. Four of these patients were HCV positive. Only one of the eight HCV-negative Toronto patients developed progression of hepatic fibrosis. The validity of this study has been questioned on grounds of adequacy of the liver biopsy samples and differences at base line of patients treated with DFO or deferiprone. Moreover, the results on the same biopsies were not confirmed by Callea[54]. None of the 25 HCV-negative patients reported collectively by Tondury[17], Hoffbrand[18] and Galanello[55] developed this complication. No evidence of fibrosis progression was found among any of the 29 patients, both HCV positive and negative, representing part of the children who participated in the LA-02 multicentre study, receiving L1 for a mean duration of 22-26 months, studied prospectively by repeated liver biopsies[55,56] or in patients studied in Italy by Stella et al[57]. Wanless et al[58] most recently evaluated serial liver biopsies from 56 TM patients participating in the Italian multicentre study. They demonstrated no evidence of deferiprone-induced progression of hepatic fibrosis during long term therapy. This analysis is of the largest collection of liver biopsies reported to date and was analysed independently by three pathologists. The effect of long-term L1 treatment on liver histology was also examined by Berdoukas et al[59]. Repeat liver biopsies were performed at an interval of about 1 year in 14 patients treated with L1 and compared with those of 22 patients receiving DFO. The authors concluded that L1 appears to stabilise liver iron and that it does not significantly increase liver fibrosis.

5.7. Other reported side effects. Early suggestions that deferiprone causes immune disturbance have not been substantiated by studies of CD4, CD8 T cell and B cell numbers, immunoglobulin levels, specific antibody titres, antinuclear factor anti-histone antibodies, rheumatoid factor complement levels in many studies (reviewed refs 1-5). Deaths in patients receiving deferiprone have mainly been due to cardiac abnormalities from iron overload, progression of myelodysplasia or infection not associated with agranulocytosis or immune deficiency.

6. SITE OF IRON CHELATION:SHUTTLE EFFECT

Grady et al[60] have shown that when DFO and deferiprone are given together to humans, there is a synergistic increase in urine and faecal iron excretion to 2.4-3.4 times that when DFO is given alone. They proposed a "sink and shuttle" effect. Deferiprone has a partition coefficient twenty times greater than that of DFO and thus can chelate iron from body compartments inaccessible to DFO. Studies of Shalev et al[61,62] showed that deferiprone chelates iron from membranes of thalassaemic and sickle cell red cells in vitro and in vivo, the free drug rapidly entering these cells but the chelate, of three times the M.W. of the free drug, being released more slowly from the cells. Desferrioxamine has a M.W. three times greater than deferiprone and is confined to plasma and the hepato-biliary tract. Grady et al[60] suggested deferiprone could chelate iron from inside cells and make this available to DFO in plasma for excretion in urine or stools.

Breuer et al[63] have further explored this shuttle effect They developed a new assay for DFO chelatable iron (DCI) in plasma. They consider this to be a component of non-transferrin bound iron since iron bound to transferrin is not available to DFO. The method uses a fluorescein-DFO probe whose fluorescence is stoichiometrically quenched by iron. DCI was present in the serum of most thalassaemia major patients tested (21 of 27) but not in 48 healthy controls and only in 8 of 95 samples from 39 patients with genetic haemochromatosis. After giving an oral dose of deferiprone to patients with TM, they found a rapid and large rise in plasma DCI, up to 10μM within 30-60 minutes. When DFO was given to the patients at the same time as deferiprone, there was no rise in plasma DCI suggesting that iron had been transferred from deferiprone to DFO in vivo. This transfer of DCI from deferiprone to DFO was observed in vitro from preformed deferiprone iron complexes, even when the deferiprone/iron ratios exceeded 3:1. Deferiprone was also shown to transfer iron to apotransferrin in vitro suggesting it may shuttle iron to apotransferrin in vivo. This contrasts with previous observations that deferiprone is capable of removing iron from transferrin in vitro and in vivo. Breuer and colleagues suggest deferiprone may enter cells, chelate iron and shuttle this to the cell surface and extracellular fluid where it is transferred to DFO and then excreted in urine.

7. COMBINATION THERAPY

Three clinical studies have been published in which patients received deferiprone and DFO on the same day for several months. The first study in London combined deferiprone 75-100mg/kg/day on seven days a week with DFO by continuous subcutaneous infusion 2g per day on 2 to 6 days[40]. Five patients were treated for 6 to 15 months. Urine iron excretion was additive in 3 patients and in two was synergistic. In all five there was a substantial fall in serum ferritin. In the second study[64] 10 thalassaemia patients in Malaysia were treated with deferiprone at a dose of 75-85mg/kg daily for 1 year. In seven patients this was combined with DFO given subcutaneously, 1-4 gram over 8 hours on two successive days. This resulted in a rise in urine iron excretion from a mean of 13.7mg/day to 27.4mg/day. There was a significant reduction in serum ferritin in these 7 patients. Three patients also showed a significant reduction in liver iron. The third study in Lebanon showed that combination therapy was significantly more effective than either drug alone at lowering serum ferritin over a 12 month period[65]. Longer term studies of the combined therapy are in progress and have not shown any new or increased incidence of toxicity of either drug compared with the drug given alone. In the gerbil model, Hershko and colleagues have shown that combined DFO and deferiprone was more efficient at lowering liver iron concentrations and restoring mitochondrial respiratory enzyme activity than either drug alone[52].

8. CONCLUSIONS AND FUTURE PROSPECTS

Much is now known about the efficacy and toxicity of deferiprone. Used at a dose of 75mg/kg/day it appears, on average, to be about 65% as effective as DFO with a wide individual patient variation. Compliance with deferiprone is better and, for patients already committed to regular transfusions for whom deferiprone is equally effective as DFO, it is clearly preferable to take an orally active drug rather than subcutaneous infusions. Furthermore, with the emergence of advanced cardiac magnetic resonance imaging techniques, it is now possible to accurately and reproducibly assess tissue iron loading. This will enable physicians to introduce better and innovative chelation programmes, either by giving DFO alone to its maximal therapeutic dose intravenously or subcutaneously without causing toxic side effects, or using oral iron chelation with deferiprone at doses up to 100mg/kg/weight, or the combination of both. Preliminary data from our centre suggests that when severe cardiac iron loading is present (low $T2^*$) with reduced left and right ventricular ejection fractions, the best treatment is not continuous intravenous DFO but continuous subcutaneous DFO using the elastometric delivery device[66] combined with daily oral deferiprone 75mg/kg/weight (unpublished data). Improved treatment options and better techniques for assessment of tissue iron overload will improve the quality of life of TM patients and prolong their survival worldwide. It is to be hoped that other orally active chelators will become available for clinical use and thus further improve iron chelation treatment of TM and other transfusion dependent patients.

REFERENCES

1. Addis A, Loebstein R, Koren G, Einarson TR. Meta-analytic review of the clinical effectiveness of oral deferiprone (L1), *Eur J Clin Pharmacol* 1999; **55**: 1-6.
2. Barman Balfour JA and Foster RH. Deferiprone. A review of its clinical potential in iron overload in β-thalassaemia major and other transfusion-dependent diseases. *Drugs* 1999; **58**: 553-578.
3. Diav-O, Koren G. Oral iron chelation with deferiprone. *Pediatr Clin N Am* 1997; **44**: 235-247.
4. Hershko C and Hoffbrand AV. Iron Chelation Therapy. *Rev in Clin and Exp Hematol* 2000; **4**: 337-361.
5. Olivieri NF. Long-term therapy with deferiprone. *Acta Haematol* 1996; **95**: 37-48.
6. Al-Refaie FN, Hoffbrand AV. Oral iron chelation therapy. *Recent Advances in Haematology* 1993; **7**: Eds. M.K. Brenner, A.V. Hoffbrand. Churchill Livingstone, Edinburgh.
7. Al-Refaie FN, Hoffbrand AV. Oral iron chelating therapy: the L_1 experience. *Bailliere's Clinical Haematology* 1994; **77**: 941-963.
8. Kontoghiorghes GJ, Aldouri MA, Sheppard L, Hoffbrand AV. (1987) 1,2-Dimethyl-3- hydroxypyrid-4-one, an orally active chelator for treatment of iron overload. *Lancet* 1989; **i**, 1294-1295.
9. Kontoghiorghes GJ, Aldouri MA, Hoffbrand AV, Barr J, Wonke B, Kourouclaris T. Sheppard L. Effective chelation of iron in beta-thalassaemia with the oral chelator 1,2-dimethyl-3-hydroxypyrid-4-one. *Br Med J* 1987; **295**, 1509-1512.
10. Kontoghiorghes GJ, Goddard JG, Bartlett AN, Sheppard L. Pharmacokinetics studies in humans with the oral iron chelator 1,2-dimethyl-3-hydroxypyrid-4-one. *Clin Pharmacol* 1990; **488**, 255-261.
11. Matsui D, Klein J, Grunau V, McLelland R, Chung D, St-Louis P, Olivieri N, Koren G, Hermann C. Relationship between the pharmacokinetics and iron excretion pharmacodynamics of the new oral iron chelator 1,2-dimethyl-3-hydroxy-pyrid-4-one in patients with thalassemia. *Clin Pharmacol Ther* 1991; **50**, 294-298.
12. Al-Refaie FN, Sheppard LN, Nortey P, Wonke B, Hoffbrand AV. Pharmacokinetics of the oral iron chelator deferiprone (L1). *Br J Haematol* 1995; **89**, 403-408.
13. Agarwal MB, Gupte SS, Viswanathan C, Vasandano D, Ramanathan J, Desai N, Puniyani RR, Chaablani AT. Long-term assessment of efficacy and safety of L1, an oral iron chelator, in transfusion dependent thalassaemia: Indian trial. *Br J Haematol* 1992; **82**, 460-466.
14. Al-Refaie FN, Wonke B, Hoffbrand AV, Wickens DG, Nortey P, Kontoghiorghes GJ. Efficacy and possible adverse effects of the oral iron chelator 1,2-dimethyl-3-hydroxypyrid-4-one (L_1) in thalassemia major. *Blood* 1992: **80**, 593-599.
15. Al-Refaie FN, Hershko C, Hoffbrand A, Olivieri NF, Tondury P, Wonke B. Results of long-term deferiprone (L_1) therapy: a report by the International Study Group on Oral Iron Chelators. *Br J Haematol* 1995; **91**, 224-229.
16. Tondury P, Kontoghiorghes GJ, Ridolfi-Luthy A, Hirt A, Hoffbrand AV, Lottenbach AM, Sonderegger T, Wagner HP. L1 (1,2-dimethyl-3-hydroxypyrid-4-one) for oral iron chelation in patients with beta-thalassaemia major. *Br J Haematol* 1990; **76**, 550-553.
17. Tondury P, Zimmermann A, Nielsen P, Hirt A. Liver iron and fibrosis during long-term treatment with deferiprone in Swiss thalassaemic patients. *Br J Haematol* 1998; **101**, 413-415.
18. Hoffbrand AV, Al-Refaie FN, Davis B, Siritanaratkul N, Jackson BFA, Cochrane J, Prescott E, Wonke B. Long term trial of deferiprone in 51 transfusion dependent iron overloaded patients. *Blood* 1998; **91**, 295-300.
19. Kersten MJ, Lange R, Smeets ME, Vreugdentill G, Roozendaal KJ, Lameijer W, Goudsmit R. Long-term treatment of transfusional iron overload with the oral iron chelator deferiprone (L1): a Dutch multicenter trial. *Ann Hematol* 1996; 247-252.
20. Longo F, Fischer R, Engelhardt R, et al. Iron balance in thalassemia patients treated with deferiprone (Abstract). *Blood* 1998; **92**, Suppl. i, 325a.
21. Mazza P, Ammuri B, Lazzari G, Masi C, Palazzo G, Spartera MA, Giuia R, Sebastio AM, Suma V, DeMarco S, Semeraro F, Moscoqiuri R. Oral iron chelating therapy. A single centre interim report on deferiprone (L1) in thalassemia. *Haematologica* 1998; **83**, 496-501.
22. Olivieri NF, Brittenham JM, Matsui D, Berrovitch M, Blendis LM, Cameron RG, McClelland RA, Liu PP, Templeton DM, Koren G. Iron-chelation therapy with oral deferiprone in patients with thalassemia major. *N Engl J Med* 1993; **332**, 918-922.
23. Olivieri NF, Brittenham GM, Armstrong SAM, Basran RKm Daneman R, Daneman N, Iwanchko RM, Talbot AL, Koren G. First prospective randomised trial of the iron chelator deferiprone (L1) and deferoxamine. *Blood* 1995; **86**, Suppl. 249a.

24. Olivieri NF, Brittenham GM, McLaren CE, Templeton DM, Cameron RG, McClelland RA, Bust AD, Fleming KA. Long-term safety and effectiveness of iron- chelation therapy with deferiprone for thalassemia major. *N Engl J Med* 1998: **339**, 417-423.
25. Cohen AR, Galanello R, Piga A, Di Palma A, Vullo C, Tricta F. Safety profile of the oral iron chelator deferiprone: a multicentre study. *Br J Haematol* 2000; **108**, 305- 312.
26. Taher A, Chamoun FM, Koussa Saad MA, Khoriaty AI, Neeman R, Mourad FH. Efficacy and side effects of deferiprone (L1) in thalassemia patients not compliant with desferrioxamine. *Acta Haematol* 1999; **101**, 173-177.
27. Taher A, Sheikh-Taha M, Koussa S, Inati A, Neeman P, Mourad F. Comparison between deferoxamine and deferiprone (L1) in iron-loaded thalassemia patients. *Europ J Haemat* 2001; (in press).
28. Del Vecchio GC, Crollo E, Schettini F, Schettini, F, Fischer R, De Mattia D. Factors influencing effectiveness of deferiprone in a thalassaemia major clinical setting. *Acta Haematol* 2000; **104**, 99-102.
29. Al-Refaie FN, Wickens DG, Wonke B, Kontoghiorghes GJ, Hoffbrand AV. Serum non-transferrin-bound iron in beta-thalassemia major patients treated with desferrioxamine and (L_1). *Br J Haematol* 1992; **82**, 431-436.
30. Wolfe L, Olivieri N, Sallan D, Collan S, Roe V, Propper R, Freedman MH, Nathan DG. Prevention of cardiac disease by subcutaneous deferoxamine in patients with thalassemia major. *N Engl J. Med* 1985; **312**, 1600-
31. Aldouri MA, Wonke B, Hoffbrand AV, Flynn DM, Ward SE, Agnew JE, Hilson AJW. (1990) High incidence of cardiomyopathy in beta-thalassaemia patients receiving regular transfusion and iron chelation: Reversal by intensified chelation. *Acta Haematol* 1990; **84**, 113-117.
32. Modell B, Khan M, Darlison M. Survival of beta thalassemia major in the United Kingdom: data from the UK Thalassaemia Register. *Lancet* 2000; **355**, 2051-2052.
33. Olivieri NF, Nathan DG, MacMillan JH, Wayne AD, Martin M, McGee A, Koren C, Liu PP, Cohen AR. Survival in medically treated patients with homozygous beta-thalassemia. *N Engl J Med* 1994; **331**, 574-578.
34. Wonke B, Hoffbrand AV, Brown D, Dusheiko G. Antibody to hepatitis C virus in multiply transfused patients with thalassaemia major. *J Clin Path* 1990; **43**, 638-640.
35. Chapman, RWG, Hussain MAM, Gorman A, Laulicht M, Politis D, Flynn DM, Sherlock S, Hoffbrand AV. Effect of ascorbic acid deficiency on serum ferritin concentrations in patients with beta-thalassaemia major and iron overload. *J Clin Path* 1982; **35**, 487-491.
36. Olivieri NF, Brittenham GM. Iron-chelating therapy and the treatment of thalassemia. *Blood* 1997; **89**, 739-761.
37. Anderson L, Holden S, Davis B, Prescott E, Charrier C, Bunce N, Firmin D, Wonke B, Porter J, Walker JM, Pennell D. Cardiovascular T2-star (T2*) magnetic resonance for the early diagnosis of myocardial iron overload. *Europ Heart J* (in press).
38. Anderson L, Porter JB, Wonke B, Walker JM, Holden S, Davies BA, Prescott E, Channier C, Firmin DN, Pennel DJ. Relationship of myocardial iron overload to right and left bventricular function in homozygous beta thalassaemia. *Blood* 2000; Supplement Abstract **2597**, 605a.
39. Anderson L, Wonke B, Prescott E, Holden S, Walker M, Panel DJ. Improved cardioprotection with the oral iron chelator L1 over standard therapy with subcutaneous deferoxamine in iron overload cardiomyopathy. *J Amer Cardiac College* 2001; **37(2)**, Abst.
40. Wonke B, Wright C, Hoffbrand AV. Combined therapy with deferiprone and desferrioxamine. Br J Haematol 1998; **103**, 361-364.
41. Diav-Citrin O, Atanackovic G, Koren G. An investigation into the variability in the therapeutic response to deferiprone in patients with thalassemia major. *Therapeutic Drug Monitoring* 1999; **21**, 74-81.
42. Pippard MJ, Weatherall DJ. Oral iron chelation in thalassaemia: an uncertain scene. *Br J Haematol* 2000; **111**, 2-5.
43. Olivieri NF, Koren G, Matsui D, Liu PP, Blendis L, Cameron, R, McLelland R, Templeton DM. Reduction of tissue iron stores and normalization of serum ferritin during treatment with the oral iron chelator in thalassemia intermedia. *Blood* 1992; **79**, 2741-2748.
44. Cragg L, Hebbel RP, Miller W, Solovey A, Selby S, Enright H. The iron chelator L1 potentiates oxidative damage in iron-loaded liver cells. *Blood* 1998; **92**. 632-638.
45. Link G, Konijn AM, Hershko C. Cardio-protective effect of alpha-tocopherol, ascorbate, deferoxamine, and deferiprone: mitochondrial function in cultured, iron-laded heart cells. *J Lab Clin Med* 1999; **133**, 179-188.
46. Bartlett AN, Hoffbrand AV, Kontoghiorghes GJ. Long-term trial with the oral iron chelator 1,2-dimethyl-4-hydroxypyrid-4-one (L_1). II. Clinical observations. *Br J Haematol* 1990; **76**, 301-304.
47. Hoffbrand AV, Bartlett A, Veys PA, O'Connor NTJ, Kontoghiorghes GJ. Agranulocytosis and thrombocytopenia in patient a with Blackfan-Diamond anaemia during oral chelator trial. *Lancet* 1989; **ii**, 57.
48. Al-Refaie FN, Wonke B, Wickens DG, Aydinok Y, Hoffbrand AV. Zinc concentrations in patients with iron overload receiving oral iron chelator 1,2- dimethyl-hydroxyprid-4-one or desferrioxamine. *J Clin Path* 1994; **46**, 657-660.

49. Mangiagli A, De Sanctis V, Campisi S, Di Silvestro G, Urso L. Treatment with deferiprone (L1) in a thalassaemia patient with bone lesions due to desferrioxamine. *J Paed Endocrinology* 2000; **13**, 677-680.
50. Cunningham JM, Al-Refaie FN, Hunter AE, Sheppard LN, Hofffbrand AV. Differential toxicity of alpha-keto hydroxypyridine iron chelators and desferrioxamine to human haemopoietic precursors in vitro. *Europ J Haemat* 1994; **52**, 176-179.
51. Al-Refaie FN, Wonke B, Hoffbrand AV. Deferiprone-associated myelotixicity. *Eur J Haemat* 1994; **53**, 398-
52. Carthew P, Smith AG, Hider RC, Dorman B, Edwards RE, Francis JC. Potentiation of iron accukmulation in cardiac myocytes during the treatment of iron overload in gerbils with the hydroxypyridnone iron chelator CP94. *Biometals* 1994; 7, 267-271.
53. Hershko C, Konijn AM, Huerta M, Rosenmann E, Reinus C, Link G. Safety and efficacy of iron chelating therapy with DFO and L1 in the gerbil model of hemosiderosis. *J Lab Clin Med* 2001; (in press).
54. Callea F. Iron chelation with oral deferiprone in patients with thalassemia. (Letter) N Engl J Med 1998; **339**, 1711.
55. Galanello R, De Virgiliis S, Agus A. Sequential liver fibrosis grading during deferiprone treatment in patients with thalassemia major. *9th International Conference on Oral Chelation in the Treatment of Thalassaemia and Other Diseases,* Hamburg, March 14 25-28 1999; Abstract 50.
56. Piga A, Facello S, Caglioti C, Pucci A, Pietribiasi F. No progression of liver fibrosis in thalassemia major during deferiprone or desferrioxamine iron chelation. *Blood* 1998; **92**, Suppl. 2, 216.
57. Stella M, Pinzello G, Maggio A. Iron chelation therapy with oral iron deferiprone in patients with thalassaemia (Letters) *N Engl J Med* 1998; **339**, 1712.
58. Wanless JR, Sweeney G, Dhillon AP, Guido M, Piga A, Galanello R, De Sanctis V, Schwartz E, Cohen AR. Absence of deferiprone induced hepatic fibrosis@ a milti centre study. *Blood* 2000; **96**, Suppl. 1, 2600, Abstraccrt 606a.
59. Berdoukas VA, Lindeman R, Eagle C. Liver assessment in a group of transfusion dependent patients treated with deferiprone and compared to a group of patients treated with desferrioxamine. Forty-First Annual Meeting, American Society of Hematology, December 3-7, New Orleans, LA. *Blood* 1999; **94**, Suppl. 1, Abstract3283.
60. Grady RW, Berdoukas VA, Giardina PJ. Iron chelators: combined therapy could be a better approach. Blood 1998; Suppl. 1, Pt. 2:16b.
61. Shalev O, Repka T, Goldfarb A, Grinberg L, Abramov A, Olivieri NF, Rachmilewitz EA, Hebbel RP. Deferiprone (L1) chelated pathologic iron deposits from membranes of intact thalassemic and sickle red blood cells both in vitro and in vivo. *Blood* 1995; **86**, 2008-2013.
62. Shalev O, Hileti, D, Nortey P, Hebbel RP, Hoffbrand AV. Transport of ^{14}C-deferiprone in normal, thalassaemic and sickle red blood cells. *Br J Haematol* 1999; **105**, 1081-1083.
63. Breuer W, Erness MJJ, Pootrakul P, Abramov A, Hershko C, Cabantenik ZI. Desferrioxamine-chelatable iron a component of serum iron-transferrin-bound iron, used for assessing chelation therapy. *Blood* 2001; **97**, 792-798.
64. Balveer K, Pyar K, Wonke B. Combined oral and parenteral iron chelation in beta thalassaemia major. *Med J Malaysia* 2000; **55**, 493-497.
65. Taher A, Mourad FH, Sheikh-Taha M, Koussa S, Inati A, Neeman R, Hoffbrand AV. Comparison of desferrioxamine, deferiprone and a combination therapy in iron overloaded thalassemia patients. *11th International Conference on Oral Chelation -Catania* 2000; p.144.
66. Arauju A, Kosaryan M, MacDowell A, Wickens D, Puni S, Wonke B, Hoffbrand AV. A novel delivery system for continuous desferrioxamine infusion in transfusional iron overload. *British J Haemat* 1996; **93**, 835-837.

IRON CHELATOR CHEMISTRY

Zu D. Liu, Ding Y. Liu and Robert C. Hider[*]

1. INTRODUCTION

Iron overload is a serious clinical condition which can be largely prevented by the use of iron-specific chelating agents. Desferrioxamine-B (**1**), the most widely used iron chelator in haematology over the past thirty years, has a major disadvantage of being orally inactive[1]. Consequently, the successful design of an orally active, non-toxic, selective iron chelator has become a much sought after goal.

When designing iron chelators for clinical application, the properties governing metal selectivity and ligand-metal complex stability are paramount. In theory, chelating agents can be designed for either the iron(II) (ferrous) or the iron(III) (ferric) oxidation state. Ligands that prefer iron(II) contain "soft" donor atoms, exemplified by nitrogen-containing ligands such as 2,2'-bipyridyl. Although these compounds are selective for iron(II) over iron(III), they retain an appreciable affinity for other biologically important bivalent metals such as copper(II) and zinc(II) ions.[2] Thus the design of a nontoxic iron(II)-selective ligand is extremely difficult and indeed may not be possible. In contrast iron(III)-selective ligands, typically oxyanions and notably hydroxamates and catecholates, are generally more selective for tribasic metal cations over dibasic cations.[2] Many tribasic cations, for instance aluminium(III) and gallium(III), are not essential for living cells and thus in practice iron(III) is a suitable target for "clinical chelator" design. An additional advantage of high-affinity iron(III) chelators is that, under aerobic conditions, they will chelate iron(II) and facilitate autoxidation to iron(III)[3]. Thus, high-affinity iron(III)-selective ligands bind both iron(III) and iron(II) under most physiological conditions.

- Robert C Hider, Zu D Liu and Ding Y Liu, Department of Pharmacy, King's College London, Franklin-Wilkins Building, 150 Stamford Street, London SE1 9NN, U.K.

2. DESIGN FEATURES OF IRON(III) CHELATOR

2.1. Thermodynamic stability of iron(III) complexes

The coordination requirements of iron(III) are best satisfied by six donor atoms ligating in an octahedral fashion to the metal centre. The stability of a metal complex is governed by the number of chelate rings formed in the resultant ligand-metal complex, the more rings, the more stable the complex. The number of chelating rings can be enhanced by increasing the number of donor atoms attached to a single chelator; for example, a metal ion with co-ordination number six may form three rings with a bidentate ligand or five rings with a hexadentate ligand (**Figure 1**). Thus in order to maximise the thermodynamic stability of the iron(III) complex it is necessary to incorporate all six donors into a single molecular structure thereby creating a hexadentate ligand. This increase in stability is largely associated with the entropic changes that occur on going from free ligand and solvated free metal to the ligand-metal complex.

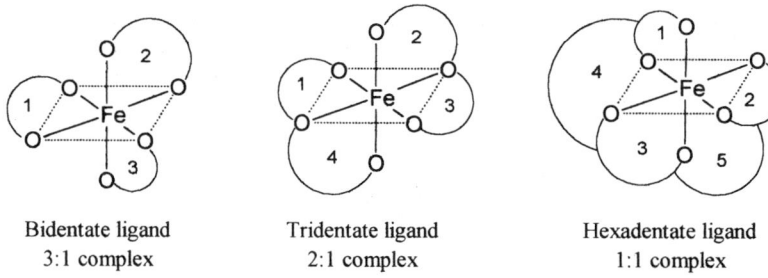

Figure 1: Schematic representation of chelate ring formation in metal-ligand complexes

Under biological conditions, a parameter which is generally more useful than the conventional stability constant for comparison of chelators is the pM value or specifically for iron(III) the pFe^{3+} value[4,5]. pFe^{3+} is defined as the negative logarithm of the concentration of the free iron(III) in solution. For clinically relevant conditions, pFe^{3+} values are typically calculated for total [ligand] = 10^{-5}M and total [iron] = 10^{-6}M at pH 7.4. The comparison of ligands under these conditions is useful, as the pFe^{3+} value, unlike the stability constant logK or $log\beta_3$, takes into account the effects of ligand protonation and denticity as well as differences in metal-ligand stoichiometries. The higher the pFe^{3+} value, the stronger the chelator.

The formation of a complex will also be dependent on both free metal and free ligand concentration and as such will be sensitive to concentration changes. Thus the degree of dissociation for a *tris*-bidentate ligand-metal complex is dependent on $[ligand]^3$ whilst the hexadentate ligand-metal complex dissociation is only dependent on [ligand]. Hence the dilution sensitivity of complex dissociation follows the order hexadentate < tridentate < bidentate.

2.2. Iron(III) ligand selection

By virtue of its high charge density, iron(III) forms most stable bonds with ligands containing weakly polarizable atoms, such as oxygen, indeed the affinity of such compounds for iron(III) reflects the pKa values of the chelating oxygen atoms, the higher the affinity for iron(III), the higher the pKa value (**Table 1**).

Catechols: Catechol moieties possess a high affinity for iron(III). This extremely strong interaction with tripositive metal cations results from the high electron density of both oxygen atoms. However this high charge density is also associated with the high affinity for protons (pKa values, 12.1 and 8.4). Thus the binding of cations by catechol has marked pH sensitivity[6]. For simple bidentate catechols, the 2:1 complex is the dominant form in the pH range 5.5-7.5 (**Figure 2A**). Such complexes bear a net charge and therefore are unlikely to permeate membranes by simple diffusion[7]. An additional problem with catechol-based ligands is their susceptibility towards oxidation[6].

Hydroxamates: The hydroxamate moiety possesses a lower affinity for iron than catechol. However it has the advantage of forming neutral *tris*-complexes with iron(III) which are in principle able to permeate membranes by non-facilitated diffusion[7]. The selectivity of hydroxamates, like catechols, favours tribasic cations over dibasic cations. Because of the relatively low protonation constant (pKa ~ 9), hydrogen ion interference at physiological pH is less pronounced than for that of catechol ligands, consequently the 3:1 complex predominates at pH 7.0 when sufficient ligand is present (**Figure 2B**). However the affinity of a simple bidentate hydroxamate ligand for iron is insufficient to solubilise iron(III) at pH 7.4 at clinically achievable concentrations (**Figure 2C**), thus only hexadentate hydroxamates are likely to be effective iron(III) scavengers under such conditions.

Hydroxypyridinones: Hydroxypyridinones (HPOs) combine characteristics of both hydroxamate and catechol groups, forming 5-membered chelate rings in which the metal is coordinated by two vicinal oxygen atoms. The hydroxypyridinones are monoprotic acids at pH 7.0 and thus form neutral *tris*-iron(III) complexes. The affinity of such compounds for iron(III) reflects the pKa values of the chelating oxygen atoms, the higher the affinity for iron(III), the higher the pKa value (**Table 1**). The surprisingly high pKa value of the carbonyl function of 3-hydroxypyridin-4-one results from extensive delocalisation of the lone pair associated with the ring nitrogen atom. The pyridin-4-ones form a neutral 3:1 complex with iron(III), which is stable over a wide range of pH values (**Figure 2D**). Although catechol derivatives possess higher β_3 values than that of 3-hydroxypyridin-4-one, the corresponding pFe^{3+} values are lower (**Table 1**). This difference is due to the relatively higher affinity of catechol for protons. Of all dioxygen ligand classes, 3-hydroxypyridin-4-ones possess the greatest affinity for iron(III) in the physiological pH range (**Table 1**).

Aminocarboxylates: Aminocarboxylate ligands are excellent iron(III) chelating agents. Several polycarboxylate ligands such as ethylenediaminetetraacetic acid (EDTA) (**8**) and diethylenetriaminepentaacetic acid (DTPA) (**9**), have been widely investigated for their iron

Table 1: pKa values and affinity constants of dioxobidentate ligands for iron(III)

Ligand	Structure	pKa_1	pKa_2	$Log\beta_3$	pFe^{3+}
N,N-Dimethyl-2,3-dihydroxybenzamide (DMB) (2)		8.4	12.1	40.2	15
Acetohydroxamic acid (3)		-	9.4	28.3	13
2-Methyl-3-hydroxypyran-4-one (maltol) (4)		-	8.7	28.5	15
1-Hydroxypyridin-2-one (5)		-	5.8	27	16
1-Methyl-3-hydroxypyridin-2-one (6)		0.2	8.6	32	16
1,2-Dimethyl-3-hydroxypyridin-4-one (deferiprone) (7)		3.6	9.9	37.2	19

chelating abilities. However, the selectivity of these molecules for iron(III) is relatively poor. This lack of selectivity leads to zinc depletion in patients receiving aminocarboxylate-based ligands such as DTPA[8].

Hydroxycarboxylates: Hydroxycarboxylate ligands are strong chelating agents, which are more selective for iron(III) than the corresponding aminocarboxylates, due to all the coordinating atoms being oxygen. The interaction between iron(III) and citrate (10) has been well characterised[9], but by virtue of its tridentate nature, a large number of complexes have been identified[10] including iron/citrate polymers[11,12]. In contrast hexadentate hydroxycarboxylate ligands for instance staphloferrin (11)[13] and rhizoferrin (12)[14] have iron(III) complex chemistries dominated by the formation of 1:1 complexes.

(11) R = H
(12) R = COOH

IRON CHELATOR CHEMISTRY

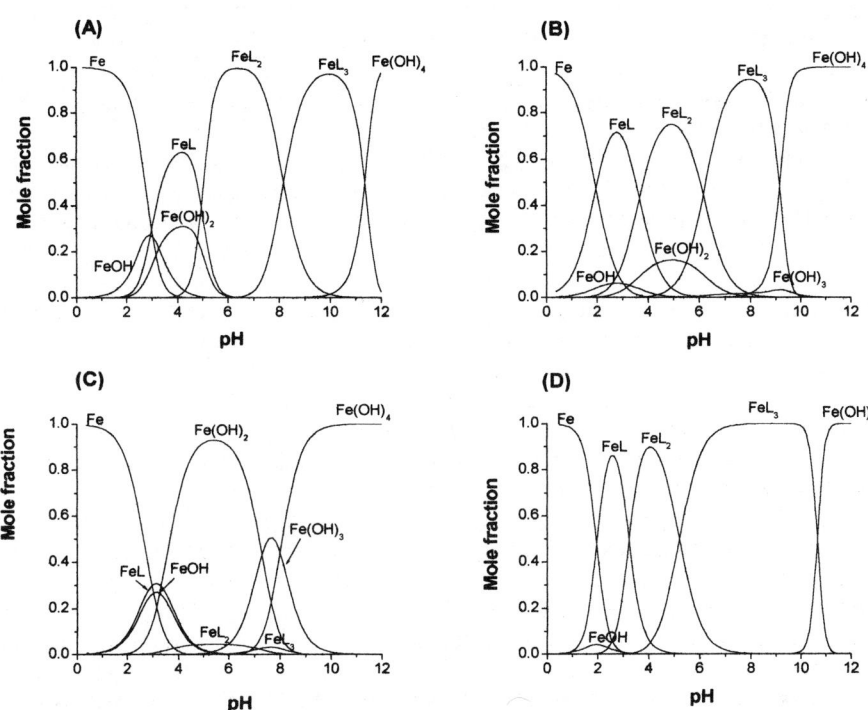

Figure 2: Speciation plot of iron(III) in the presence of (A) N,N-dimethyl-2,3-dihydroxybenzamide (2), $[Fe^{3+}]_{total} = 1 \times 10^{-6}M$; [Ligand] = $1 \times 10^{-5}M$; (B) acetohydroxamic acid (3), $[Fe^{3+}]_{total} = 1 \times 10^{-6}M$; [Ligand] = $1 \times 10^{-4}M$; (C) acetohydroxamic acid (3), $[Fe^{3+}]_{total} = 1 \times 10^{-6}M$; [Ligand] = $1 \times 10^{-5}M$; (D) 1,2-dimethyl-3-hydroxypyridin-4-one (7), $[Fe^{3+}]_{total} = 1 \times 10^{-6}M$; [Ligand] = $1 \times 10^{-5}M$.

3. CRITICAL FEATURES FOR CLINICAL APPLICATION

3.1. Lipophilicity and molecular weight

In order for a chelating agent to exert its pharmacological effect, a drug must be able to reach the target sites at sufficient concentration. Hence the key property for an orally active iron chelator is its ability to be efficiently absorbed from the gastrointestinal tract and to cross biological membranes thereby gaining access to the desired target sites such as the liver. There are three major factors which influence the ability of a compound to freely permeate a lipid membrane, namely, lipophilicity, ionisation state and molecular size.

Lipophilicity is a measure of the affinity of a molecule for a lipophilic environment and is commonly assessed by measuring the distribution ratio of the solute (partition coefficient) between two phases, typically *n*-octanol and water. In order to achieve efficient oral absorption, the chelator should possess appreciable lipid solubility which may facilitates penetration of the gastrointestinal tract (partition coefficient greater than 0.2)[15]. However, highly lipid-soluble chelators can also penetrate most cells and critical barriers such as the blood-brain and placental barriers, thereby enhancing possible toxic side effects. Membrane permeability can also be affected by the ionic state of the compound. Uncharged molecules penetrate cell membranes more rapidly than charged molecules[16]. It is for this reason that aminocarboxylate-containing ligands are unlikely to possess high oral activity.

Molecular size is yet another factor which influences the rate of non-facilitated drug absorption[17,18]. The transcellular route involves diffusion into the enterocyte and thus utilises some 95% of the surface of the small intestine. In contrast, the paracellular route only utilises a small fraction of the total surface area and the corresponding flux *via* this route is much smaller. The "cut-off" molecular weight for the paracellular route in the human small intestine is approximately 400[19] and this route is unlikely to be quantitatively important for molecules with molecular weights > 200[18]. There is no clear "cut off" value for the transcellular route, but as judged by PEG permeability, penetration falls off rapidly with molecular weights > 500[20,21]. Thus in order to achieve greater than 70% absorption, subsequent to oral application, the chelator molecular weight probably needs to be less than 300[20]. This molecular-weight limit provides a considerable restriction on the choice of chelator and may effectively exclude hexadentate ligands from consideration, most siderophores, for instance DFO, have molecular weights in the range 500-900. Although EDTA has a molecular weight of only 292, it is too small to fully encompass the chelated iron, thereby facilitating the potential toxicity of the metal[15]. Bidentate and tridentate ligands, by virtue of their much lower molecular weights, are predicted to possess higher rates of absorption. The fraction of the absorbed dose for a range of bidentate 3-hydroxypyridin-4-ones has typically been found to fall between 50-70%, as assessed in the rabbit[22].

3.2. Chelator disposition

The metabolic properties of chelating agents play a critical role in determining both their efficacy and toxicity. It is important to ensure that the chelator is not degraded to metabolites which lack the ability to bind iron. Such properties will inevitably require the use of higher chelator levels thereby increasing the risk of inducing toxicity. Chelators are likely to be more resistant to metabolism if their backbones lack ester and amide links[23]. The catechol function is also a disadvantage with respect to metabolism as there are numerous enzymes specifically designed to modify the catechol entity, for instance catechol-*O*-methyl transferase and tyrosinase[23]. Ideally for maximal scavenging effect, a chelator must be present within the extracellular fluids in the range 10-25μM and for a sufficient length of time to ensure interception of iron from both the extracellular and intracellular pools. Compounds with short-plasma half lives are thus likely to be less effective due to the limited pool of chelatable iron present within the body at any one time. DFO possesses a short plasma half life and it is for this reason that the molecule is administered *via* an infusion pump[24].

3.3. Toxicity

The toxicity associated with iron chelators originates from a number of factors; including inhibition of iron-containing metalloenzymes; lack of metal selectivity, which may lead to the deficiency of other physiologically important metals such as zinc(II); redox cycling of iron-complexes between iron(II) and iron(III), thereby generating free radicals and the kinetic lability of the iron-complex leading to iron redistribution. Comparative properties of iron(III) chelators are summarised in **Table 2**.

Metal selectivity: An ideal iron chelator should be highly selective for iron(III) in order to minimise chelation of other biologically essential metal ions which could lead to deficiency with prolonged usage. Unfortunately, many ligands that possess a high affinity for iron(III) may also have appreciable affinities for other metals such as zinc(II), this being especially so with carboxylate- and nitrogen-containing ligands. Selectivity is less of a problem with the oxygen-containing bidentate catechol, hydroxamate and hydroxypyridinone ligand families, each of which possesses a strong preference for tribasic over dibasic cations. Although in principle competition with copper(II) could be a problem, under most biological conditions it is not so, as copper is extremely tightly bound to proteins and the unbound fraction is reported to be less than $10^{-20}M$[25]. Copper is exchanged between proteins *via* specialised high affinity chaperone molecules[26].

Table 2: *Comparative properties of bidentate, tridentate and hexadentate ligands*

Bidentate ligands	Tridentate ligands	Hexadentate ligands
Possible for all coordinating atoms to be "hard" oxygen centres which renders ligands highly selective for iron(III)	Very difficult for all coordinating atoms to be "hard" oxygen centres which may lead to poor metal selectivity	Possible for all coordinating atoms to be "hard" oxygen centres which renders ligands highly selective for iron(III)
Affinity for iron is concentration dependent	Affinity for iron is concentration dependent	Affinity for iron is concentration independent
Kinetically labile – iron redistribution is possible	Kinetically labile – iron redistribution is possible	Kinetically inert – iron redistribution is unlikely
Form partially coordinated 2:1 complexes	Can form partially coordinated 1:1 complexes	Only form fully coordinated 1:1 complexes,
Can form uncharged iron(III) complexes	All iron(III) complexes are charged	Can form uncharged iron(III) complexes
Do not form polymeric complexes	Can form polymeric complexes which are likely to be trapped within cells	Do not form polymeric complexes
Penetration of BBB is dependent on lipophilicity	Penetration of BBB is dependent on lipophilicity	Generally low penetration of the BBB

Complex structure: In order to prevent free radical production, iron should be coordinated in such a manner as to avoid direct access of oxygen and hydrogen peroxide. Most hexadentate ligands such as DFO are kinetically inert and reduce hydroxyl radical production to a minimum by entirely masking the surface of the iron (**Figure 3**). However, not all hexadentate ligands are of sufficient dimensions to entirely mask the surface of the bound iron, in which case the resulting complex may enhance the ability of iron to generate free radicals. This phenomenon is particularly marked at neutral or alkaline pH values when the solubility of noncomplexed iron(III) is severely limited. The classic example of this type of behaviour is demonstrated by EDTA, where a seventh co-ordination site is occupied by a water molecule. This water molecule is kinetically labile and is capable of rapidly exchanging with oxygen, hydrogen peroxide and many other ligands present in biological media.

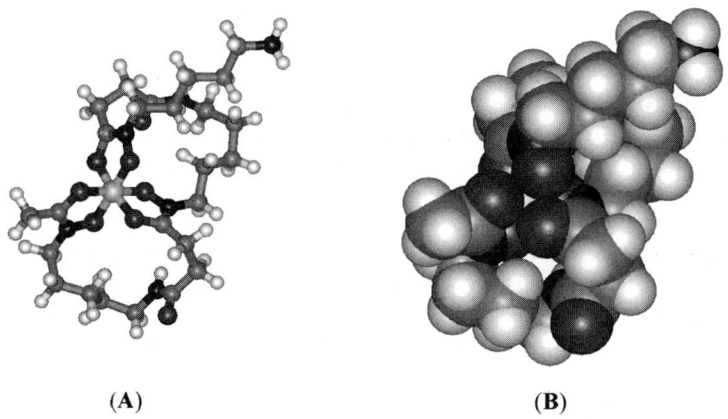

(A)   (B)

Figure 3: Energy-minimized structure of ferrioxamine B (Λ-C-trans,trans conformation) (A) Ball-stick model; (B) Space-filling model.

In contrast to the kinetically stable ferrioxamine complex, bidentate and tridentate ligands are kinetically more labile and the iron(III) complexes can dissociate at low ligand concentrations. Partial dissociation of bi- and tridentate ligand iron complexes renders the iron(III) cation surface accessible to other ligands. The concentration dependence of 3-hydroxypyridin-4-one iron complex speciation is minimal at pH 7.4 when the ligand concentration is above 1μM, due to the relatively high affinity of the ligand for iron(III). Thus bidentate 3-hydroxypyridin-4-ones behave more like hexadentate ligands, the 3:1 complex being the dominant species at pH 7.4 (**Figure 2D**). With such complexes, the iron atom is completely coordinated by the hydroxypyridinone ligands (**Figure 4**). This is in marked contrast to iron(III) complexes of most catechol species[7]. Even at low hydroxypyridinone concentrations there is minimal hydroxyl radical production[27], especially in the presence of physiological levels of citrate, which is likely to lead to the formation of a tertiary complex where the iron is completely coordinated by two hydroxypyridinone ligands and one citrate[28].

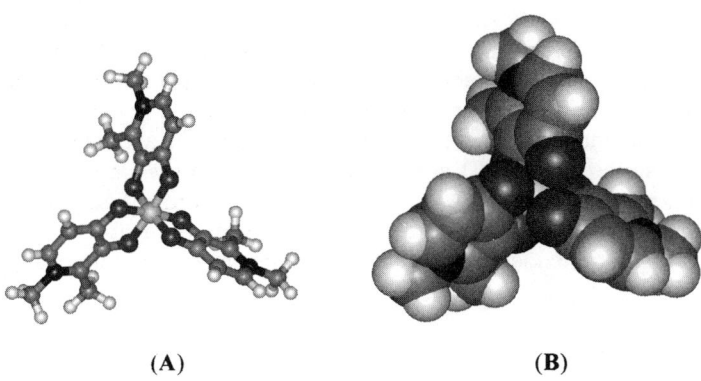

Figure 4: Energy-minimized structure of the 3:1 iron(III) complex of 1,2-dimethyl-3-hydroxypyridin-4-one (7). (A) Ball-stick model; (B) Space-filling model.

Redox activity: Chelators that bind both iron(II) and iron(III) are capable of redox cycling, a property that has been utilised by a wide range of enzymes[29] and industrial catalysts[30]. However this is an undesirable property for iron scavenging molecules, as redox cycling can also lead to the production of hydroxyl radicals. Significantly the high selectivity of catechol and hydroxamate siderophores for iron(III) over iron(II) renders redox cycling under biological conditions unlikely. Chelators containing nitrogen ligands tend to possess lower redox potentials and the coordinated iron can be reduced enzymatically under biological conditions. Such complexes may redox cycle and under aerobic conditions generate oxygen radicals.

Enzyme inhibition: In general, iron chelators do not directly inhibit haem-containing enzymes due to the inaccessibility of porphyrin-bound iron to chelating agents. In contrast many non-haem iron-containing enzymes such as the lipoxygenase and aromatic hydroxylase families and ribonucleotide reductase are susceptible to chelator-induced inhibition[31,32]. Lipoxygenases are generally inhibited by hydrophobic chelators, therefore the introduction of hydrophilic characteristics into a chelator tend to minimize such inhibitory potential[33]. Although this relationship holds with hydroxypyridinones, where the size of the alkyl substitution is increased in the 1-position of the pyridinone ring, in an essentially linear manner (**Figure 5A**), it is less evident for compounds with large substituents in the 2-position[34]. In fact, the variation in the inhibitory properties of the HPO chelators possessing different R_2 substituents is more dependent on the size of the substituent than the lipophilicity of the chelator (**Figure 5B**). The introduction of a hydrophilic substitutent at the 2-position of hydroxypyridinones markedly reduces the inhibitory properties, presumably due to steric

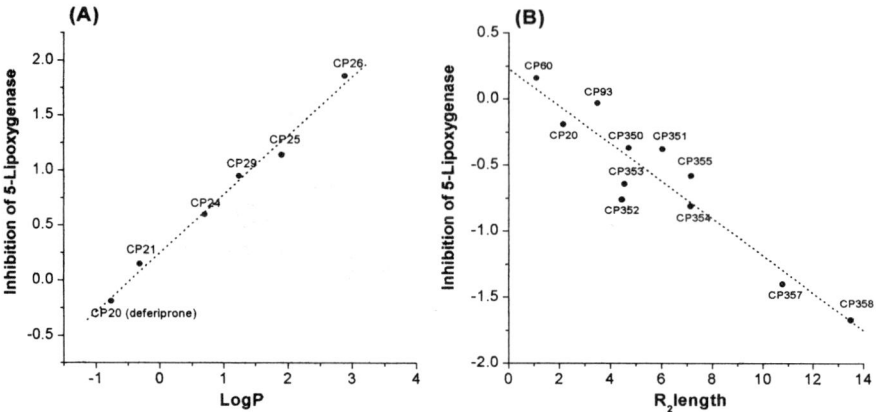

Figure 5: Relationship between the chemical nature of the ligand and the inhibition of 5-lipoxygenase; (A) HPOs with different R_1 substituents and (B) HPOs with different R_2 substituents.

interference of the chelation process at the enzyme active site[34]. In contrast lipophilicity is reported to be the dominant factor in controlling the ability of HPO chelators to inhibit mammalian tyrosine hydroxylase[35], hydrophilic chelators (LogP ≤ -1.0) tending to be relatively weak inhibitors of this enzyme. Clearly by careful modification of their physicochemical properties, iron chelators can be designed which exert minimal inhibitory influence on many metalloenzymes[33-36].

Hydrophilicity: Although bidentate and tridentate ligands possess a clear advantage over hexadentate ligands with respect to oral bioavailability, their enhanced ability to permeate membranes renders them potentially more toxic. Thus the penetration of the blood-brain barrier (BBB) is one of the likely side-effects associated with bidentate and tridentate ligands. The ability of a compound to penetrate the BBB is critically dependent on the partition coefficient as well as the molecular weight[37]. BBB permeability is predicted to be low for most hexadentate compounds, by virtue of their higher molecular weight. With low molecular weight molecules (M.W. < 300), penetration is largely dependent on the lipophilicity and molecules with partition coefficients < 0.05 tend to penetrate the BBB inefficiently[38]. Thus, chelators with partition coefficients lower than this critical value are predicted to show poor entry into the central nervous system. Indeed brain penetration of 3-hydroxypyridin-4-ones is strongly dependent on their lipophilicity[39], a clear correlation being observed between BBB permeability and the percentage polar surface area of the molecule (**Figure 6**). Thus 1,ω-hydroxyalkylhydroxypyridinones penetrate the BBB much more slowly than the simple 1-alkyl derivatives[39]. These results suggest that the biological distribution pattern of the HPOs can be significantly altered by simple modification of chemical structure without compromising their pharmacological function (selective iron chelation).

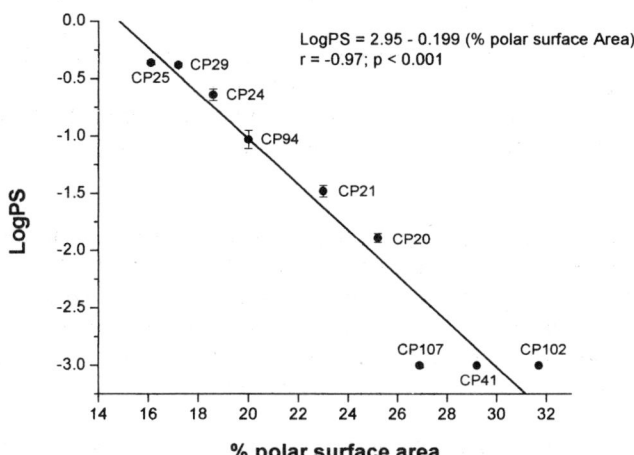

Figure 6: Relationship between the percentage polar surface area of the HPO chelators and their blood-brain barrier permeability (LogPS).

4. IRON(III) CHELATORS WITH POTENTIAL THERAPEUTIC APPLICATION

4.1. Hexadentate chelators

Desferrioxamine (See Chapters 6 & 9 of this volume): Naturally occurring siderophores provide excellent models for the development of therapeutic useful iron chelators. Indeed, desferrioxamine (DFO) (**1**), a growth-promoting agent secreted by the microorganism *Streptomyces pilosus*, is presently the therapeutic agent of choice for the clinical treatment of chronic iron overload. DFO is a *tris*-hydroxamic acid derivative and chelates ferric iron in an 1:1 molar ratio. It possesses an extremely high affinity for iron(III) and a much lower affinity for other metal ions present in biological fluids, such as zinc, calcium and magnesium. Although DFO is a large and a highly hydrophilic molecule ($D_{7.4}$ = 0.01), it gains entry into the liver *via* a facilitated transport system. It can therefore interact with both hepatocellular and extracellular iron, promoting urinary and biliary iron excretion[40]. Ferrioxamine, the DFO-iron complex, is kinetically inert and possesses a relatively low lipophilicity and thus is unlikely to enter cells. This property reduces the potential of iron redistribution. However, DFO is far from being an ideal therapeutic agent due to its oral inactivity and rapid renal clearance (plasma half-life of 5-10 minutes[41]). In order to achieve sufficient iron excretion, it has to be administered subcutaneously or intravenously for 8-12 hours a day, 5-7 days a week[24]. Consequently, patient compliance with this expensive and cumbersome regimen is often poor. Moreover, although DFO has been demonstrated to be a safe drug when administered in the presence of an elevated body iron burden, intensive therapy in young patients with relatively lower body iron stores may result in serious neurotoxicity, abnormalities of cartilage formation and other serious adverse effects[42-45].

In an attempt to improve the oral bioavailability of this chelator, a range of DFO prodrugs obtained *via* esterification of the labile hydroxamate functions has been investigated[46]. However, only marginal improvement in oral activity was found with tetra-acyl derivatives and none have been identified which possess comparable activity to that of subcutaneous DFO[47]. Several strategies centred on modification of the DFO backbone have also been pursued[48,49]. Unfortunately, no lead compound has yet emerged for further development.

Aminocarboxylates: DTPA (**9**) is an aminocarboxylate hexadentate ligand and has been used in patients who develop toxic side-effects with DFO[50]. Due to its net charge at neutral pH, DTPA is largely confined to extracellular compartments *in vivo* and is excreted in the urine within 24 h of administration[8,51]. DTPA is not orally active and due to its relative lack of selectivity for iron(III) leads to zinc depletion[8]. Consequently zinc supplementation is required to prevent the toxic sequelae of such depletion.

In order to enhance the selectivity of the aminocarboxylate ligands for iron(III), several analogues which contain both carboxyl and phenolic ligands have been designed[52,53]. A particularly useful compound is *N,N'*-bis(2-hydroxybenzyl)-ethylenediamine-*N,N'*-diacetic acid (HBED) (**13**) which is significantly more effective than DFO when given intramuscularly to iron overloaded rats[54,55]. It binds ferric iron strongly with an overall stability constant (logK_1) of 40 and a pFe^{3+} value of 31, rendering this molecule a potent ligand for chelation of iron *in vivo*. Unfortunately, HBED is not efficiently absorbed *via* the oral route in either primate[55,56] or man[57] because of the zwitterionic nature of the molecule. Consequently a considerable effort has been placed into the design of HBED ester prodrugs[58,59]. Pitt and co-workers demonstrated oral activity for a number of HBED diester derivatives[58], however the rate of hydrolysis of simple alkyl esters was found to be slow particularly in primate[56] and the efficacy of the compounds disappointing. In addition many of the compounds were found to be neurotoxic, the esters apparently crossing the blood brain barrier.

(13) $R_1 = R_2 = H$
(14) $R_1 = R_2 = CH(CH_3)_2$
(15) $R_1 = C_2H_5; R_2 = H$

The optimum compound was reported to be the di-isopropyl ester of HBED (**14**)[58]. Significantly this series of compounds possess a relatively small molecular weight (MW < 500) as compared with the majority of hexadentate ligands. More recently these investigations have been extended to include a wide range of ester and amide derivatives of HBED[59]. A compound of particular interest was found to be the mono ethyl ester of HBED (**15**), which possesses good oral availability[60]. The reasons for the efficacy of this compound is probably manyfold including, a distruption of intramolecular H-bonding thereby improving water solubility; an enhancement of partition constant, particularly in media of low dielectric constant; the ability of the mono ester to bind iron(III) and the activation of ester hydrolysis of the resulting iron(III) complex[61]. Unfortunately, this compound unlike the parent HBED induces liver toxicity when administered to mammals orally. HBED has a relatively high affinity for zinc (Log K_1 = 18.4) due to the presence of the two nitrogen ligands, and is capable of removing zinc from the zinc finger protein MTF-1[62].

Catechols: Hexadentate tricatechols are iron(III) chelators per excellence, enterobactin (**16**) typifying the group. However such molecules possess a relatively high molecular weight and therefore their oral bioavailability is poor, particularly at clinically useful doses. A further complication with such molecules is that they adopt a high net negative charge when coordinated to iron(III), which tends to minimise their rate of efflux by non-facilitated diffusion. A number of synthetic analogues have been prepared which retain the high affinity for iron(III) typical of enterobactin and yet are more stable under biological conditions, for instance the tripodal molecule (**17**) and MECAM (**18**). Unfortunately many of these hexadentate catechols bind to the enterobactin receptor expressed by many pathogenic organisms and hence will supply iron to such bacteria, an undesirable feature for clinical use[63].

Hydroxypyridinones: Hexadentate siderophore analogues can be constructed by derivatizing prototype bidentate hydroxypyridinones and attaching them to suitable molecular frameworks. Although enterobactin has an extremely high stability constant for iron(III), the effectiveness of this molecule and its analogues under acid conditions is limited by their weak acid nature and the required loss of six protons on binding iron(III). In contrast hydroxypyridinones are stronger acids than catechols, and since they are monoprotic, hexadentate ligands formed from three such units only loose three protons on formation of a six-coordinated complex. Thus hexadentate HPOs compete well with hexadentate catechols at neutral pH values. Another potential advantage of these molecules is that they may not be recognised by siderophore receptors and are therefore less likely to donate iron to pathogenic organisms. Several hexadentate ligands based on the hydroxypyridinone moiety have been investigated, such as (**19**)[64] and (**20**)[65]. Although the pFe^{3+} values of the hexadentate ligands were significantly higher (approximately 7 and 8 log units) than those of the corresponding bidentate ligands, a clear decrease in the formation constants of up to 2 log units was observed with the hexadentate ligands when compared to the bidentate analogues, indicating the lack of ligand predisposition for metal binding. Hexadentate pyridinone molecules are likely to be less toxic than their bidentate analogues because of a more restricted biodistribution. However, by virtue of their higher molecular weight, such molecules, like the analogous siderophores, generally possess low oral bioavailability.

(19)

(20)

4.2. Tridentate chelators

A potential problem associated with all tridentate ligands is that, unlike bidentate and most hexadentate compounds, there is a possibility of polymeric structure formation (21). Such structures (21) are difficult to clear *via* the kidney and are likely to become trapped within cells. For structural reasons, tridentate ligands with a high affinity for iron(III) all include at least one nitrogen atom as a ligand, and thus the selectivity of iron(III) over iron(II) is compromised. Another unique feature of tridentate ligands is that, without exception, they form a charged complex with iron(III). Although this is an advantage for enhancing hydrophilicity, it may also lead to associated intracellular trapping.

(21)

Desferrithiocins: Desferrithiocin (DFT) (22) is a siderophore isolated from *Streptomyces antibioticus*. It forms a 2:1 complex with iron(III) at neutral pH using a phenolate oxygen, a carboxylate oxygen and a nitrogen atom as ligands[67]. It possesses a high affinity for ferric iron (Log β_2 = 29.6), however by virtue of the presence of the nitrogen and carboxylate ligands, it also binds zinc with relatively high affinity (Log β_2 = 15.3)[68].

(22)

Long term studies of DFT in normal rodents and dogs at low doses have shown toxic side effects, such as reduced body weight and neurotoxicity[69]. A range of synthetic analogues of DFT have been prepared[70], however, to date no suitable candidates have been identified for

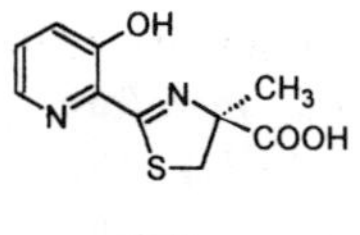

IRON CHELATOR CHEMISTRY

(22)

Triazoles (See Chapter 14 of this volume): Recently, triazoles have been investigated as ligands by Novartis[72]. These molecules chelate iron(III) with two phenolate oxygens and one triazolyl nitrogen. The lead compound ICL670A (**23**) is extremely hydrophobic, with a logP value of 3.8 and a $logD_{7.4}$ value of 1.0[73]. As a result it can penetrate membranes easily and possesses good oral availability. It is highly effective at removing iron from both the iron-loaded rat and marmoset and is the current lead orally active iron chelator of Novartis[73]. The triazole (**23**) forms a 2:1 iron complex which possesses a net charge of 3⁻ and a molecular weight over 800[74]. Should such a complex form intracellularly, with the exception of hepatocytes, it is likely that the complex will remain trapped within the cell. The triazoles can exist in two conformations: one a tridentate structure (**23A**) and the alternate a tetradentate structure (**23B**) with a strong tendency to form polymeric complexes (**Figure 7**). The latter conformation (**23B**) favours zinc(II) since 50% of the coordinating ligands are nitrogen[74,75].

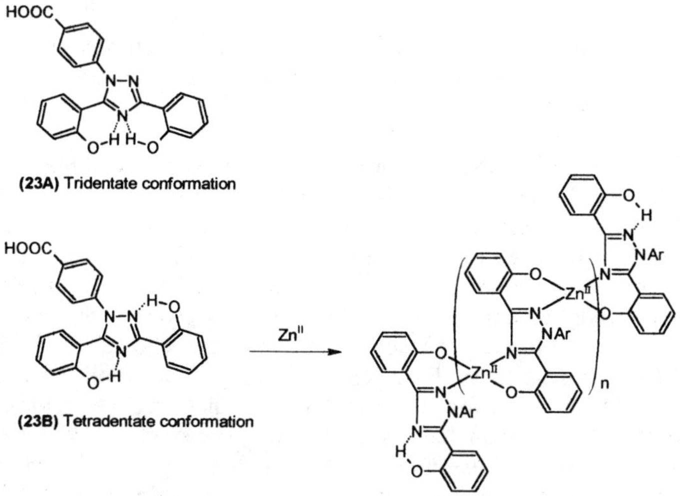

(23A) Tridentate conformation

(23B) Tetradentate conformation

Figure 7: *Two possible conformations of the triazole derivatives: (A) the tridentate conformation and (B) the tetradentate conformation. The tetradentate structure has a strong tendency to form polymeric complexes.*

PIH Analogues (See Chapter 13 of this volume): Pyridoxal isonicotinoyl hydrazone (PIH) (**24**), together with a wide range of analogues have been subjected to extensive evaluation as iron(III) chelators[76-79]. Many have been demonstrated to be orally active in rodents. The efficiency of *in vivo* iron removal increases with lipophilicity of both the ligand and the iron complex[80], but as more lipophilic chelators are likely to be associated with enhanced toxicity, log P values close to unity are preferred[80]. Many of the PIH analogues are uncharged at neutral pH values and therefore gain ready access to cells, indeed salicylaldehyde isonicotinoyl hydrazone (SIH) (**25**) gains entry to a range of cell types more rapidly than most other chelators, including the smaller hydroxypyridinones[81]. The binding of iron(III) by PIH and related ligands is complicated because of the existence of a number of dissociatable protons both in the free and coordinated states. Never-the-less the 2:1 complex is the favoured species over the pH range 4-8 and the affinity for iron(III) compares well with that of other tridentate ligands. The two nitrogen atoms in the coordinating sphere of the complex[82] endow PIH analogues with the ability to bind iron(II) with appreciable affinity[83].

(24)   (25)

4.3. Bidentate chelators

On the basis of selectivity and affinity, particularly considering the pFe^{3+} value, 3-hydroxypyridin-4-one is the optimal bidentate ligand for the chelation of iron(III) over the pH range of 6.0 – 9.0 (**Table 1**).

Dialkylhydroxypyridinones (See Chapters 7, 8, 9 & 10 of this volume) The 1,2-dimethyl derivative (deferiprone, L1, CP20) (**7**) is the only orally active iron chelator currently available for clinical use (marketed by Apotex Inc. Toronto, Canada as Ferriprox™). Unfortunately, the dose required to keep a previously well chelated patient in negative iron balance with Ferriprox™ is relatively high, in the region of 75 to 100 mg/kg/day[84] and side effects have been observed in some patients receiving deferiprone in this dose range[85,86]. One of the major reasons for the limited efficacy of deferiprone in clinical use is that it undergoes extensive phase II metabolism in the liver. The 3-hydroxyl functionality, which is crucial for scavenging iron, is a prime target for glucuronidation. Urinary recovery studies conducted on deferiprone in both rats and man have shown that respectively >44% and >85% of the administered dose is recovered in the urine as the non-chelating 3-*O*-glucuronide conjugate[87]. The use of deferiprone for the treatment of iron overload remains a hotly debated subject at the present time[86, 88-93], however until a superior orally active chelator becomes available, deferiprone remains in the unique position of being the only orally active chelator available for clinical use.

The 1,2-diethyl analogue CP94 (**26**) has also been extensively investigated[94-97]. This chelator has been found to be more efficient at iron removal than deferiprone in several mammalian species e.g. rat[94] and cebus monkey[97]. The presumed reason for the greater efficacy of CP94 in the rat is its unusual phase I metabolic pathway which leads to the formation of the 2-(1'-hydroxyethyl) metabolite CP365 (**27**) (**Figure 8**)[87,98]. This metabolite does not undergo further phase II metabolism to form a glucuronide conjugate and hence

IRON CHELATOR CHEMISTRY

retains the ability to chelate iron. Promising results obtained in rat models, led to the limited clinical evaluation of CP94 in thalassaemic patients[96]. Unfortunately, the metabolism of CP94 in man does not parallel that of the rat. The main urinary metabolite of CP94 in man is the 3-O-glucuronide conjugate (>85%)[95]. Extensive conversion to this metabolite was found to severely limit clinical efficacy.

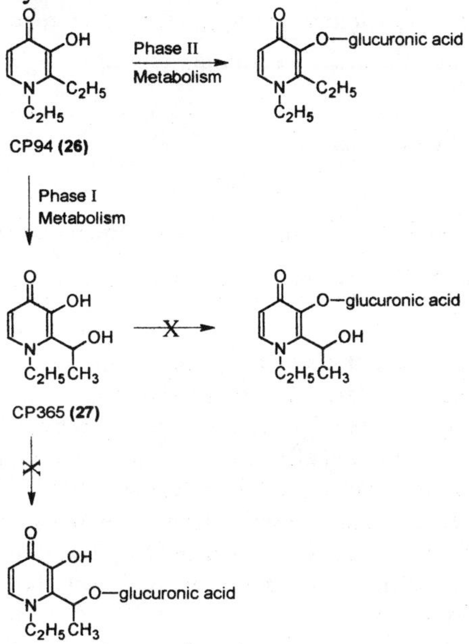

Figure 8: *Metabolism of CP94 (26) via phase I and phase II metabolic pathways. The 2-1'-hydroxylated metabolite CP365 (27) is the major metabolite in the rat, whereas phase II glucuronidation is the major metabolic route of CP94 in man.*

Hydroxypyridinone prodrugs: The critical dependence of chelator efficacy on metabolic behaviour led to a concept of ligand design which minimises conjugation reactions with glucuronic acid. Despite the limited efficacy of CP94 in man, the superior extracellular and intracellular iron mobilisation ability in the rat provided important information for chelator design. The lack of glucuronidation of the 2-(1'-hydroxyethyl) metabolite of CP94 led to the investigation of the possibility of developing structurally related compounds. Indeed, 1-hydroxyalkyl derivatives of HPOs such as CP40 (**28**) and CP102 (**29**), which are not extensively metabolised *via* phase II reactions, have been identified[99,100]. Although the use of 1-hydroxyalkyl derivatives of HPO may offer a significant improvement over previously evaluated HPOs[101-103], a disadvantage of these compounds, especially with some of the more hydrophilic analogues is their reduced liver extraction.

Another potential problem associated with orally active bidentate HPOs is that, by virtue of their relatively low molecular weight and favourable distribution coefficients, they rapidly penetrate most cells and critical barriers such as the blood-brain barrier and the placental barrier. Ideally, the distribution of iron chelators developed for the treatment of general iron overload, such as β-thalassaemia major, is best limited to the extracellular space and the liver. The anticipated distribution coefficient ($D_{7.4}$) requirements for an ideal iron chelator are outlined in **Table 3**. Clearly, there is no single compound fulfilling these requirements since

Table 3: Anticipated optimal distribution coefficients of an ideal iron chelator

	$D_{7.4}$
Good absorption from the gastrointestinal tract	> 0.2
Efficient liver extraction	> 1.0
Poor entry into peripheral cells (thymus, muscle, heart, bone marrow)	< 0.001
Poor ability to penetrate the blood-brain and maternal/placental barriers	< 0.001

the optimal distribution coefficient for absorption from the gastrointestinal tract is quite different from that necessary to limit access to the brain and placenta. In principle this problem can be overcome by the use of prodrugs, whereby a hydrophobic prodrug is absorbed from the gastrointestinal tract and then efficiently extracted by the liver during the "first pass". Once in the hepatocyte, the prodrug ester link is cleaved to yield a much more hydrophilic chelator (**Figure 9**). This chelator can scavenge iron in the hepatocyte, but also efflux into the systemic circulation, thereby scavenging the extracellular iron pool. Because of its hydrophilic nature, the ability to cross critical membrane barriers will be markedly reduced, thereby minimising toxicity problems.

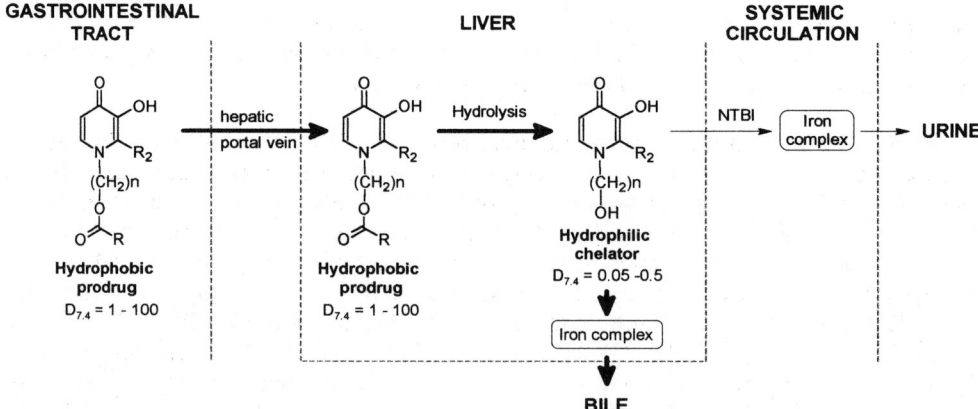

Figure 9: Schematic representation of the use of ester prodrugs of 1-hydroxyalkyl HPOs to enhance both the absorption from the gastrointestinal tract and liver extraction from the hepatic portal vein. In the hepatocyte, the prodrug is rapidly converted to a hydrophilic

chelator which will scavenge iron in the hepatocyte. This hydrophilic chelator can efflux into the systemic circulation, thereby scavenging the extracellular iron pool. Due to its hydrophilic nature, the molecule will not be expected to readily penetrate critical membrane barriers such as the blood-brain barrier.

A wide range of ester prodrugs of 1-hydroxyalkyl HPOs has been investigated[104-106]. *In vitro* esterase studies indicate that the pivaloyl esters and the aromatic ester analogues partially fulfil the requirements for relatively efficient liver extraction[104-108]. Preliminary pharmacokinetic and absorption studies in the rat have demonstrated that the prodrugs are rapidly absorbed from the gastrointestinal tract in the intact form and subsequently undergo extensive first pass metabolism[107,108]. In many cases the ester prodrug leads to superior iron excretion *via* the bile than the corresponding alcohol (**Table 4**), indicating selective delivery to the liver. However, not all prodrugs provide increased efficacy, as for instance with the benzoyl ester of CP102 (CP183) (**30**) where the increased hydrophobicity of the ester leads to rapid conjugation of the prodrug.

Table 4: *Selected ester prodrugs of 1-hydroxyalkyl HPOs.*

Chelator	R_2	R	$D_{7.4}$	Efficacy (%)[1]
CP40 (**28**)	CH_3	H	0.08	3.7
CP162	CH_3	$COC(CH_3)_3$	4.70	32.0
CP280	CH_3	COC_6H_5	10.90	13.0
CP102 (**29**)	C_2H_5	H	0.22	12.9
CP117	C_2H_5	$COC(CH_3)_3$	14.50	19.2
CP183 (**30**)	C_2H_5	COC_6H_5	32.80	9.6

[1] Iron mobilisation efficacy of chelators (450μmol/kg) was measured using the ^{59}Fe-ferritin loaded rat model[103].

"**High pFe^{3+}**" **hydroxypyridinones**: In order to further improve chelation efficacy and minimise drug-induced toxicity, considerable effort has been applied to the design of novel hydroxypyridinones with enhanced pFe^{3+} values[109,110]. Recently we have demonstrated that the introduction of a 1'-hydroxyalkyl group (**31**)[111] or an amido function (**32**)[112] at the 2-position of 3-hydroxypyridin-4-ones enhances the affinity for iron(III) in the pH range 5-8. This effect results from stabilising the ionised species due to the combined effect of intramolecular hydrogen bonding between the 2-(1'-hydroxyl) group or the 2-amido substitutent with the adjacent 3-hydroxyl function and electron withdrawal from the pyridinone ring. Although such an effect reduces the overall iron(III) stability constant, it also reduces the pKa values of the chelating function. These combined changes result in an increase in the corresponding pFe^{3+} values[110-112]. Interestingly a recent Novartis lead compound (**33**)[113] also possesses an 1'-hydroxyl group at the 2-position and this is also responsible for the observed enhanced pFe^{3+} value of this ligand.

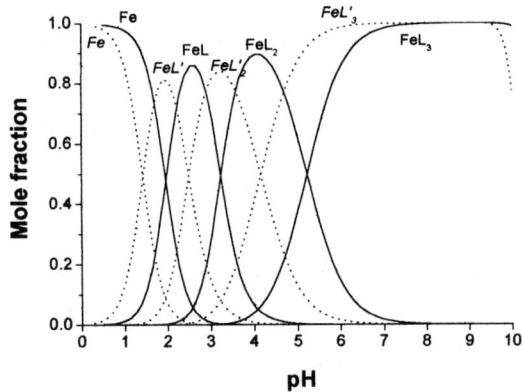

(31)   (32)   (33)

The enhancement of pFe^{3+} values has a dramatic effect on the speciation plot of the corresponding iron(III) complexes of these ligands. Thus the complex of the N-methyl amido derivative CP502 (**34**) dissociates less readily than the dialkyl derivatives, leading to lower concentrations of the L$_2$Fe$^+$ complex when compared for instance with deferiprone (**Figure 10**). These novel high pFe^{3+} HPOs show great promise in their ability to remove iron under *in vivo* conditions[110-112] (**Table 5**). Furthermore the low lipophilicity of many of these molecules will severely limit the distribution of the chelator, which in turn is likely to minimise the potential toxicity. A small number of related compounds have been selected for pre-clinical investigation.

Figure 10: Comparison of the speciation plots of iron(III) in the presence of deferiprone (**7**) and CP502 (**34**); L = deferiprone; L' = CP502; [Fe^{3+}]$_{total}$ = 1 x 10^{-6} M and [Ligand] = 1 x 10^{-5} M.

5. CONCLUSIONS

Over the past 30 years many attempts have been directed towards the design of nontoxic orally active iron chelators, but only one clinically useful compound has emerged to date, deferiprone. Since 1995 a number of significant advances have been made and the authors are confident that other more efficacious orally active chelators will soon join deferiprone. The successful introduction of such compounds will impact considerably on the therapeutic

Table 5: *Physicochemical properties and iron mobilisation efficacies of novel high pFe^{3+} 3-hydroxypyridin-4-ones.*

Chelator	R_1	R_2	R_6	$D_{7.4}$	pKa	$Log\beta_3$	pFe^{3+}	Efficacy (%) [1]
CP20 (7)	CH_3	CH_3	H	0.17	3.56, 9.64	36.3	19.4	9.5
CP361	CH_3	$CH(CH_3)OH$	CH_3	0.25	3.55, 8.97	35.5	21.5	44.5
CP363	CH_3	$CH(CH_3)OCH_3$	CH_3	1.13	3.22, 9.55	36.1	21.0	69.7
CP365 (27)	C_2H_5	$CH(CH_3)OH$	H	0.27	3.11, 8.74	34.8	21.3	50.6
CP375	CH_3	$CH(C_2H_5)OCH_3$	CH_3	3.85	N.D.	N.D.	N.D.	68.1
CP502 (34)	CH_3	$CONHCH_3$	CH_3	0.04	2.77, 8.44	34.3	21.7	54.8

[1] Iron mobilisation efficacy of chelators (450μmol/kg) was measured using the ^{59}Fe-ferritin loaded rat model[103].

outcome and quality of life for the thalassemic population. There is a potential for iron chelation in a wider range of clinical situations and once an orally active chelator has been clinically proven in thalassemic patients, such compounds will almost certainly find application for the treatment of other disease states, for instance sickle cell anaemia.

Acknowledgements. The authors thank Apotex Research Inc. Canada and Medical Research Council of United Kingdom for financial support.

REFERENCE:

1. Hershko C, Konijn AM, Link G. Iron chelators for thalasaemia. *Br J Haematol* 1998; **101**: 399-406.
2. Martell AE, Smith RM. Critical stability constant. Vol. 1-6. London: Plenum Press, 1974-1989.
3. Harris DC, Aisen P. Facilitation of Fe(II) autoxidation by Fe(III) complexing agents. *Biochim Biophys Acta* 1973; **329**: 156-158.
4. Raymond KN, Muller G, Matzanke BF. Complexation of iron by siderophores: A review of their solution and structural chemistry and biological function. *Top Curr Chem* 1984; **58**: 49-102.
5. Hider RC, Liu ZD, Piyamongkol S. The design and properties of 3-hydroxypyridin-4-one iron chelators with high pFe^{3+} values. *Transfus Sci* 2000; **23**: 201-209.
6. Hider RC, Mohd-Nor AR, Silver J, Morrison IEG, Rees LVC. Model compounds for microbial iron-transport compounds. Part 1. Solution chemistry and mössbauer study of iron(II) and iron(III). Complexes from phenolic and catecholic system. *J Chem Soc Dalton Trans* 1981; **2**: 609-622.

7. Hider RC, Hall AD. Clinically useful chelators of tripositive elements. *Prog Med Chem* 1991; **28**: 41-173.
8. Pippard MJ, Jackson MJ, Hoffman K, Petrou M, Model CB. Iron chelation using subcutaneous infusion of diethyl triaminopentaacetic acid (DTPA). *Scand J Haem.* 1986; **36**: 466-472.
9. Martin RB. Citrate binding of Al^{3+} and Fe^{3+}. *J Inorg Biochem* 1986; **28**: 181-187.
10. Pierre JL, Gautier-Luneau I. Iron and citric acid: A fuzzy chemistry of ubiquitous biological relevance. *Biometals* 2000; **13**: 91-96.
11. Spiro TG, Bates G, Saltman P. The hydrolytic polymerisation of ferric citrate II. The influence of excess citrate. *J Am Chem Soc* 1967; **89**: 5559-5562.
12. Strouse J, Layten SW, Strouse CE. Structural studies of transition metal complexes of triionized and tetraionized citrate. Models or the coordination of the citrate ion to transition metal ions in solutions and at the active site of aconitase. *J Am Chem Soc* 1977; **99**: 562-572.
13. Meiwes J, Fiedler HP, Haag H, Zahner H, Konetschnyrapp S, Jung G. Isolation and characterization of Staphyloferrin A: A compound with siderophore activity from S*taphylococcus hyicus* DSM-20459. *FEMS Microbiol Lett* 1990; **67**: 201-205.
14. Thieken A, Winkelmann G. Rhizoferrin - A complexone type siderophore of the Mucorales and Entomophthorales (Zygomycetes). *FEMS Microbiol Lett* 1992; **94**: 37-42.
15. Tilbrook GS, Hider RC. Iron Chelators for clinical use. In: Sigel A, Sigel H, eds. *Metal Irons in Biological Systems. Vol. 35: Iron Transport and Storage in Microorganisms, Plants and Animals.* New York: Marcel Dekker, 1998:691-730.
16. Florence AT, Attwood D. Physicochemical Principles of Pharmacy. 2nd. ed. London: Macmillan, 1988
17. Holander D, Ricketts D, Boyd CAR. Importance of probe molecular geometry in determining intestinal permeability. *Can J Gastroenterol* 1988; **2**: 35A-38A.
18. Fagerholm U; Nilsson D; Knutson L; Lennernas H. Jejunal permeability in humans in vivo and rats in situ: Investigation of molecular size selectivity and solvent drag. *Acta Physiol Scand* 1999; **165**: 315-324.
19. Travis S, Menzies I. Intestinal permeability: functional assessment and significance. *Clin Sci* 1992; **82**: 471-488.
20. Maxton DG, Bjarnason I, Reynolds AP, Catt SD, Peters TJ, Menzies IS. Lactulose ^{51}Cr-labelled ethylenediaminetetra-acetate, L-rhamnose and polyethyleneglycol 500 as probe markers for assessment in vivo of human intestinal permeability. *Clin Sci* 1986; **71**: 71-80.
21. Kim M. Absorption of polyethylene glycol oligomers (330-1122 Da) is greater in the jejunum than in the ileum of rats. *J Nutr* 1996; **126**: 2172-2178.
22. Yokel RA, Fredenburg AM, Meurer KA, Skinner TL. Influence of lipophilicity on the bioavailability and disposition of orally active 3-hydroxypyridin-4-one metal chelators. *Drug Metab Dispo* 1995; **23**: 1178-1180.
23. Porter JB, Huehns ER, Hider RC. The development of iron chelating drugs. *Bailliere's Clin Haematol* 1989; **2**: 257-292.
24. Pippard MJ, Callender ST, Weatherall DJ. Intensive iron-chelation therapy with desferrioxamine in iron-loading anaemias. *Clin Sci Mol Med* 1978; **54**: 99-106.
25. O'Halloran TV. Transition-metals in control of gene-expression. *Science* 1993; **261**: 715-725.
26. Rae TD; Schmidt PJ; Pufahl RA; Culotta VC; O'Halloran TV. Undetectable intracellular free copper: The requirement of a copper chaperone for superoxide dismutase. *Science* 1999; **284**: 805-808.
27. Singh S, Khodr H, Taylor MI, Hider RC. Therapeutic iron chelators and their potential side-effects. *Biochem Soc Symp* 1995; **61**: 127-137.
28. Hider RC, Kayyli R, Evans P. The production of hydroxyl radicals by Deferiprone-iron compounds under physiological conditions. *Blood* 1999; **94**(suppl):406A-406A.
29. Lippard SJ, Berg JM. Principles of Bioinorganic Chemistry. California: University Science Books, 1994.
30. Britovsek GJP, Gibson VC, Wass DF. The search for new-generation olefin polymerization catalysts: Life beyond metallocenes. *Angew Chem Int Ed* 1999; **38**: 428-447.
31. Hider RC. Potential protection from toxicity by oral iron chelators. *Toxicol Lett* 1995; **82-3**: 961-967.
32. Hider RC, Singh S, Porter JB. Iron chelating agents with clinical potential. *Proc Roy Soc Edin* 1992; **99B**: 137-168.
33. Abeysinghe RD, Robert PJ, Cooper CE, MacLean KH, Hider RC, Porter JB. The environment of the lipoxygenase iron binding site explored with novel hydroxypyridinone iron chelators. *J Biol Chem* 1996; **271**: 7965-7972.

34. Liu ZD, Kayyali RS, Hider RC, Theobald AE. Design, synthesis, and evaluation of novel 2-substituted 3-hydroxypyridin-4-ones: Structure-activity investigation of metalloenzyme inhibition by iron chelators. *J Med Chem*, in press.
35. Liu ZD, Lockwood M, Rose S, Theobald AE, Hider RC. Structure-activity investigation of the inhibition of 3-hydroxypyridin-4-ones on mammalian tyrosine hydroxylase. *Biochem Pharmacol*. 2001; **61**: 285-290.
36. Cooper CE, Lynagh GR, Hoyes KP, Hider RC, Cammack R, Porter JB. The relationship of intracellular iron chelation to the inhibition and regeneration of human ribonucleotide reductase. *J Biol Chem* 1996; **271**: 20291-20299.
37. Levin VA. Relationship of octanol/water partition coefficient and molecular weight to rat brain capillary permeability. *J Med Chem* 1980; **23**: 682-684.
38. Oldendorf WH. Lipid solubility and drug penetration of the blood brain barrier. *Proc Soc Expt Bio Med* 1974; **147**: 813-816.
39. Habgood MD, Liu ZD, Dehkordi LS, Khodr HH, Abbott J, Hider RC. Investigation into the correlation between the structure of hydroxypyridinones and blood-brain barrier permeability. *Biochem Pharmacol* 1999; **57**: 1305-1310.
40. Hershko C, Grady RW, Cerami A. Mechanism of iron chelation in the hypertransfused rat: definition of two alternative pathways of iron mobilisation. *J Lab Clin Med* 1978; **92**: 144-149.
41. Summers MR, Jacobs A, Tudway D, Perera P, Ricketts C. Studies in desferrioxamine and ferrioxamine metabolism in normal and iron-loaded subjects. *Br J Haematol* 1979; **42**: 547-555.
42. Porter JB, Jawson MC, Huehns ER, East CA, Hazell JWP. Desferrioxamine toxicity: Evaluation of risk factors in thalassaemic patients and guidelines for safe dosage. *Br J Haematol* 1989; **73**: 403-405.
43. Freedman MH, Grisaru D, Olivieri NF, MacLusky I, Thorner PS. Pulmonary syndrome in patients with thalassemia major receiving intravenous deferoxamine infusions. *Am J Dis Child* 1990; **144**: 565-569.
44. Olivieri NF, Buncic JR, Chew E, Gallant T, Harrison RV, Keenan N, Logan W, Mitchell D, Ricci G, Skarf B. Visual and auditory neurotoxicity in patients receiving subcutaneous desferrioxamine infusions. *New Engl J Med* 1986; **314**: 869-873.
45. Olivieri NF, Harris J, Koren G, Khattak S, Freedman MH, Templeton DM, Bailey JD, Reilly BJ. Growth failure and bony changes induced by deferoxamine. *Am J Pediatr Hematol Oncol* 1992; **14**: 48-56.
46. Peter HH. Industrial aspects of iron chelators: Pharmaceutical application. In: Spik G, Montreuil J, Crichton RR, Mazurier J, eds. *Proteins of Iron Storage and Transport*. Amsterdam: Elsevier, 1985:293-303.
47. Fechtig B, Peter H. New O-acylhydroxamic acid derivatives. *International Patent WO 8603745*, 1984.
48. Bergeron RJ, Wiegand J, McManis JS, Perumal PT. Synthesis and biological evaluation of hydroxamate-based iron chelators. *J Med Chem* 1991; **34**:3182-3187.
49. Bergeron RJ, Liu Z, McManis JS, Wiegand J. Structural alterations in desferrioxamine compatible with iron clearance in animals. *J Med Chem* 1992; **35**:4739-4744.
50. Jackson MJ, Brenton DP, Modell B. DTPA in the management of iron overload in thalassaemia. *J Inher Met Dis* 1983; **6**(suppl 2):97-98.
51. Bannerman RM, Callender ST, Williams DL. Effect of desferrioxamine and DTPA in iron overload. *Br Med J* 1962; **2**:1573-1577.
52. L'Eplattenier F, Murase I, Martell AE. New multidentate ligands. IV. Chelating tendencies of N,N'-di(2-hydroxybenzyl)ethylenediamine-N,N'-diacetic acid. *J Am Chem Soc* 1967; **89**: 837-843.
53. Martell AE, Motekaitis RJ, Clarke ET. Synthesis of N,N'-di(2-hydroxybenzyl)ethylenediamine-N,N'-diacetic (HBED) derivatives. *Can J Chem* 1986; **64**:449-456.
54. Lau EH, Cerny EA, Wright BJ, Rahman YE. Improvement of iron removal from the reticuloendothelial system by liposome encapsulation of N,N'-bis[2-hydroxybenzyl]-ethylenediamine-N,N'-diacetic acid (HBED) - Comparison with desferrioxamine. *J Lab Clin Med* 1983; **101**: 806-816.
55. Bergeron RJ, Wiegand J, Brittenham GM. HBED: A potential alternative to deferoxamine for iron-chelating therapy. *Blood* 1998; **91**:1446-1452.
56. Peter HH, Bergeron RJ, Streiff RR, Wiegand J. A comparative evaluation of iron chelators in a primate model. In: Bergeron RJ, Brittenham GM, eds. *The Development of Iron Chelators for Clinical Use*. London: CRC Press, 1993:373-394.
57. Grady RW, Salbe AD, Hilgartner MW, Giardina PJ. Results from phase I clinical trial of HBED. *Adv Exp Med Biol* 1994; **356**: 351-359.
58. Pitt CG, Bao Y, Thompson J, Wani MC, Rosenkrantz H, Metterville J. Esters and lactones of phenolic amino carboxylic acids: Prodrugs for iron chelation. *J Med Chem* 1986; **29**:1231-1237.

59. Gasparini F, Leutert T, Farley DL. N,N'-bis(2-hydroxybenzyl)ethylene-diamine-N,N'-diacetic acid derivatives as chelating agents. *International Patent WO 95/16663*, 1995.
60. Lowther N, Tomlinson B, Fox R, Faller B, Sergejew T, Donnelly H. Caco-2 cell permeability of a new (hydroxybenzyl)ethylenediamine oral iron chelator: Correlation with physicochemical properties and oral activity. *J Pharm Sci* 1998; **87**: 1041-1045.
61. Faller B, Spanka C, Sergejew T, Tschinke V. Improving the oral bioavailability of the iron chelator HBED by breaking the symmetry of the intramolecular H-bond network. *J Med Chem.* 2000; **43**:1467-1475.
62. Hider RC, Bittel D, Andrews GK. Competition between iron(III)-selective chelators and zinc-finger domains for zinc(II). *Biochem Pharmacol* 1999; **57**:1031-1035.
63. Guterman SK, Morris PM, Tannenberg WJK. Feasibility of enterochelin as an iron-chelating drug: Studies with human serum and a mouse model system. *Gen Pharmac* 1978; **9**:123-127.
64. White DL, Durbin PW, Jeung N, Raymond KN. Specific sequestering agents for the actinides. 26: Synthesis and initial biological testing of polydentate oxohydroxypyridinecarboxylate ligands. *J Med Chem* 1988; **31**:11-18.
65. Xu JD, Kullgren B, Durbin PW, Raymond KN. Specific sequestering agents for the actinides. 28: Synthesis and initial evaluation of multidentate 4-carbamoyl-3-hydroxy-1-methyl-2(1H)-pyridinone ligands for in vivo plutonium(IV) chelation. *J Med Chem* 1995; **38**:2606-2614.
66. Rai BL, Khodr H, Hider RC. Synthesis, physico-chemical and iron(III)-chelating properties of novel hexadentate 3-hydroxy-2(1H)pyridinone ligands. *Tetrahedron* 1999; **55**:1129-1142.
67. Hahn FN, McMurry TJ, Hugi A, Raymond KN. Coordination chemistry of microbial iron transport. 42: Structural and spectroscopic characterisation of diastereometric Cr(III) and Co(III) complexes of desferrithiocin. *J Am Chem Soc* 1990; **112**:1854-1860.
68. Anderegg G, Raber M. Metal complex formation of a new siderphore desferrithiocin and of three related ligands. *J Chem Soc Chem Commun* 1990; 1194-1196.
69. Wolfe LC. Desferrithiocin. *Sem Haematol* 1990; **27**:117-120.
70. Bergeron RJ, Wiegand J, Dionis JB, Egli-Karmakka M, Frei J, Huxley-Tencer A, Peter HH. Evaluation of desferrithiocin and its synthetic analogues as orally effective iron chelators. *J Med Chem* 1991; **34**:2072-2078.
71. Baker E, Peter WH, Jacobs A. Desferrithiocin is an effective iron chelator in vivo and in vitro but ferrithiocin is toxic. *Br J Haematol* 1992; **81**:424-431.
72. Lattmann R, Acklin P. Substituted 3,5-diphenyl-1,2,4-triazoles and their use as pharmaceutical metal chelators. *International Patent WO 97/49395*, 1997.
73. Nick HP, Acklin P, Faller B, Jin Y, Lattmann R, Rouan M-C, Sergejew T, Thomas H, Wiegand H, Schnebli HP. A new, potent, orally active iron chelator. In: Badman DG, Bergeron RJ, Brittenham GM, eds. *Iron Chelators: New development strategies*. Florida: The Saratoga Group. 2000; 311-331.
74. Heinz U, Hegetschweiler K, Acklin P, Faller B, Lattmann R, Schnebli HP. 4-[3,5-bis(2-hydroxyphenyl)-1,2,4-triazol-1-yl]benzoic acid: A novel efficient and selective iron(III) complexing agent. *Angew Chem Int Edit* 1999; **38**:2568-2570.
75. Ryabukhin YI, Shibaeva NV, Kuzharov AS, Korobkova VG, Khokhlov AV, Garnovskii AD. Synthesis and investigation of complex compounds of transition metals with di(o-hydroxyphenyl)-1,2,4-oxadiazole and its 1,2,4-triazole analogs. *Sov J Coordinat Chem* 1988; 493-499.
76. Hershko C, Avramovici-Grisaru S, Link G, Gelfand L, Sarel S. Mechanism of in vivo iron chelation by pyridoxal isonicotinoyl hydrazone and other imino derivatives of pyridoxal. *J Lab Clin Med* 1981; **98**:99-108.
77. Ponka P, Borova J, Neuwirt J, Fuchs O. Mobilization of iron from reticulocytes. *FEBS Lett* 1979; **97**:317-321.
78. Johnson DK, Pippard MJ, Murphy TB, Rose NJ. An in vivo evaluation of iron-chelating drugs derived from pyridoxal and its analogs. *J Pharmacol Exp Ther* 1982; **221**:399-403.
79. Cikrt M, Ponka P, Necas E, Neuwirt J. Biliary iron excretion in rats following pyridoxal isonicotinoyl hydrazone. *Br J Haematol* 1980; **45**:275-283.
80. Edward JT, Ponka P, Richardson DR. Partition-coefficients of the iron(III) complexes of pyridoxal isonicotinoyl hydrazone and its analogs and the correlation to iron chelation efficacy. *Biometals* 1995; **8**:209-217.
81. Zanninelli G, Glickstein H, Breuer W, Milgram P, Brissot P, Hider RC, Konijn AM, Libman J, Shanzer A, Cabantchik ZI. Chelation and mobilization of cellular iron by different classes of chelators. *Mol Pharmacol* 1997; **51**:842-852.

82. Webb J, Vitolo LM. Pyridoxal isonicotinoyl hydrazone (PIH): A promising new iron chelator. *Birth Defects Original article Ser* 1988; **23**: 63-70.
83. Avramovicigrisaru S, Sarel S, Cohen S, Bauminger RE. The synthesis, crystal and molecular-structure, and oxidation-state of iron complex from pyridoxal isonicotinoyl hydrazone and ferrous sulfate. *Israel J Chem* 1985; **25**:288-292.
84. Balfour JAB, Foster RH. Deferiprone - A review of its clinical potential in iron overload in beta-thalassaemia major and other transfusion-dependent diseases. *Drugs* 1999; **58**:553-578.
85. Brittenham GM. Development of iron-chelating agents for clinical use. *Blood* 1992; **80**:569-574.
86. Olivieri NF, Brittenham GM, McLaren CE, Templeton DM, Cameron RG, McClelland RA, Burt AD, Fleming KA. Long-term safety and effectiveness of iron-chelation therapy with deferiprone for thalassemia major. *New Engl J Med* 1998; **339**:417-423.
87. Singh S, Epemolu O, Dobbin PS, Tilbrook GS, Ellis BL, Damani LA, Hider RC. Urinary metabolic profiles in man and rat of 1,2-dimethyl- and 1,2-diethyl substituted 3-hydroxypyridin-4-ones. *Drug Metab Dispo* 1992; **20**:256-261.
88. Nathan DG. An orally active iron chelator. *New Engl J Med* 1995; **332**:953-954.
89. Hoffbrand AV, Al-Refaie F, Davis B, Siritanakatkul N, Jackson BFA, Cochrane J, Prescott E, Wonke B. Long-term trial of deferiprone in 51 transfusion-dependent iron overloaded patients. *Blood* 1998; **91**:295-300.
90. Kowdley KV, Kaplan MM. Iron-chelation therapy with oral deferiprone - Toxicity or lack of efficacy? *New Engl J Med* 1998; **339**:468-469.
91. Tricta F, Spino M. Iron chelation with oral deferiprone in patients with thalassemia. *New Engl J Med* 1998; **339**:1710.
92. Maggio A, Capra M, Ciaccio C, Magnano C, Rizzo M, et al. Evaluation of efficacy of L1 versus desferrioxamine by clinical randomized multicentric study. *Blood* 1999; **94**(suppl): 34B-34B.
93. Cohen AR, Galanello R, Piga A, Gamberini R, DeSanctis V, Tricta F. Deferiprone and neutropenia: Incidence and characteristics in a long-term safety study. *Blood* 1999; **94**(suppl):406A-406A.
94. Porter JB, Morgan J, Hoyes KP, Burke LC, Huehns ER, Hider RC. Relative oral efficacy and acute toxicity of hydroxypyridin-4-one iron chelators in mice. *Blood* 1990; **76**:2389-2396
95. Porter JB, Abeysinghe RD, Hoyes KP, Barra C, Huehns ER, Brooks PN, Blackwell MP, Araneta M, Britenham G, Singh S, Dobbin P, Hider RC. Contrasting interspecies efficacy and toxicology of 1,2-diethyl-3-hydroxypyridin-4-one CP94, relates to differing metabolism of the iron chelating site. *Br J Haematol* 1993; **85**:159-168.
96. Porter JB, Singh S, Katherine PH, Epemolu O, Abeysinghe RD, Hider RC. Lessons from preclinical and clinical studies with 1,2-diethyl-3-hydroxypyridin-4-one, CP94 and related compounds. *Adv Expt Med Biol* 1994; **356**:361-370
97. Bergeron RJ, Streiff RR, Wiegand J, Luchetta G, Creary EA, Peter HH. A comparison of the iron-clearing properties of 1,2-dimethyl-3-hydroxypyrid-4-one, 1,2-diethyl-3-hydroxypyrid-4-one, and Deferoxamine. *Blood* 1992; **79**:1882-1890.
98. Lu SL, Gosriwatana I, Liu DY, Liu ZD, Mallet AI, Hider RC. Biliary and urinary metabolic profiles of 1,2-diethyl-3-hydroxypyridin-4-one (CP94) in the rat. *Drug Metab Dispos* 2000; **28**:873-879.
99. Singh S, Epemolu RO, Ackerman R, Porter JB, Hider RC. Development of 3-hydroxypyridin-4-ones which do not undergo extensive phase II metabolism. 3rd NIH-Sponsored Symposium on The Development of Iron Chelators for Clinical Use. Gainsville, FL, 1992, Abstract: 52.
100. Singh S, Choudhury R, Epemolu RO, Hider RC. Metabolism and pharmacokinetics of 1-(2'-hydroxyethyl)- and 1-(3'-hydroxypropyl)-2-ethyl-3-hydroxypyridin-4-ones in the rat. *Eur J Drug Metab Pharmacokinet* 1996; **21**:33-41
101. Zanninelli G, Choudury R, Loreal O, Guyader D, Lescoat G, Arnaud J, Verna R, Cosson B, Singh S, Hider RC, Brissot P. Novel orally active iron chelators (3-hydroxypyridin-4-ones) enhance the biliary excretion of plasma non-transferrin-bound iron in rats. *J Hepatol* 1997; **27**:176-184.
102. Rai BL, Dekhordi LS, Khodr H, Jin Y, Liu Z, Hider RC. Synthesis, physicochemical properties and evaluation of N-substituted-2-alkyl-3-hydroxy-4(1H)-pyridinones. *J Med Chem* 1998; **41**:3347-3359
103. Liu ZD, Lu SL, Hider RC. In vivo iron mobilisation evaluation of hydroxypyridinones in ^{59}Fe-ferritin loaded rat model. *Biochem Pharmacol* 1999; **57**:559-566.
104. Rai BL, Liu ZD, Liu DY, Lu SL, Hider RC. Synthesis, physicochemical properties and biological evaluation of ester prodrugs of 3-hydroxypyridin-4-ones: Design of orally active chelators with clinical potential. *Eur J Med Chem* 1999; **34**:475-485.

105. Liu ZD, Liu DY, Lu SL, Hider RC. Synthesis, physicochemical properties and biological evaluation of aromatic ester prodrugs of 1-(2'-hydroxyethyl)-2-ethyl-3-hydroxypyridin-4-one (CP102): Orally active iron chelators with clinical potential. *J Pharm Pharmacol* 1999; **51**:555-564.
106. Liu ZD, Liu DY, Lu SL, Hider RC. Design, synthesis, and biological evaluation of aromatic ester prodrugs of 1-(3'-hydroxypropyl)-2-methyl-3-hydroxypyridin-4-one (CP41) as orally active iron chelators. *Arzneim-Forsch/Drug Res.* 2000; **50** (I):461-470.
107. Choudhury R, Epemolu RO, Rai BL, Hider RC, Singh S. (1997) Metabolism and pharmacokinetics of 1-(2'-trimethylacetoxyethyl)-2-ethyl-3-hydroxypyridin-4-one (CP117) in the rat. *Drug Metab Dispo* 1997; **25**:332-339.
108. Liu DY, Liu ZD, Lu SL, Hider RC. Hydrolytic and metabolic characteristics of the esters of 1-(3'-hydroxypropyl)-2-methyl-3-hydroxypyridin-4-one (CP41), potentially useful iron chelators. *Pharmacol Toxicol* 2000; **86**:228-233.
109. Zbinden P. Hydroxypyridinones. *US Patent 5,688,815,* 1997.
110. Hider RC, Tilbrook GS, Liu ZD. Novel Orally Active Iron(III) Chelators. *International Patent WO 98/54138*, 1998.
111. Liu ZD, Khodr HH, Liu DY, Lu SL, Hider RC. Synthesis, physicochemical characterisation and biological evaluation of 2-(1'-hydroxyalkyl)-3-hydroxypyridin-4-ones: Novel iron chelators with enhanced pFe^{3+} values. *J Med Chem* 1999; **42**:4814-4823.
112. Liu ZD, Piyamongkol S, Liu DY, Khodr HH, Lu SL, Hider RC. Synthesis of 2-amido-3-hydroxypyridin-4(1*H*)-ones: Novel iron chelators with enhanced pFe^{3+} values. *Bioorgan. Med. Chem.*, 2001; **9**:563-573.
113. Lowther N, Fox P, Faller B, Nick H, Jin Y, Sergejew T, Hirschberg Y, Oberle R, Donnelly H. In vitro and in situ permeability of a 'second generation' hydroxypyridinone oral iron chelator: Correlation with physico-chemical properties and oral activity. *Pharmaceut Res* 1999; **16**:434-440.

STRUCTURE–ACTIVITY RELATIONSHIPS AMONG DESAZADESFERRITHIOCIN ANALOGUES

Raymond J. Bergeron*, Jan Wiegand, James S. McManis, William R. Weimar and Guangfei Huang

1. INTRODUCTION

Iron, a transition metal, represents 5% of the earth's crust. It exists in a variety of oxidation states ranging from the zero-valent metal itself to Fe(VI), occurring in such diverse forms as iron disulfide (FeS_2, "fool's gold"), iron oxides, including magnetite (Fe_3O_4), and hemoglobin. However, the Fe(II) and Fe(III) oxidation states (Equation 1) are of the most relevance in the present discussion.

$$Fe(II) \rightleftharpoons Fe(III) + e^- \qquad (1)$$

Iron is central to molecules which are responsible for oxygen processing and electron transport. It is critical to the function of metalloenzymes, including oxidases, hydrogenases, reductases, and the like[1,2]. Among eukaryotes, life without this metal is virtually unknown.

The equilibrium shown in Equation 1 is very sensitive to both pH and the nature of the donor groups surrounding the metal. Thus, iron can behave as an electron donor (i.e., a reducing agent) or an electron acceptor (i.e., an oxidizing agent). At lower pHs, iron tends to exist as Fe(II), whereas at physiological and higher pH values iron is principally in the form of the highly insoluble Fe(III) hydroxides[3]. Although iron is an essential micronutrient, its "excess" can cause significant cellular, tissue, and organ damage, a problem which arises out of iron's relationship with oxygen[4].

---

* Raymond J. Bergeron, Author, Department of Medicinal Chemistry, University of Florida, Gainesville, Florida, 32610, USA, Jan Wiegand, James S. McManis, William R. Weimar and Guangfei Huang, Department of Medicinal Chemistry, University of Florida, Gainesville, Florida, 32610, USA

Iron Chelation Therapy.
Edited by Chaim Hershko, Kluwer Academic/Plenum Publishers, 2002.

The nature of the bonding between the two atoms in molecular oxygen (O_2) precludes its reactivity with most organic molecules. That is, the triplet ground state of O_2, with its two unpaired electrons, renders "spin-forbidden" its reaction with the paired electrons seen in most organic molecules. However, molecular oxygen does react with transition metals to generate superoxide anion (Equation 2), a radical anion which is far more reactive than is O_2 itself.

$$M(n) + O_2 \rightleftharpoons M(n+1) + O_2^{\cdot -} \qquad (2)$$

This reaction can occur between oxygen and various metals [e.g., Fe(II)] either directly or as a by-product of enzyme-mediated reactions. For example, xanthine oxidase produces uric acid from xanthine, liberating superoxide anion in the process[5]. Although superoxide anion *per se* may be toxic, it is the subsequent reactions of this species' product, hydrogen peroxide (H_2O_2, Equation 3), which is likely the primary source of iron's toxicity.

$$2 O_2^{\cdot -} + 2 H^+ \rightleftharpoons H_2O_2 + O_2 \qquad (3)$$

Under normal circumstances, this conversion occurs primarily via superoxide dismutase (SOD), which is a copper–zinc dependent enzyme in the cytosol and a manganese form in the mitochondria. Of less significance is Fe(II)-driven dismutation of superoxide anion. Normally, various "housekeeping" enzymes, such as glutathione peroxidase and catalase, manage hydrogen peroxide by converting it to water or to water and molecular oxygen respectively (Equations 4 and 5). Problems arise when hydrogen peroxide and Fe(II) are present in too large a concentration for these enzymes to manage[4] or when these species are present in the absence of the enzymes, either extracellularly or intracellularly.

$$\begin{array}{c} 2 \text{ Glutathione–SH} \quad (\text{Glutathione–S})_2 \\ H_2O_2 \longrightarrow 2 H_2O \end{array} \qquad (4)$$

$$2 H_2O_2 \xrightarrow{\text{Catalase}} 2 H_2O + O_2 \qquad (5)$$

On reaction with Fe(II), hydrogen peroxide produces the hydroxyl radical and the hydroxide anion (Equation 6); the radical is the offending species.

$$H_2O_2 + Fe(II) \rightleftharpoons HO^{\cdot} + HO^- + Fe(III) \qquad (6)$$

The hydroxyl radical is highly reactive, disrupting membranes and DNA structure, producing hypochlorous acid, and initiating other chain reactions, all of which result in damage to cells, tissues, and organs. Unfortunately, the event shown in Equation 6 is not a single occurrence; Fe(III) is reduced to Fe(II) readily via a number of physiological reductants, for example, ascorbic acid. Even superoxide anion can reduce Fe(III) to Fe(II), thus re-initiating the cycle if sufficient hydrogen peroxide is available.

Although the mechanisms of iron-mediated cellular disruption are manifold and remain incompletely elucidated, the removal of "excess" iron ameliorates the problem[4].

Thus, the solution to iron-mediated damage becomes one of identifying a chelator that can remove and sequester excess iron.

2. FAMILIES OF IRON CHELATORS

2.1. Siderophores, natural product iron chelators: An iron accessibility problem arose with the conversion of soluble Fe(II) in the biosphere to insoluble Fe(III) hydroxide when molecular oxygen was produced by blue-green algae. In response, bacteria assembled a group of low molecular weight, virtually Fe(III)-specific ligands, siderophores. When secreted into the extracellular milieu, these chelators sequester the metal and render it utilizable by the microorganism.

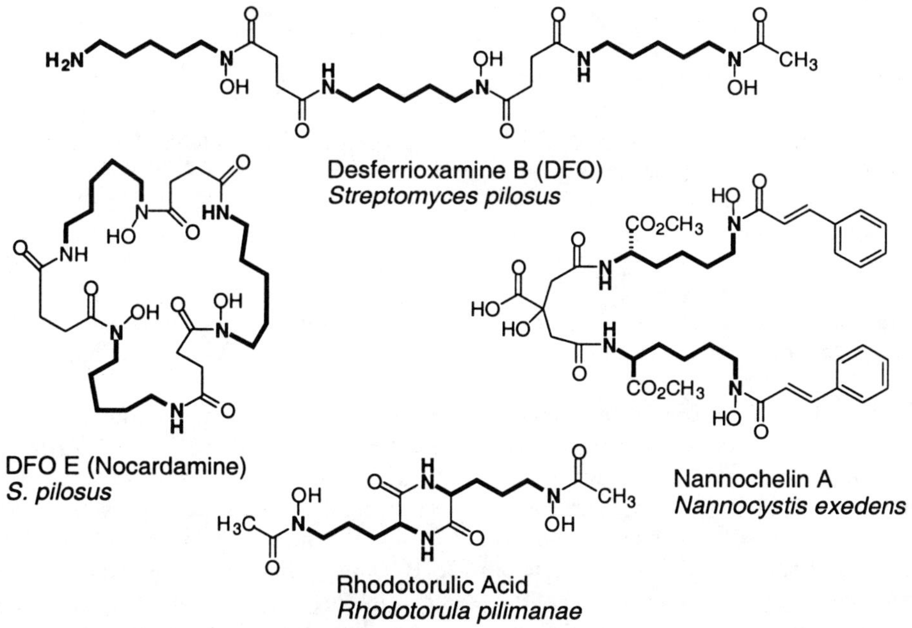

Figure 1. Natural iron chelators, siderophores: Representative hydroxamates. Polyamine backbones are depicted by darkened bonds.

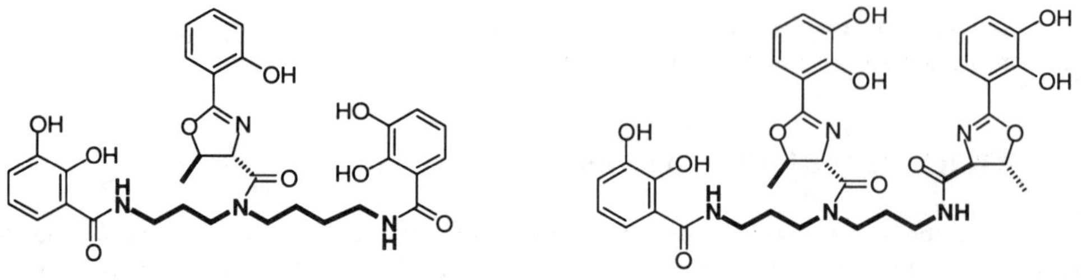

L-Parabactin
Paracoccus denitrificans

Vibriobactin
Vibrio cholerae

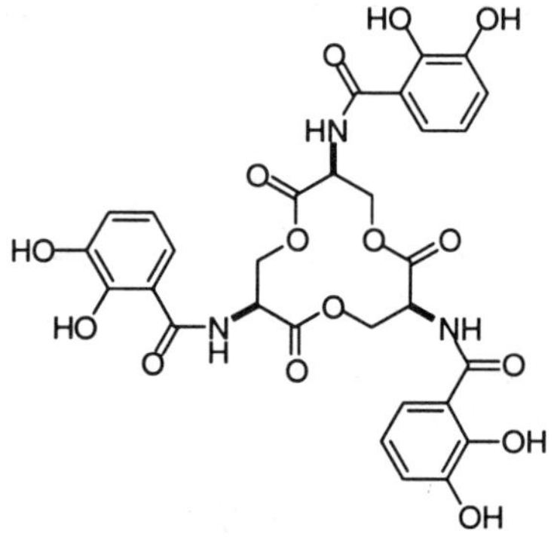

Enterobactin
Escherichia coli

Figure 2. Natural iron chelators, siderophores: Representative catecholamides. Polyamine backbones are depicted by darkened bonds. Note that enterobactin is an exception.

The siderophores can be separated on the basis of the compounds' donor groups into two major classes, the hydroxamates and the catecholamides (Figures 1 and 2)[6]. Many of these compounds also share common structural denominators; their backbones are predicated on polyamines or their precursor amino acids. These fragments are derived from relatively small molecules in the case of the hydroxamates. For instance, cadaverine is the backbone on which desferrioxamine B (DFO, Figure 1)[7] is assembled by *Streptomyces pilosus*. Interestingly, this same microorganism produces several other analogous hydroxamates, not only linear molecules, but also a macrocycle, nocardamine (DFO E, Figure 1)[8], which is also predicated on cadaverine. The cadaverine precursor, lysine, is the scaffolding for the product of *Nannocystis exedens*, nannochelin[9]; the putrescine precursor, ornithine, is the basis for rhodotorulic acid, which is secreted by *Rhodotorula pilimanae*[10]. Among the catecholamides (Figure 2), parabactin, isolated from *Paracoccus denitrificans*[11,12], is predicated on a spermidine backbone; vibriobactin, a product of *Vibrio cholerae*[13], on a norspermidine backbone.

As shown in Table 1, many siderophores, including enterobactin and DFO, are hexacoordinate ligands, forming 1:1 complexes with Fe(III). The hydroxamate Fe(III) chelates have formation constants on the order of 10^{30} M^{-1}, whereas the catecholamides, having stability constants of > 10^{42} M^{-1}, form stronger complexes with Fe(III).

Table 1. Characteristics of Selected Natural and Synthetic Ligands

Ligand	Natural/ Synthetic	Class	Stoichiometry of Fe(III) Complex	log K or β_3	Reference(s)
Desferrioxamine B	Natural	Hydroxamate	1:1	30.99	14
Desferrioxamine E	Natural	Hydroxamate	1:1	32.5	15,16
DTPA	Synthetic	Aminopolycarboxylate	1:1	28.0	14
HBED	Synthetic	Phenolic aminopolycarboxylate	1:1	36.74	14
1,2-dimethyl-3-hydroxypyridin-4-one (L1)	Synthetic	Hydroxypyridinone	3:1	35.92	17,18
Parabactin	Natural	Catecholamide	1:1	48	19
Enterobactin	Natural	Catecholamide	1:1	52	16,20,21
Desferrithiocin	Natural	Miscellaneous	2:1	28	22,23

2.2. Synthetic iron chelators: There are many more synthetic ligands to choose from than there are natural products. As with the siderophores, the synthetic chelators can be classified on the basis of their donor groups. Figure 3 shows representative compounds from these classes. These compounds include diethylenetriamine pentaacetic acid (DTPA), an aminopolycarboxylate; *N,N'*-bis(2-hydroxybenzyl)ethylenediamine-*N,N'*-diacetic acid (HBED)[24,25], a phenolate analogue of the aminopolycarboxylates; 1,2-dimethyl-3-hydroxypyridin-4-one (deferiprone, L1)[26], a hydroxypyridinone; an amideless desferrioxamine[27], a hydroxamate; 1,3,5-*N,N',N''*-tris(2,3-dihydroxybenzoyl)-triaminomethylbenzene (MECAM)[28], a catecholamide; and *N,N*-bis(pyridoxyl)-ethylenediamine-*N,N'*-diacetic acid (PLED)[29], a hydroxypyridine. However, the first three ligands have been the most thoroughly studied. Both HBED and DTPA are also hexacoordinate ligands, forming 1:1 complexes with Fe(III); L1 is a bidentate chelator and forms a 3:1 complex. The formation constants of these three compounds range from 10^{28} to 10^{37} M^{-1} (Table 1).

DTPA   HBED   L1

Amideless DFO

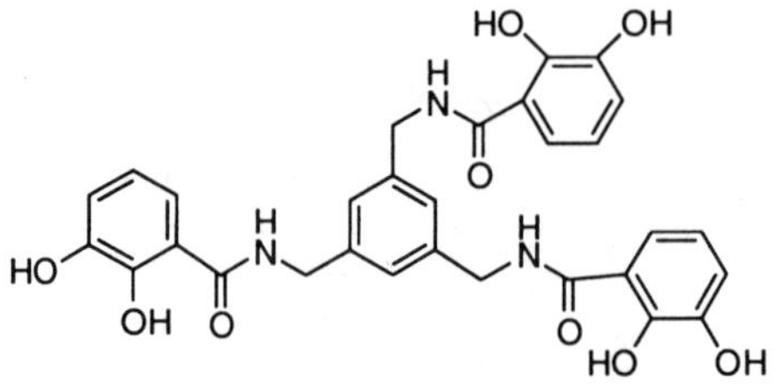

MECAM

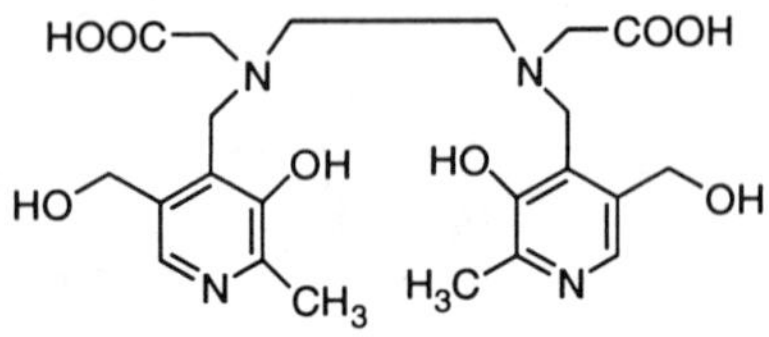

PLED

Figure 3. (continued from previous page) Representative synthetic iron chelators.

3. SELECTING A THERAPEUTIC IRON CHELATOR

3.1. Natural vs. synthetic iron chelators: The question is whether there is any implicit advantage of natural products over synthetic chelators, or *vice versa*. A comparison of the siderophore DFO (Figure 1) with the synthetic ligand DTPA (Figure 3) provides some answers. Both ligands were discovered and first tested in humans at about the same time[30-32]; when administered subcutaneously, both clear iron from humans with about the same efficacy[33]. However, unlike DFO, DTPA also effectively clears other metals, particularly zinc. This lack of selectivity of DTPA is responsible for the observed toxic effects at therapeutic dose levels. These side effects were ameliorated in patients who received zinc supplements concurrently with DTPA[33,34]. On the other hand, DFO is far more selective in clearing iron. At therapeutic doses, the side effects of DFO are unrelated to the removal of metals. The metal selectivity of DFO should not be surprising in view of the fact that microorganisms have had hundreds of millions of years to perfect a chelator that would provide them with Fe(III).

3.2. Desferrioxamine B, the current chelator of choice: Although hundreds of synthetic iron chelators have been assembled over the last 25 years in hopes of identifying an alternative to DFO for the treatment of iron overload, DFO remains the ligand of choice in the clinic[35]. Nevertheless, DFO has many shortcomings, the first of which is its efficiency. The chelator must be administered subcutaneously or intravenously for prolonged periods of time[36]. Further, intravenous administration of the drug must be done very cautiously, as serious hypotensive reactions can occur[37]. Frequently, DFO causes significant and often painful reactions at the site of injection[38-41], can affect hearing[42], and can promote *Yersinia* infections[43,44]. The combination of these side effects has led to patient compliance problems.

Efficiency, a measure of how much iron excretion is promoted by a ligand relative to the iron excretion that should be elicited on a theoretical basis, is probably the most profound shortcoming of DFO. Because DFO forms a very tight 1:1 complex with Fe(III), a millimole of Desferal (657 mg), the methanesulfonic acid salt of DFO, given to a patient should cause the excretion of a mg-atom of Fe(III) (56 mg) if the drug were 100% efficient. Unfortunately, the efficiency of DFO is only between 5 and 7%[33,36]. Owing in part to this poor efficiency, the drug is generally administered by subcutaneous infusion 8–12 hours daily, 4–6 days a week[35,36]. If this chelator were ten times as efficient, the dosing regimen would be very different; many of the compliance problems could be ameliorated.

4. CHELATOR DESIGN CONSIDERATIONS

In the search for clinically effective iron chelators, there are three principal considerations: the efficiency, the toxicity profile, and how the chelator is to be administered. Obviously, the more efficient a chelator is, the more likely it is to be a desirable candidate. This parameter affects both the toxicity profile and modes of administration.

Drugs derive their toxicity from two sources: that which is implicit in the structure and unrelated to its intended use and that which is associated with the drug's pharmacological function. Iron chelators are no different in this sense. Obviously, there is a limit as to how much iron can be removed from a patient without the occurrence of a lethal event. Although this issue can be of great significance in preclinical toxicity trials of iron chelators, this is an unlikely concern in iron–overloaded patients. The problem of toxicity implicit in the structure of the ligand itself is always of concern. However, the more efficient a ligand is, the less of the drug is required, thus minimizing this latter problem.

Again, the difficulty of how to administer a compound is tied to that compound's efficiency. A highly efficient parenteral device will certainly be met with better patient compliance than is DFO. Although an orally effective iron chelator is certainly the most desirable device, it, too, must be very efficient. It is known from the HIV arena that compliance is poor when patients are asked to take multiple daily medications, even orally.

5. DESFERRITHIOCIN – A PLACE TO BEGIN

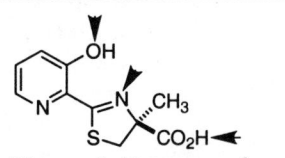

Figure 4. Structure of desferrithiocin (DFT). The chelating centers are depicted by the arrows.

Desferrithiocin (DFT, Figure 4) is a tridentate siderophore isolated from *Streptomyces antibioticus*[45]. It forms a 2:1 complex with Fe(III) and has a formation constant of 4×10^{29} M^{-1} (Table 1)[22,23]. The donor groups include a phenolic oxygen, a thiazoline nitrogen, and a carboxyl.

Desferrithiocin was one of the first iron chelators which was shown to be orally active in both the bile duct-cannulated rodent model, in which the efficiency was 5.5%[46], and the iron-overloaded *Cebus apella* primate model, in which the efficiency was 16%[47,48]. The latter figure is three times the efficiency of DFO given subcutaneously in the primate model; the drawback is that DFT elicits severe nephrotoxicity[48]. However, because of this remarkable oral activity DFT was chosen as a platform from which to launch structure–activity studies[46,48-52]. These studies were intended to assemble DFT analogues which are orally active iron chelators without the toxicity of the parent molecule. The initial structural modifications of DFT were (1) stripping the thiazoline methyl to produce (S)-4,5-dihydro-2-(3-hydroxy-2-pyridinyl)-4-thiazolecarboxylic acid [(S)-DMDFT], (2) removal of the aromatic nitrogen to yield (S)-4,5-dihydro-2-(2-hydroxyphenyl)-4-methyl-4-thiazolecarboxylic acid [(S)-DADFT], and (3) removing both the aromatic nitrogen and the thiazoline methyl to produce (S)-4,5-dihydro-2-(2-hydroxyphenyl)-4-thiazolecarboxylic acid [(S)-DADMDFT] (Figure 5)[48,52]. Abstraction of the thiazoline methyl diminished the efficiency substantially in both rodents and primates[46,48]. Although removal of the pyridine nitrogen attenuated the efficiency in rodents, the efficiency of this compound was increased in primates[52]. Finally, the absence of both the thiazoline methyl and the pyridine nitrogen left a molecule [(S)-DADMDFT] which was considerably less active in the rodent model but was still quite active in the primate[48,51].

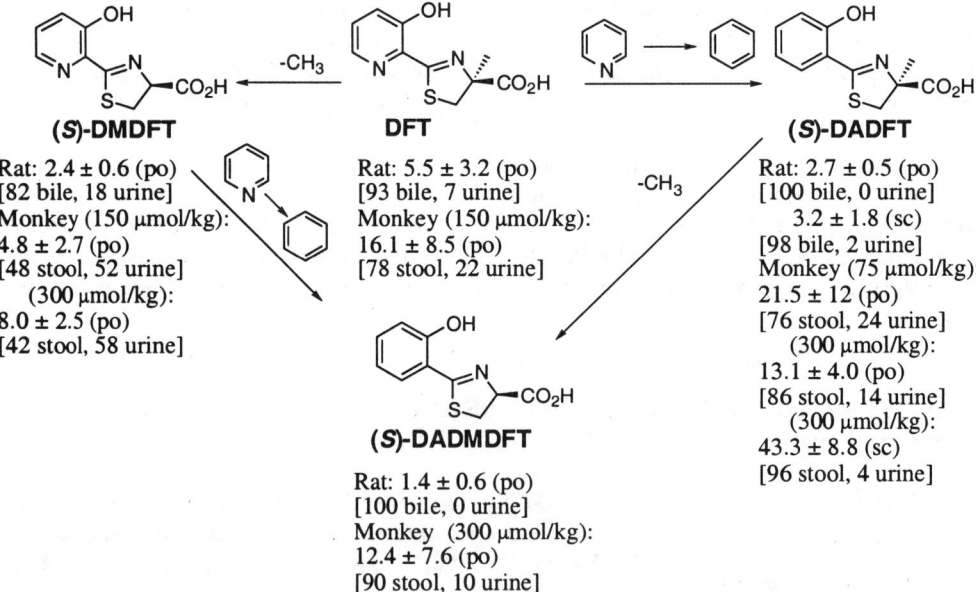

Figure 5. Structure–activity relationships of the desferrithiocins and iron clearing efficiency. The dose of DFT or analogue in the rats is 150 μmol/kg; the dose in the monkeys is as shown in parentheses for each ligand. The mode of administration is shown in parentheses next to the efficiency (%, ± standard deviation). The fraction of iron excreted in the bile or stool and urine is shown in brackets.

The impact of these modifications on the toxicity of these molecules was very profound. No toxicity was found with (S)-DMDFT (Figure 6), even in a 30-day trial in rodents[51]; unfortunately, this is the least effective of the four chelators (Figure 5). Interestingly, neither (S)-DADFT nor (S)-DADMDFT elicited nephrotoxicity; rather, the toxicity was gastrointestinal (Figure 6)[48,51,52]. These findings were critical to subsequent structure–activity studies.

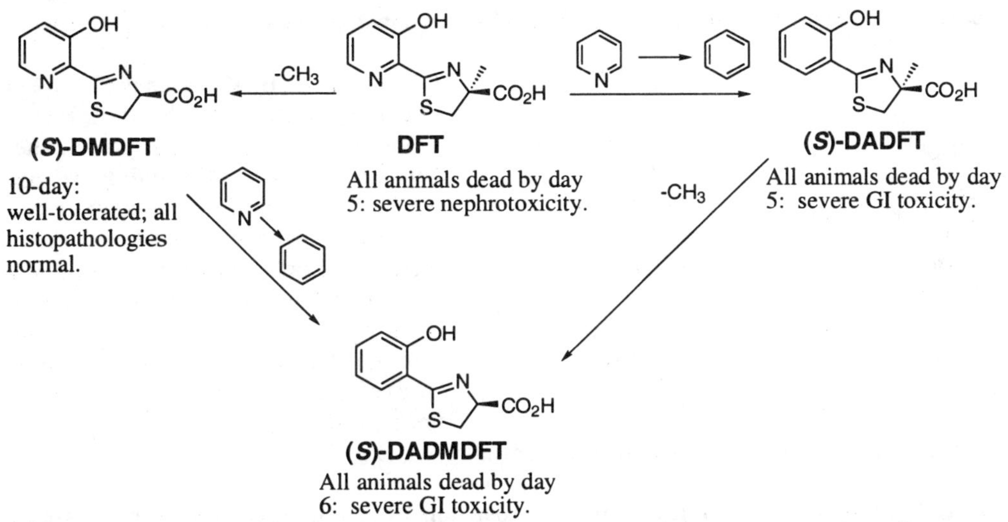

(S)-DMDFT
10-day: well-tolerated; all histopathologies normal.

DFT
All animals dead by day 5: severe nephrotoxicity.

(S)-DADFT
All animals dead by day 5: severe GI toxicity.

(S)-DADMDFT
All animals dead by day 6: severe GI toxicity.

Figure 6. Structure–activity relationship of the DFTs and toxicity. The ligands were administered orally at a dose of 384 μmol/kg/day, equivalent to 100 mg/kg/day of the sodium salt of DFT.

Systems such as (S)-DADMDFT are very easy to access synthetically. The parent drug DFT can be viewed formally as a condensation product between two different fragments: an aromatic nitrogen-containing compound, 3-hydroxypicolinic acid, and a rare amino acid, (S)-α-methylcysteine (Figure 7). Owing to the aromatic nitrogen, the picolinic acid does not lend itself easily to structural modifications. Thus, in spite of its toxicity, (S)-DADMDFT was the choice from which to carry out additional structure–activity analyses.

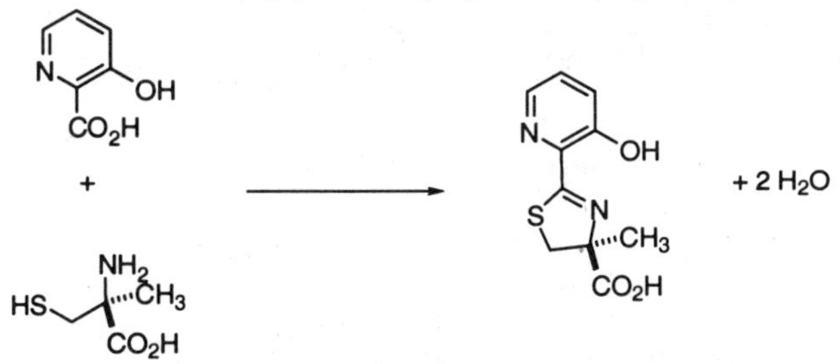

Figure 7. Schematic of the condensation between 3-hydroxypicolinic acid and (S)-α-methylcysteine to yield DFT.

6. DESAZADESMETHYLDESFERRITHIOCIN STRUCTURE–ACTIVITY RELATIONSHIPS – IRON CLEARANCE

The structural changes explored in DADMDFT were (1) alteration of the distances between the chelating centers, (2) modification of the thiazoline ring, (3) changing the stereochemistry at C-4, (4) fusion of aromatic rings, and (5) alteration of the redox potential of the aromatic ring by addition of electron-donating and –withdrawing groups.

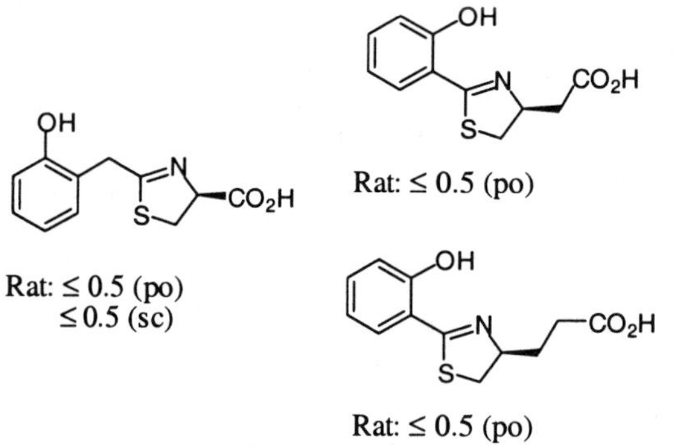

6.1. Alterations of distances between chelating centers: As shown in Figure 8, introduction of one methylene either between the aromatic and thiazoline rings or between the thiazoline ring and the carboxyl or the insertion of two methylenes between the thiazoline ring and the carboxyl abrogated the oral iron clearing activity[51]. The design implication is to leave the basic thiazoline–carboxylate fragment unchanged.

Figure 8. Alteration of distances between chelating centers.

6.2. Thiazoline ring modification: Expansion of the thiazoline ring by a single methylene, oxidation of the thiazoline to a thiazole, reduction of the thiazoline to a thiazolidine, or replacement of the sulfur with a methylene, a nitrogen, or an oxygen all resulted in ligands that were inactive when given orally to rodents (Figure 9)[49,51]. Both the (R)- and (S)-enantiomers of the oxazoline were tested in rodents and primates. Pharmacokinetic studies of the (R)-enantiomer indicated that it was well-absorbed orally[53]; thus, bioavailability was not the key issue regarding the lack of oral activity of these ligands. The design cue here is to leave the thiazoline ring intact.

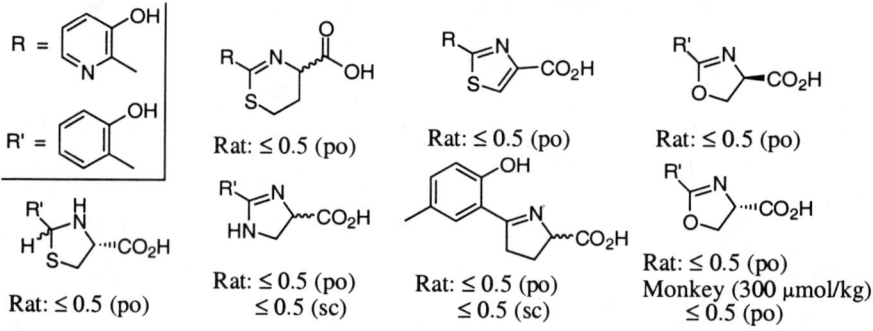

Figure 9. Thiazoline ring modifications.

6.3. Alteration of C-4 Stereochemistry:
One of the most interesting findings was realated to the impact of configurational changes at C-4 on the ligands' activity (Figures 10 and 11). Although the difference in activity between the (R)- and (S)-enantiomers is not consistent in the rodent model, the results clearly suggest the importance of the stereochemistry at C-4 in the compounds' iron-clearing properties as measured in the primate model[50,52]. The (S)-enantiomers are generally more active than their corresponding (R)-enantiomers in the latter animal model. This is apparent for (R)- and

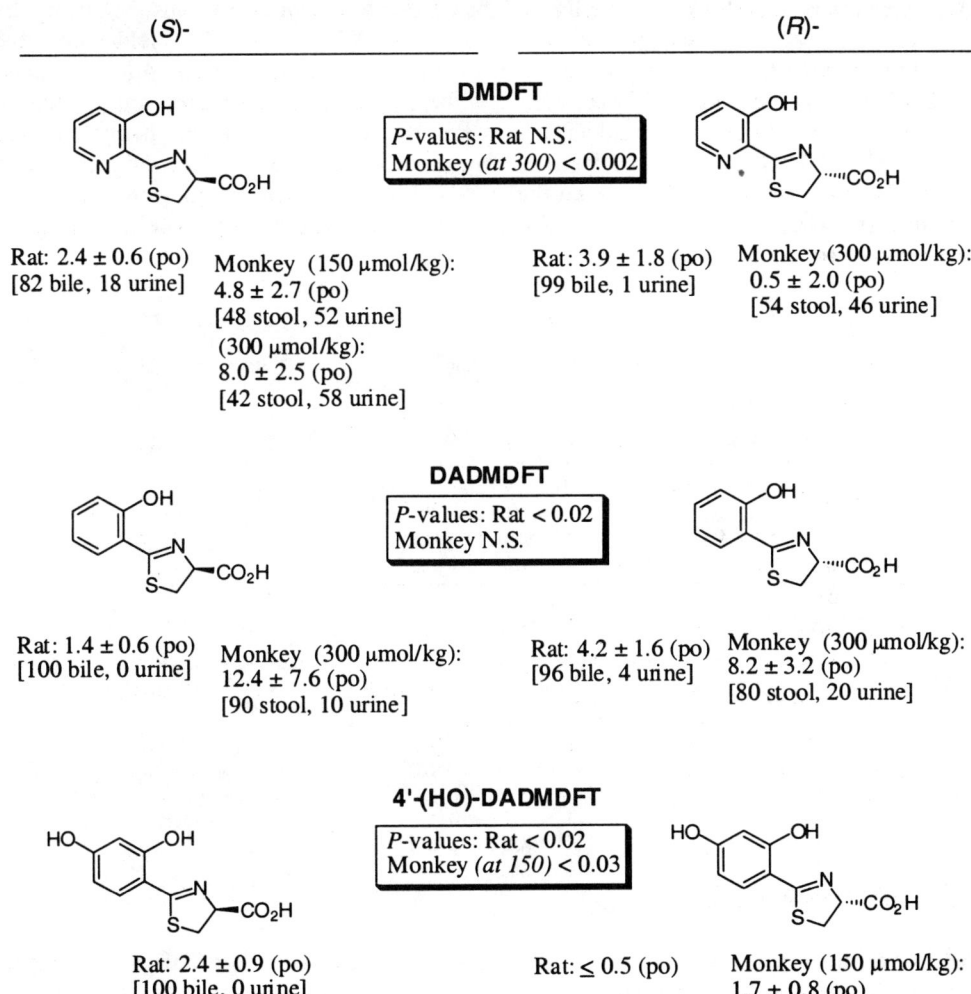

Figure 10. C-4 Stereochemistry of **DMDFT**, **DADMDFT**, and **4'-(HO)-DADMDFT**. N.S., $P > 0.05$.

(S)-DMDFT and (R)- and (S)-4'-(HO)-DADMDFT in Figure 10 and for the second pair of (R)- and (S)-naphthyl analogues and the (R)- and (S)-quinoline analogues in Figure 11. The difference is not as obvious for (R)- and (S)-DADMDFT. The design implication is that (S)-enantiomers are more likely to have a high index of success.

6.4. Aromatic ring fusion: Both (R)- and (S)-DMDFT and (R)- and (S)-DADMDFT were benz-fused (Figure 11); the rationale was that by increasing the lipophilicity of these analogues the residence time of the ligands would be more protracted, thus increasing the efficiency. In fact, 5,6-benz-fusion of both (R)- and (S)-DADMDFT yielded inactive chelators in both rodents and primates[50]. However, the situation for the 4,5-benz-fused DADMDFTs is somewhat different. Although the (S)-enantiomer was the more active in both the rodent and primate models, it was still not as effective in the primate as the parent (S)-DADMDFT[50]. A similar situation was encountered with the 4,5-benz-fused DMDFTs in the primate, that is, even the activity of the more active (S)-enantiomer was less than that observed for its parent, (S)-DMDFT[51]. The suggestion for chelator design is that benz-fusion is of no value.

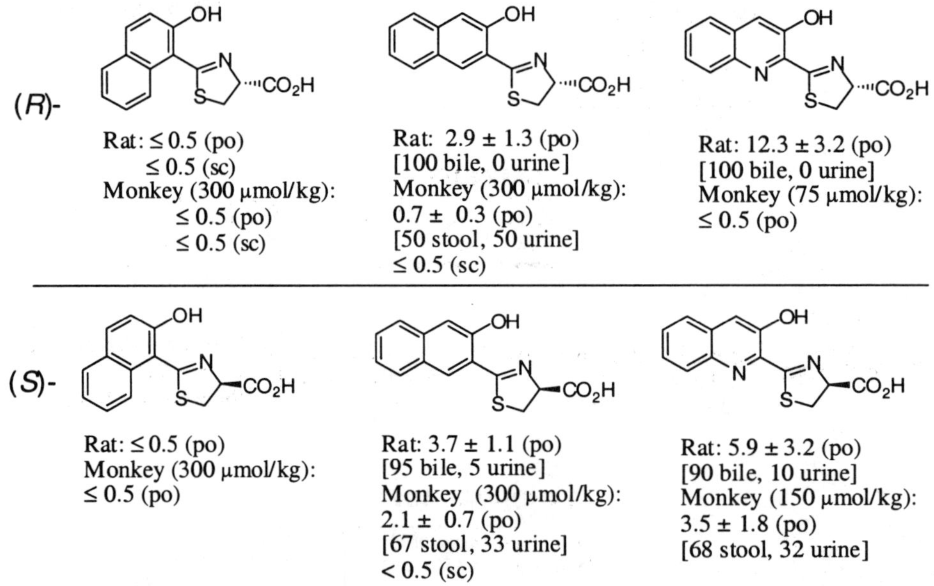

(R)-

Rat: ≤ 0.5 (po)
≤ 0.5 (sc)
Monkey (300 μmol/kg):
≤ 0.5 (po)
≤ 0.5 (sc)

Rat: 2.9 ± 1.3 (po)
[100 bile, 0 urine]
Monkey (300 μmol/kg):
0.7 ± 0.3 (po)
[50 stool, 50 urine]
≤ 0.5 (sc)

Rat: 12.3 ± 3.2 (po)
[100 bile, 0 urine]
Monkey (75 μmol/kg):
≤ 0.5 (po)

(S)-

Rat: ≤ 0.5 (po)
Monkey (300 μmol/kg):
≤ 0.5 (po)

Rat: 3.7 ± 1.1 (po)
[95 bile, 5 urine]
Monkey (300 μmol/kg):
2.1 ± 0.7 (po)
[67 stool, 33 urine]
< 0.5 (sc)

Rat: 5.9 ± 3.2 (po)
[90 bile, 10 urine]
Monkey (150 μmol/kg):
3.5 ± 1.8 (po)
[68 stool, 32 urine]

Figure 11. Fusion of aromatic rings.

6.5. Alteration of the redox characteristics of the aromatic ring: This SAR analysis was predicated on the idea that by altering the redox potential of the aromatic ring the metabolic profile of the drug, and thereby its toxicity, could change. The importance of the metabolic issue as it affects analogue activity is underscored by our observation that DADMDFT is metabolized to 5'-(HO)-DADMDFT, which is inactive as an iron chelator.

The redox potential of the aromatic rings was changed in two different ways (Figure 12), by introduction of either electron-donating or electron-withdrawing groups as considered relative to the thiazoline ring[51]. The electron-withdrawing groups that were attached to the ring included a 3'-(OH), a 3'-(OCH$_3$), and a 4'-(CO$_2$H); the electron-donating group was a 4'-(OH). In each instance, introduction of electron-withdrawing groups into the aromatic ring of (S)-DADMDFT resulted in chelators with little, if any, iron-clearing activity in rodents (Figure 12). However, introduction of a 4'-(OH) into the aromatic ring of (S)-DADMDFT was compatible with iron clearance in both the rodent and primate models (Figure 12)[51], although the hydroxylated compound was less active than the parent drug in the primate. When DADFT[52] and a β,β-dimethyl cysteine analogue of DADMDFT (DM)[48] were 4'- hydroxylated, both ligands were very efficient iron chelators when given orally (Figure 12); the activity was similar to those of their respective parent drugs.

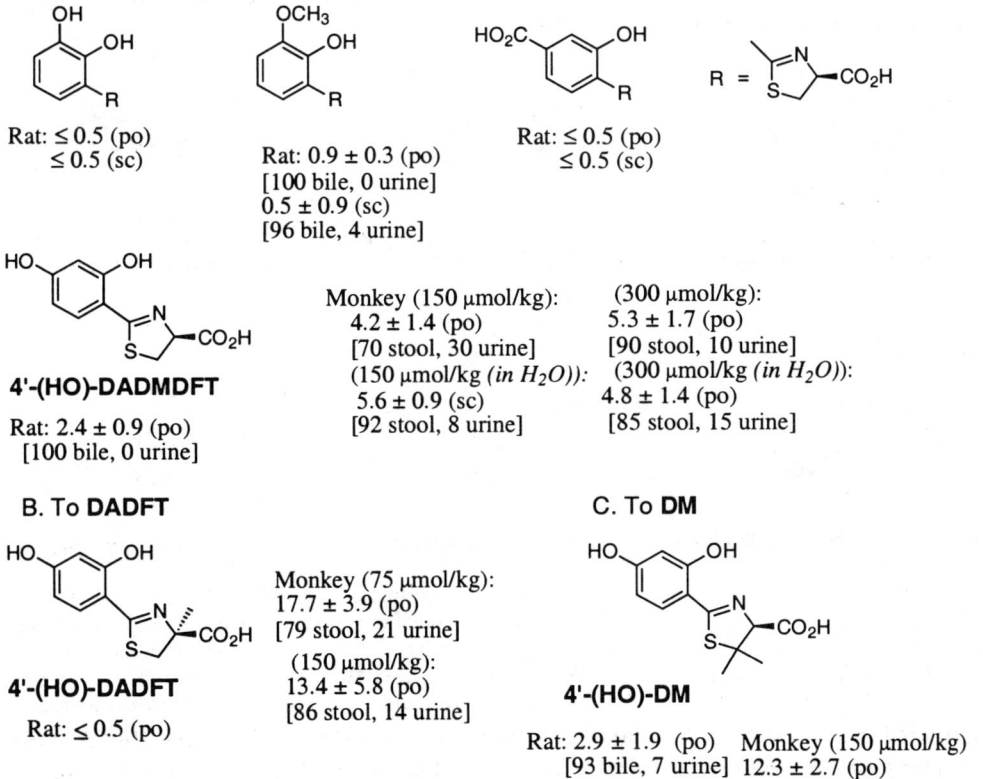

Figure 12. Addition of electron-donating and –withdrawing groups.

7. DESAZADESMETHYLDESFERRITHIOCIN STRUCTURE–ACTIVITY RELATIONSHIPS – TOXICITY

Since the 4'-hydroxylated compounds [4'-(HO)-DADMDFT, 4'-(HO)-DADFT, and 4'-(HO)-DM] were active in the *Cebus apella* primate model, the toxicity profiles were evaluated under two different dosing schedules in rodents (Figure 13). One schedule involved a 10-day trial of each of these ligands at a uniform dose of 384 μmol/kg/day p.o. This dose is equivalent to a dose of 100 mg/kg/day of the sodium salt of DFT[48].

Under a second schedule, iron-overloaded rats were given a dose of drug which was equivalent to a dose necessary to promote the excretion of 450 μg iron/kg of body weight per day in the primate model. This 450 μg iron/kg/day number is derived from calculations as to how much iron would need to be removed on a daily basis to put a significantly iron-overloaded patient in negative iron balance[54]. Thus, the dose of the chelators given to the rodents was an equivalent iron-clearing dose in each case (Figure 13). The trials were carried out for 30 days.

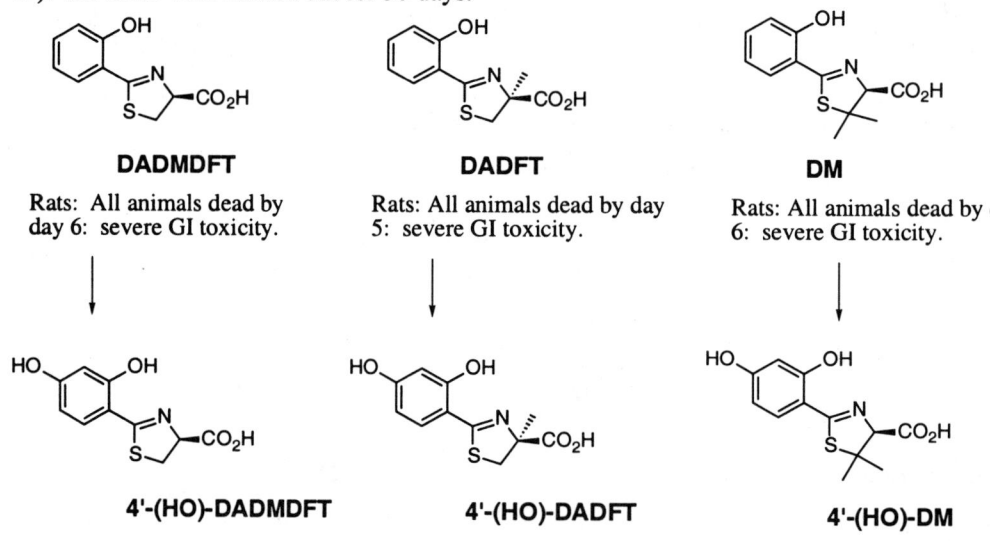

Figure 13. Structure–activity relationship of the DFTs and toxicity. In the 10-day trials, the ligands were administered orally at a dose of 384 μmol/kg/day, equivalent to 100 mg/kg/day of the sodium salt of DFT. In the 30-day studies, the dose was based on the dose which clears 450 μg Fe/kg of body weight in the primate model: 300 μmol/kg/day (72.6 mg/kg/day) for **4'-(HO)-DADMDFT**, 119 μmol/kg/day (30.1 mg/kg/day) for **4'-(HO)-DADFT**, and 130 μmol/kg/day (34.8 mg/kg/day) for **4'-(HO)-DM**.

The difference in toxicity between the parent and their respective hydroxylated analogues is profound (Figure 13). In the case of the parent drugs (DADMDFT, DADFT, and DM) all of the rats were dead within six days from severe gastrointestinal toxicity[48,51,52]. In contrast, all three of the hydroxylated analogues were well tolerated in both the 10- and 30-day trials; nearly all histopathologies on these animals were normal (Figure 13).

8. METABOLISM OF DESAZADESFERRITHIOCINS

Gentisic acid, a principal metabolite of aspirin, is derived from hydroxylation at the 5-position of the aromatic ring[55,56]. To find whether this was the fate of (S)-DADMDFT and how this hydroxylation might affect its activity, a search for 5'-hydroxydesazadesmethylDFT in the urine of rats treated with (S)-DADMDFT was conducted. The putative metabolite was synthesized; HPLC methodologies[57] for quantitation of DFTs in tissue and fluids worked well for this compound. Analysis of 24-hour urine specimens from animals treated with (S)-DADMDFT indicated that this ligand was indeed 5'-hydroxylated; furthermore, this metabolite represents about 18% of the total concentration of drug and metabolites in the urine. This metabolite has not been measured in bile or other tissues. Interestingly, when the synthetic metabolite was given to rodents in an iron clearance experiment, it showed no observable clearance activity; thus, 5'-hydroxylation inactivates the drug (R.J.B., J.M., W.R.W., and J.W., unpublished observations). This raises some very interesting questions regarding other DADFT analogues. Is oxidative deactivation occurring with the DADFTs or with their 4'-hydroxylated counterparts, and to what extent? Does this deactivation correlate with iron clearing efficiency? If so, will blocking this position by introduction of a different atom or functional group at this position increase the chelators' efficiency? Furthermore, what role does oxidative deactivation play in primates, if any? The toxicity of the metabolites remains to be evaluated thoroughly.

9. SUMMARY AND CONCLUSIONS

Desferrithiocin, a natural product iron chelator (siderophore), offers an excellent platform from which to construct orally active iron chelators which have a good therapeutic window.

A systematic structure–activity study on desferrithiocin identified the structural fragments necessary for the compound's oral iron-clearing activity. There are strict requirements regarding the distance between the ligating centers; they cannot be altered without loss of efficacy. The thiazoline ring must remain intact. Benz-fusions, which were designed to improve the ligands' tissue residence time and possibly iron-clearing efficiency, are ineffective. The maintenance of an (S)-configured C-4 carbon is optimal in the design of desferrithiocin-based iron chelators. With this information in hand, alteration of the redox potential of the aromatic ring was initiated. Introduction of a hydroxy in the 4'-position of at least three different desazadesferrithiocin analogues resulted in moderate to small changes in iron clearing efficacy yet dramatic reductions in the toxicity of the compounds were observed. Although the toxicity studies of these desferrithiocin analogues are continuing, it is clear that it is possible to alter a siderophore in such a way as to ameliorate its toxicity profile while maintaining its iron-clearing properties.

ACKNOWLEDGMENT

We thank Dr. Eileen Eiler-McManis for her assistance in the organization and editing of this manuscript. The U.S. National Institutes of Health (Grant No. R01-DK49108) is gratefully acknowledged for its support of the research from the Bergeron laboratory.

REFERENCES

1. Ortiz de Montellano PR ed Cytochrome p450 - structure, metabolism, and biochemistry. New York: Plenum; 1986.
2. Sahlin M, Petersson L, Graslund A, et al. Magnetic interaction between the tyrosyl free radical and the antiferromagnetically coupled iron center in ribonucleotide reductase. *Biochemistry* 1987;**26**:5541-5548.
3. Raymond KN, Carrano CJ. Coordination chemistry and microbial iron transport. *Acc Chem Res* 1979;**12**:183-190.
4. Halliwell B. Iron, oxidative damage, and chelating agents. Bergeron RJ, Brittenham GM, eds. *The development of iron chelators for clinical use*. Boca Raton, CRC, 1994,33-56.
5. Britigan BE, Pou S, Rosen GM, Lilleg DM, Buettner GR. Hydroxyl radical is not a product of the reaction of xanthine oxidase and xanthine. The confounding problem of adventitious iron bound to xanthine oxidase. *J Biol Chem* 1990;**265**:17533-17538.
6. Bergeron RJ. Iron: A controlling nutrient in proliferative processes. *Trends Biochem Sci* 1986;**11**:133-136.
7. Bickel H, Hall GE, Keller-Schierlein W, et al. Metabolic products of actinomycetes. XXVII. Constitutional formula of ferrioxamine B. *Helv Chim Acta* 1960;**43**:2129-2138.
8. Bickel H, Bosshardt R, Gäumann E, et al. Metabolic products of actinomycetes. XXVI. Isolation and properties of ferrioxamines A to F, representing new sideramine compounds. *Helv Chim Acta* 1960;**43**:2118-2128.
9. Kunze B, Trowitzsch-Kienast W, Höfle G, Reichenbach H. Nannochelins A, B, and C, new iron-chelating compounds from *Nannocystis exedens* (myxobacteria). *J Antibiot* 1992;**45**:147-150.
10. Atkin CL, Neilands JB. Rhodotorulic acid, a diketopiperazine dihydroxamic acid with growth-factor activity. I. Isolation and characterization. *Biochemistry* 1968;**7**:3734-3739.
11. Tait GH. The identification and biosynthesis of siderochromes formed by *Micrococcus denitrificans*. *Biochem J* 1975;**146**:191-204.
12. Peterson T, Neilands JB. Revised structure of a catecholamide spermidine siderophore from *Paracoccus denitrificans*. *Tetrahedron Lett* 1979:4805-4808.
13. Griffiths GL, Sigel SP, Payne SM, Neilands JB. Vibriobactin, a siderophore from *Vibrio cholerae*. *J Biol Chem* 1984; **259**:383-385.
14. Martell AE, Motekaitis RJ, Sun Y, Clarke ET. Ligand design of chelating agents effective in the coordination of Fe(III) and for the removal of iron in cases of iron overload. Bergeron RJ, Brittenham GM, eds. *The development of iron chelators for clinical use*. Boca Raton, CRC, 1994,329-351.
15. Anderegg G, L'Eplattenier F, Schwarzenbach G. Hydroxamate complexes. II. Application of the pH method. *Helv Chim Acta* 1963;**46**:1400-1408.
16. Harris WR, Carrano CJ, Cooper SR, et al. Coordination chemistry of microbial iron transport compounds. XIX. Stability constants and electrochemical behavior of ferric enterobactin and model complexes. *J Am Chem Soc* 1979;**101**:6097-6104.

17. Motekaitis RJ, Martell AE. Stabilities of the iron(III) chelates of 1,2-dimethyl-3-hydroxy-4-pyridinone and related ligands. *Inorg Chim Acta* 1991;**183**:71-80.
18. Clarke ET, Martell AE. Stabilities of 1,2-dimethyl-3-hydroxy-4-pyridinone chelates of divalent and trivalent metal ions. *Inorg Chim Acta* 1992;**191**:56-63.
19. Neilands JB, Peterson T, Leong SA. High affinity iron transport in microorganisms. Iron (III) coordination compounds of the siderophores agrobactin and parabactin. Martell AE, ed. *Inorganic chemistry in biology and medicine*. Vol. 140. American Chemical Society symposia. Washington, D.C., American Chemical Society, 1980,263-278.
20. Harris WR, Carrano CJ, Raymond KN. Spectrophotometric determination of the proton-dependent stability constant of ferric enterobactin. *J Am Chem Soc* 1979;**101**:2213-2214.
21. Avdeef A, Sofen SR, Bregante TL, Raymond KN. Coordination chemistry of microbial iron transport compounds. 9. Stability constants for catechol models of enterobactin. *J Am Chem Soc* 1978;**100**:5362-5370.
22. Anderegg G, Räber M. Metal complex formation of a new siderophore desferrithiocin and of three related ligands. *J Chem Soc, Chem Commun* 1990:1194-1196.
23. Hahn FE, McMurry TJ, Hugi A, Raymond KN. Coordination chemistry of microbial iron transport. 42. Structural and spectroscopic characterization of diastereomeric Cr(III) and Co(III) complexes of desferriferrithiocin. *J Am Chem Soc* 1990;**112**:1854-1860.
24. L'Eplattenier F, Murase I, Martell AE. New multidentate ligands. VI. Chelating tendencies of N,N'-di(2-hydroxybenzyl)ethylenediamine-N,N'-diacetic acid. *J Am Chem Soc* 1967; **89**:837-843.
25. Pitt CG, Gupta G, Estes WE, et al. The selection and evaluation of new chelating agents for the treatment of iron overload. *J Pharmacol Exp Ther* 1979;**208**:12-18.
26. Kontoghiorghes GJ, Sheppard L, Chambers S. New synthetic approach and iron chelating studies of 1-alkyl-2-methyl-3-hydroxypyrid-4-ones. *Arzneimittelforschung* 1987;**37**:1099-1102.
27. Bergeron RJ, Liu Z-R, McManis JS, Wiegand J. Structural alterations in desferrioxamine compatible with iron clearance in animals. *J Med Chem* 1992;**35**:4739-4744.
28. Harris WH, Raymond KN. Ferric ion sequestering agents. 3. The spectrophotometric and potentiometric evaluation of two new enterobactin analogs: 1,5,9-N,N',N''-tris(2,3-dihydroxybenzoyl)cyclotriazatridecane and 1,3,5-N,N',N''-tris(2,3-dihydroxybenzoyl)triaminomethylbenzene. *J Am Chem Soc* 1979;**101**:6534-6541.
29. Motekaitis RJ, Sun Y, Martell AE. N,N'-bis(pyridoxyl)ethylenediamine-N,N'-diacetic acid (PLED) and N,N'-bis(2-hydroxy-5-sulfobenzyl)ethylenediamine-N,N'-diacetic acid (SHBED). *Inorg Chim Acta* 1989;**159**:29-39.
30. Fahey JL, Rath CE, Princiotto JV, Brick IB, Rubin M. Evaluation of trisodium calcium diethylenetriaminepentaacetate in iron storage disease. *J Lab Clin Med* 1961;**57**:436-449.
31. Smith RS. Iron excretion in thalassemia major after administration of chelating agents. *Br Med J* 1962;**2**:1577-1580.
32. Wöhler F. The treatment of haemochromatosis with desferrioxamine. *Acta Haematol* 1963;**30**:65-87.
33. Pippard MJ, Jackson MJ, Hoffman K, Petrou M, Modell CB. Iron chelation using subcutaneous infusions of diethylene triamine penta-acetic acid (DTPA). *Scand J Haematol* 1986;**36**:466-472.
34. Wonke B, Hoffbrand AV, Aldouri M, et al. Reversal of desferrioxamine induced auditory neurotoxicity during treatment with Ca-DTPA. *Arch Dis Child* 1989;**64**:77-82.
35. Olivieri NF, Brittenham GM. Iron-chelating therapy and the treatment of thalassemia. *Blood* 1997;**89**:739-761.
36. Pippard MJ. Desferrioxamine-induced iron excretion in humans. *Bailleres Clin Haematol* 1989;**2**:323-343.
37. Whitten CF, Gibson GW, Good MH, Goodwin JF, Brough AJ. Studies in acute iron poisoning. I. Desferrioxamine in the treatment of acute iron poisoning: Clinical observations, experimental studies, and theoretical considerations. *Pediatrics* 1965;**36**:322-335.
38. Athanasiou A, Shepp MA, Necheles TF. Anaphylactic reaction to desferrioxamine. *Lancet* 1977;**2**:616.
39. Shalit M, Tedeschi A, Miadonna A, Levi-Schaffer F. Desferal (desferrioxamine)— a novel activator of tissue-type mast cells. *J Allergy Clin Immunol* 1991;**88**:854-860.
40. Bousquet J, Navarro M, Robert G, Aye P, Michel FB. Rapid desensitization for desferrioxamine anaphylactoid reactions. *Lancet* 1983;**2**:859-860.
41. Miller KB, Rosenwasser LJ, Bessette JAM, Beer DJ, Rocklin RE. Rapid desensitisation for desferrioxamine anaphylactic reaction. *Lancet* 1981;**1**:1059.
42. Olivieri NF, Buncic JR, Chew E, et al. Visual and auditory neurotoxicity in patients receiving subcutaneous

deferoxamine infusions. *N Engl J Med* 1986;**314**:869-873.
43. Nouel O, Voisin PM, Vaucel J, Dartois Hoguin M, Le Bris M. [*Yersinia enterocolitica* septicemia associated with idiopathic hemochromatosis and deferoxamine therapy. A case]. *Presse Med* 1991;**20**:1494-1496.
44. Bentur Y, McGuigan M, Koren G. Deferoxamine (desferrioxamine). New toxicities for an old drug. *Drug Saf* 1991;**6**:37-46.
45. Naegeli H-U, Zähner H. Metabolites of microorganisms. Part 193. Ferrithiocin. *Helv Chim Acta* 1980;**63**:1400-1406.
46. Bergeron RJ, Wiegand J, Dionis JB, et al. Evaluation of desferrithiocin and its synthetic analogues as orally effective iron chelators. *J Med Chem* 1991;**34**:2072-2078.
47. Bergeron RJ, Streiff RR, Wiegand J, et al. A comparative evaluation of iron clearance models. *Ann N Y Acad Sci* 1990;**612**:378-393.
48. Bergeron RJ, Streiff RR, Creary EA, et al. A comparative study of the iron-clearing properties of desferrithiocin analogues with desferrioxamine B in a *Cebus* monkey model. *Blood* 1993;**81**:2166-2173.
49. Bergeron RJ, Liu CZ, McManis JS, et al. The desferrithiocin pharmacophore. *J Med Chem* 1994;**37**:1411-1417.
50. Bergeron RJ, Wiegand J, Wollenweber M, et al. Synthesis and biological evaluation of naphthyldesferrithiocin iron chelators. *J Med Chem* 1996;**39**:1575-1581.
51. Bergeron RJ, Wiegand J, Weimar WR, et al. Desazadesmethyldesferrithiocin analogues as orally effective iron chelators. *J Med Chem* 1999;**42**:95-108.
52. Bergeron RJ, Wiegand J, McManis JS, et al. Effects of C-4 stereochemistry and C-4' hydroxylation on the iron clearing efficiency and toxicity of desferrithiocin analogues. *J Med Chem* 1999;**42**:2432-2440.
53. Bergeron RJ, Weimar WR, Wiegand J. Pharmacokinetics of orally administered desferrithiocin analogs in *Cebus apella* primates. *Drug Metab Dispos* 1999;**27**:1496-1498.
54. Brittenham GM. Pyridoxal isonicotinoyl hydrazone (PIH): Effective iron chelation after oral administration. *Ann N Y Acad Sci* 1990;**612**:315-326.
55. Shetty B, Badr M, Melethil S. Evaluation of hepatic metabolism of salicylic acid in perfused rat liver. *J Pharm Sci* 1994;**83**:607-608.
56. McMahon TF, Stefanski SA, Wilson RE, et al. Comparative acute nephrotoxicity of salicylic acid, 2,3-dihydroxybenzoic acid, and 2,5-dihydroxybenzoic acid in young and aged Fischer 344 rats. *Toxicology* 1991;**66**:297-311.
57. Bergeron RJ, Wiegand J, Ratliff-Thompson K, Weimar WR. The origin of the differences in (R)- and (S)-desmethyldesferrithiocin: Iron-clearing properties. *Ann N Y Acad Sci* 1998;**850**:202-216.

ICL670A: PRECLINICAL PROFILE

Hanspeter Nick, Agnes Wong, Pierre Acklin, Bernard Faller, Yi Jin, René Lattmann, Thomas Sergejew, Suzanne Hauffe, Helmut Thomas, Hans Peter Schnebli[*]

1. INTRODUCTION

Man is unable to actively eliminate iron from the body, once it has been acquired. Toxic and eventually lethal levels of iron accumulate as a result of repeated transfusions, e.g. in ß-thalassemia major, or due to excessive dietary iron uptake in anemias and hereditary hemochromatosis. Excess iron is deposited in the form of hemosiderins (insoluble "iron cores" of ferritin) mainly in the liver, spleen, many endocrine organs and in the myocardium. The exact mechanism of iron damage to these tissues is unknown, but it is established that organ failure correlates with iron burden in these tissues. Except for infectious diseases, cardiac complications are the major cause of death in ß-thalassemia major patients.

Iron chelators slowly mobilize these iron deposits, probably by an indirect process of continuously binding those amounts of soluble iron present in the "transit pool", which are in equilibrium with the insoluble hemosiderins. Solubilized, chelated iron is excreted in the urine and/or feces.

Thirty-five years of clinical experience with Desferal® have unquestionably established the therapeutic concept of iron chelation. In ß-thalassemia patients iron chelation therapy has been shown to reduce iron-related morbidity (1) (reduces and retards liver diseases, diabetes and other endocrine failures, normalizes growth, prevents and, in some cases, reverses cardiac insufficiency) and to improve quality of life (2). Consequently, iron chelation therapy reduces mortality dramatically: A life-table analysis shows that patients with

---

[*] Hanspeter Nick, Pierre Acklin, Bernard Faller, Yi Jin, René Lattmann, Thomas Sergejew, Suzanne Hauffe, Helmut Thomas, Hans Peter Schnebli, Novartis Pharma AG, 4002 Basel, Switzerland. Agnes Wong, Department of Pathology, University of Western, Australia, Nedlands, Western Australia 6009

Iron Chelation Therapy.
Edited by Chaim Hershko, Kluwer Academic/Plenum Publishers, 2002.

good treatment compliance all reach the age 25, while survival in the poor compliance group is only 32 % (3). The poor oral bioavailability and the short plasma half-live of Desferal® necessitate its application as slow subcutaneous or intravenous infusions resulting in poor compliance in a large segment of the patient population. The need for an iron chelator which can be given orally has been recognized for a long time. Despite major investments, the inherent problem of orally available chelators, namely, the separation of the pharmacological effect from toxic effects has been very difficult to achieve. Based on molecular modelling studies a new series of iron chelators, the bis-hydroxyphenyltriazoles, was synthesized.

This paper describes important preclinical findings with the promising orally active iron chelator, ICL670A (4-[(3,5-Bis-(2-hydroxyphenyl)-1,2,4) triazol-1-yl]-benzoic acid).

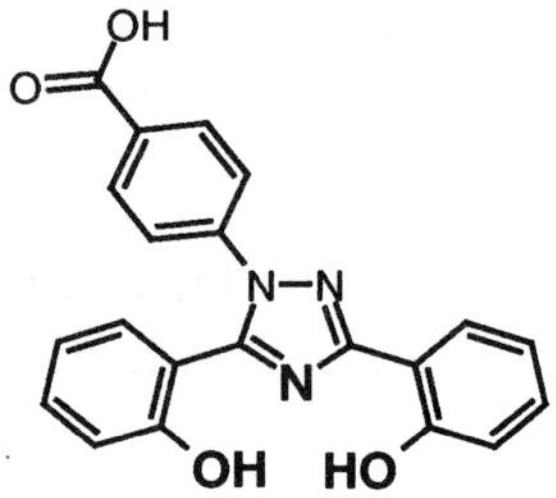

Figure 1. Structure of ICL670A with coordinating nitrogen and oxygens in bold

2. PHYSICAL CHEMISTRY

The pKa-values of ICL670A were determined by potentiometric titration. The carboxylic acid function has a pKa-value of 4.57, the pKa-values of the two hydroxyl groups are 8.71 and 10.56. This explains why solubility of ICL670A in water is strongly pH-dependent, decreasing with increasing acidity. Its solubility at pH 7.40, 25°C is 0.4 mg/mL. At pH 7.40, ICL670A distributes in octanol/ water in a 6.3:1 ratio (logD7.4= 0.8).

ICL670A is an achiral, tridentate ligand for Fe3+ (two molecules of ICL670A are needed to form a complete complex. The overall conditional complex binding constant at pH 7.4 is $10^{26.5}$ M^{-2} and the pM-value (4) is 22.5 The pM-value for ICL670A which was calculated based on a ligand concentration of 10 µM and an iron concentration of 1 µM is approximately 4 log units lower than the value reported for desferrioxamine (Desferal®, pM=26.6), but ICL670A binds iron about one thousand-fold more tightly than L1 (pM=19.5). In the millimolar to micromolar concentration range and at pH 7.4 the $(ICL670A)_2$-Fe complex with the global charge of –3 is the predominant species. The $(ICL670A)_2$-Fe complex can be protonated, for this reason its lipophilicity is increasing with increasing acidity.

The preformed $(ICL670A)_2$-Fe complex dissociates 3-4 times more slowly than that of L1 when confronted with a large excess of EDTA as competing agent.

3. IN VIVO PHARMACOLOGY

3.1. Induction of Iron Excretion in Non Iron-overloaded Rats

3.1.1. Time course of iron excretion.

Primary *in vivo* screening of ICL670A was carried out in the bile duct cannulated, non iron-overloaded rat model (5). The model allows the determination of the time course of iron excretion in the bile for 24 hours after compound administration. As can be seen in Figure 2A, considerable amounts of iron are already excreted in the first (0 to 3-hour) bile sample, demonstrating rapid absorption of the compound when administered orally.

As seen with most iron chelators, the duration of action of ICL670A increases with increasing dose, perhaps indicating a saturable excretion path. Additionally, the area under the (time) curve increases with the dose. As with all chelators tested in this model, the main route of ICL670A induced iron excretion is via the bile (> 85%).

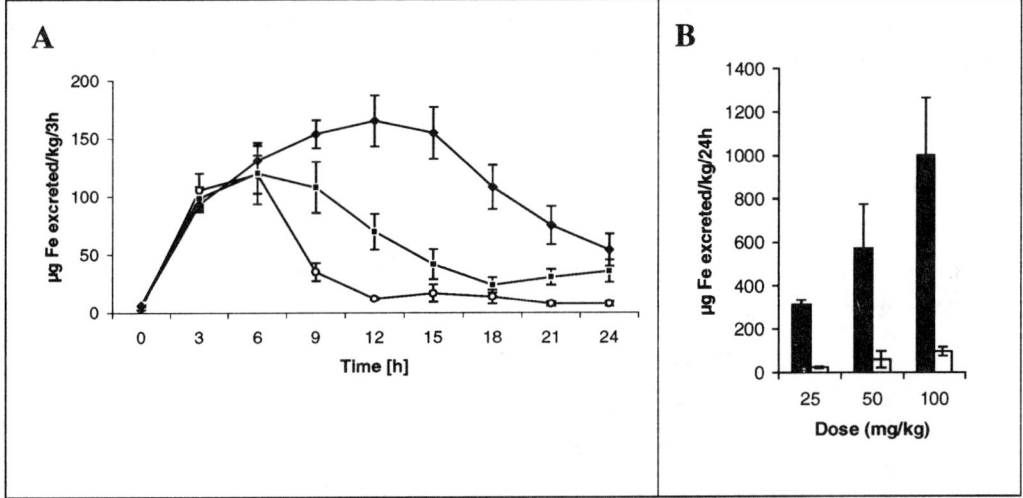

Figure 2A. Time course of biliary iron excretion induced by ICL670A in non iron-overloaded rats. Oral application of ICL670A in 0.5 % Klucel HF to Tif:RAl rats. Doses: ◆: 100 mg/kg (n = 8); ■: 50 mg/kg (n = 7); o: 25 mg/kg (n = 4). Values are means ± SEM. **Figure 2B.** Dose dependence of iron excretion into bile and urine induced by ICL670A in bile duct cannulated rats. Corresponds to the area under the curves of Figure 2A.

The dose-proportional effect of ICL670A is documented in Figure 2B, which shows the 24-hour iron excretion into bile and urine at three doses of ICL670A. The "effective oral dose" (ED) is the dose wich induces the excretion of 500 µg iron per kg body weight, the targeted daily excretion rate in ß-thalassemia patients. From the above data the ED of ICL670A was calculated to be about 32 mg/kg in Tif:RAl rats.

In the bile duct cannulated rat, ICL670A is among the most potent iron chelators ever tested. Its efficiency, i.e. the amount of excreted iron as percentage of the theoretical (stoichiometric) iron binding (IBE) of the dose, is approximately 18±5% (SD) at 50 mg/kg and 100 mg/kg p.o. By comparison, in the same model, Desferal® (s.c.) has an efficiency of about 3-4 %, while L1 has an efficiency of about 2 %.

3.1.2. Long-term Effect of ICL670A in Iron-overloaded Rats

The ability of ICL670A to reduce body iron burden was further evaluated in iron-overloaded rats. Ferrocene loading caused a more than 30-fold increase in liver iron and more modest increases in other organs, e.g. a 3-fold increase in kidney iron. Once the ferrocene loading was stopped, the rats spontaneously lost part (about one third) of their iron load, mainly during the first 4 weeks, before stabilizing.

Treatment with ICL670A at 75 µmole IBE/kg (56 mg/kg) p.o. for twelve weeks was compared to that with Desferal® and L1, both at 150 µmole IBE/kg (Figure 4). Treatment with ICL670A resulted in a continuous decrease of liver iron over the entire period. At the end of the study ICL670A had reduced non-heme liver iron concentrations by 60 % when compared to untreated, iron-overloaded control rats. ICL670A was twice as effective as s.c. Desferal® (68 % reduction, tested at twice the dose) and much more effective than L1 (29 % reduction, tested at twice the dose) .

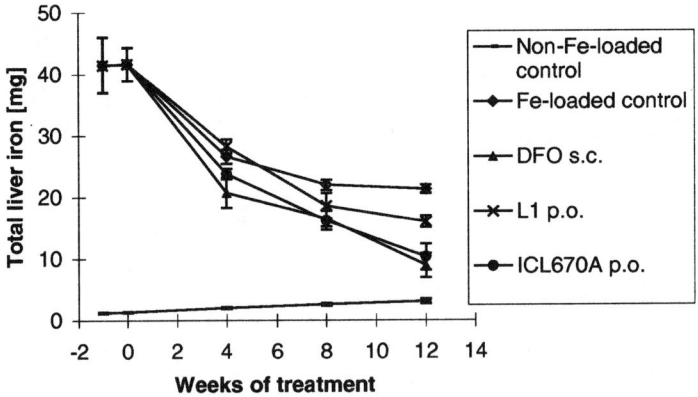

Figure 4. Long-term effect of ICL670A on liver iron in iron-overloaded male Tif:RAI rats. Doses: 150 ∟Moles IBE per kg (DFO and L1), 75 ∟Moles IBE per kg (ICL670A); vehicle 0.5% Klucel HF. Results are means of 4 animals ±.SEM.

The ferrocene induced iron-overload in the kidneys was completely reversed by ICL670A at 75 µmole IBE/kg and by Desferal® at 150 µmole IBE/kg in as little as 4 weeks. Higher doses of ICL670A rapidly depleted kidney iron to levels significantly below control (non-loaded) values. This observation may relate to the adverse effects seen in kidneys of non-loaded animals in toxicity studies.

3.1.3. Fate of the Iron Complex of ICL670A in Rat, a comparison with other chelators

In a treated animal or patient, dissociation of iron complexes prior to excretion may lead to an undesired redistribution of iron. This potential problem was studied in an experimental rat model originally proposed by M. Pippard (6) and later adopted in our laboratory with some modifications.

^{59}Fe-complexes of various chelators were preformed *in vitro* with a 20-fold excess (in IBE) of ligand. These mixtures of free ligands and iron-chelates were then injected i.v. into rats. Chelators studied were from several different classes and were studied with respect to their ability to efficiently transport iron out of the body (Table 1): DFO and HBED are hexadentate chelators, nor-DFT and ICL670A are tridentate chelators, L1, CGP65015 and DDTA belong to the bidentate family (structures and properties in Table 2).

The main reason for tridentate and bidentate chelators to be of interest lies in their relatively small size. By virtue of their lower molecular weight, tri- and bidentate molecules are more likely to be orally absorbable than hexadentates. On the other hand, these molecules (and possibly their iron complexes) can potentially penetrate tissues and cause deleterious effects.

Of the seven ligands investigated, the two hexadentate chelators DFO and HBED efficiently eliminated the metal ion since a high proportion of the injected ^{59}Fe was found in the urine or feces after 24 hours. Complexes of hexadentate chelators such as DFO and HBED are thermodynamically and kinetically more stable/inert than those of tri- or bidentate chelators. The potential for chelators to release iron from their iron complexes is higher for ligands other than hexadentates because of their greater kinetic lability and their concentration dependent affinity for the metal ion. However, the tridentate chelator ICL670A showed efficient elimination of radioiron and even with the bidentate chelator CGP65015 a substantial percentage of the injected radioiron was eliminated. This indicates that in all three chelator-classes compounds can be found which form inert iron complexes and are excreted after i.v. injection. The results further show that the Fe-chelates of nor-DFT, L1 and DDTA dissociated before excretion could take place or that these Fe-complexes are not efficiently excreted from the animal. Although nor-DFT and L1 could mobilize iron in the bile-duct cannulated rat model, the tissue distribution of ^{59}Fe injected as complexes of nor-DFT or L1 was similar to that of controls injected with unchelated radioiron. The pattern of tissue ^{59}Fe distribution by the iron-complexes of nor-DFT and L1 (initially to the skeleton / bone marrow and subsequently by red blood cells) can be explained by the release of chelated iron to transferrin which in turn transfers the iron to the erythropoietic system.

Interestingly, in the iron overloaded rats where transferrin was fully saturated with iron, nor-DFT and L1 induced incorporation of 59Fe into the skeleton and red blood cells to a similar extent as in normal rats. This is surprising, since in the control situation (injection of ^{59}Fe in the absence of chelators), the amount of radioiron incorporated into the erythron is very much reduced in the iron-loaded animals compared to the normal non iron-loaded animals. In these iron overloaded control animals the amount of ^{59}Fe retention in the liver increased to 53% of the injected dose compared to 16% in non iron-loaded controls. In normal (control) animals, the injected ^{59}Fe is rapidly bound to transferrin, thus feeding mainly into the erythron. By contrast, in the iron-loaded animals where transferrin is saturated, the injected ^{59}Fe remains as "non-transferrin bound" iron (NTBI) and thus may be quickly absorbed into the liver. The relatively high incorporation of radioiron (injected as nor-DFT or L1 complex) into the erythron on the other hand cannot be explained by feeding into the NTBI pool (through dissociation). Although there is no direct proof for this, the simplest explanation is that the ^{59}Fe from the chelates of nor-DFT and L1 rapidly exchanges with the transferrin bound iron.

In conclusion, the iron complex of ICL670A is thermodynamically/kinetically sufficiently stable/inert to be excreted to a large extent in feces after intravenous injection of the (ICL670A)2Fe-complex in rats. The fact that intravenously administered (ICL670A)2-Fe is recovered to a considerable degree is an advantageous property since in a 'real life' situation where ICL670A is administered, it can be expected that the (ICL670A)2-Fe complex once formed has a high probability to be excreted. Chelates, which in this model are poorly excreted and, above all, rather show distribution of the initially bound iron are suspected to have a higher probability to redistribute ferric ion.

Table 1. Key pharmacological properties of selected iron chelators in rats

Chelator	[a]Elimination of preformed ^{59}Fe-chelates				[a]Iron mobilization in rats		[a]Iron mobilization in marmoset monkey (4)	
	% of injected dose 24h post injection		% of injected dose 48h post injection		μg Fe/kg/24h		μg Fe/kg/48h	
	Feces	Urine	Feces	Urine	Bile	Urine	Feces	urine
[b]DFO	1.8	66			478	80	177	91
HBED	43	21			733	27	62	82
Nor-DFT	0.6	1.2			450	7	200	195
ICL670A			39	0.6	572	60	2106	306
CGP65015	20	3.7	35	6	620	20	1825	515
L1	0.4	2.7			198	22	39	145

[a] Tested at 150 μmol IBE.
[b] Given subcutaneously

Table 2. Structure and properties of chelators

Structure	Short name	Stoichiometry (Chel:Fe)	pM	Charge of complex at pH 7.4	Lipophilicity log $D_{7.4}$ of ligand
	DFO	1:1	26.6	+1	<-3
	HBED	1:1	31.1	-1	-1
	Nor-DFT	2:1	20.9	-1	-1.5
	ICL670A	2:1	22.5	-3	1.0
	CGP65015	3:1	21.3	0	0.6
	L1	3:1	19.5	0	-0.7

3.1.4. Effect on Dietary Iron Uptake

Orally administered chelators have the potential to complex dietary iron during their transit through the gastrointestinal tract and could thus promote the uptake of dietary iron. This is a safety concern as this would bring additional iron burden to the patients body instead of mobilizing and eliminating the harmful metal from the tissues.

The potential of ICL670A to increase the absorption of dietary iron was examined as follows: Groups of 4 rats were treated twice daily with oral doses of ICL670A (40 mg/ kg/ day) for two weeks; each treatment was preceded by the intragastric application of small amounts of ^{59}Fe. The control animals which received ^{59}Fe only, retained 13.2±0.8 % of the radioactivity injected. With ICL670A no increase, but a significant decrease to 8.6±1.5 % in total iron absorption was observed. The reduced retention of ^{59}Fe, most prominently seen in the liver (Table 3), not only shows that ICL670A does not promote the uptake of dietary iron, but can be taken as further evidence for its effectiveness in body iron removal.

Table 3. Radio-iron distribution in selected organs of rats after oral administration of ^{59}Fe for two weeks with and without oral ICL670A.

	Blood	Liver	Spleen	Heart	Kidneys	Femur	Carcass
Controls							
% of total	24.2±18	16.8±5.6	0.6±0.3	0.9±0.6	1.2±0.3	0.5±0.4	56±18
ICL670A							
% of total	31.9±13	12.4±3.7	1.0±0.3	1.5±0.8	1.4±0.3	0.5±0.1	51.4±7.5

Values are means ± SD.

3.2. Induction of Iron Excretion in Iron-overloaded Marmoset monkeys

Rodents differ substantially from primates in drug absorption, distribution, metabolism, excretion, and, importantly, in their iron metabolism (7) (uptake, pools, fluxes and excretion). In fact, a number of iron chelators which were active p.o. in the rat were found to be ineffective after p.o. or even s.c. administration to primates (marmoset and *Cebus* monkeys). For these reasons the pharmacology of ICL670A was also evaluated in a primate model.

In this model, marmosets (*Callithrix jacchus*) were iron-overloaded by three i.p. injections of iron-dextran, separated by a 14-day interval after each injection (200 mg/kg twice, 100 mg/kg at the third injection). Iron loading was at least 8 weeks prior to the first experiment. Seven days before and during the experiment marmosets were maintained on a low-iron diet to reduce fecal background. During the experiment the animals were kept in metabolic cages and the excretion of iron in urine and feces was followed for 2 days before (= background excretion) and 2 days after compound administration.

3.2.1. Time course of iron excretion in the marmoset

Figure 5 illustrates the type of data which is obtained with the marmoset model. Urinary background iron (pretreatment days) is negligible while the fecal background iron excretion amounts to approximately 250 µg Fe per day and kg body weight. After administration of the compound, excretion of the ICL670A-complexes is primarily through the feces and the amount of iron excreted is clearly dose dependent. Considerable amounts of

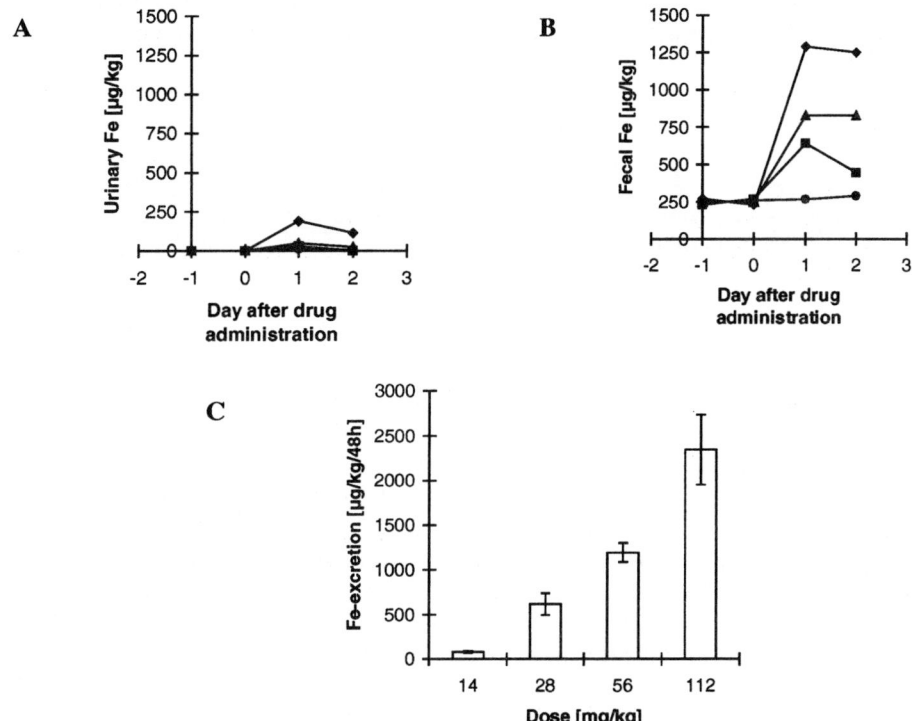

Figure 5. A and B: Time course of iron excretion in iron-overloaded marmosets induced by single doses of ICL670A. Vehicle: 40% PEG 400 + 0.45% HPMC. Doses: 150 µmol IBE/kg (112 mg/kg; n = 6); 75 µmol IBE/kg (56 mg/kg; n = 6); 37.5 µmol IBE/kg (28 mg/kg; n = 9); 18.8 µmol IBE/kg (14 mg/kg; n = 3); Error bars were omitted for reasons of clarity. C: Area under the curve calculated from Figure 5A.

iron are still excreted in the 24- to 48-hour period, particularly (as in the rat) at the higher doses. For animal welfare reasons the marmosets could only be kept in the metabolic cages for 48 hours post treatment, so that the ICL670A-induced iron excretions calculated from 48-hour data may somewhat underestimate the actual effect.

3.2.2. Dose response in the marmoset

Figure 5C illustrates the dose dependence of iron excretion (above background) induced by a single doses of ICL670A. To be able to directly relate the results of these experiments to tolerability data, the formulation used here (40% PEG 400 + 0.45% HPMC) was identical to that used in toxicity studies. ED corresponds to the dose which mobilizes 500 µg of iron per kg body weight, the targeted, daily mobilization rate for iron chelation therapy in ß-thalassemia patients. From the data in Figure 5 the ED in marmosets is estimated to be about 22 mg/kg. Note that this value is derived from 48-hour excretions after a single dose; as considerable amounts of iron are still excreted during the second day, an even lower ED is expected when daily repeated doses are given.

The *in vivo* efficiency of iron chelation calculated for ICL670A, i.e. the amount of excreted iron as percentage of the theoretical (stoichiometric) iron binding capacity (IBE) of the dose, is 29 % at the dose of 150 µmole IBE/kg. This value is among the highest ever observed with an iron chelator. For comparison, L1 has efficiencies in the marmoset and in man of about 2% and Desferal® (s.c. infusion) has an efficiency of about 5 to 10% in man. At equivalent doses (in IBE) the total iron excretion induced by oral ICL670A is more than 10-fold higher than that of L1.

3.2.3. Route of excretion in the marmoset.

As seen already in Figure 5, iron excretion induced by ICL670A is predominantly by the fecal route and only a few percent are excreted in the urine. Single doses of ICL670A were administered orally using 40% PEG 400 + 0.45% HPMC as a vehicle. At 150 µmol IBE/kg (112 mg/kg, n=6), 75 µmol IBE/kg (n=6) and 37.5 µmol IBE/kg (n=9) urinary excretion was 16.2±11.3 %, 6.0±3.6 % and 5.5±4.7 %, respectively. There is a considerable interindividual variability in the urinary/fecal ratios with a trend to increased urinary excretion at the highest dose tested (= approx. 5-fold the effective dose).

3.2.4. Specificity of iron excretion in the marmoset.

In addition to its strong affinity for iron, ICL670A also displays modest affinities for other physiologically important metals such as copper and zinc. However, in marmosets, ICL670A at 150 µmol IBE/kg did not promote any excretion of copper and zinc above background.

4. TOXICOLOGY

The aim of this program was to identify orally active iron chelators with acceptable tolerability. However, it was clear from the beginning that a very wide therapeutic margin was not likely to be achieved: Past experience has shown that most of the orally active iron chelators are toxic at, or close to, doses needed for a pharmacological action.

As many chelator toxicities are related to the pharmacological effect of the compound, i.e. iron deprivation or iron depletion, the frequency and severity of **adverse effects is influenced by the iron status** of the animals. This relationship is well established in the clinic, where the dose of Desferal® (or other chelators) must be tailored to each individual patient, based on his or her iron status and transfusion regimen. In contrast to thalassemia patients, experimental animals are usually not iron overloaded. For this reason, very potent chelators like ICL670A will render these animals iron deficient very quickly. As will be shown below, most observed adverse effects do relate to the pharmacological effectiveness of the iron chelator treatment, as at the end of the toxicity studies the liver-iron contents of the animals were reduced by up to 90 %.

Nephrotoxicity (i.e alterations in the proximal tubular epithelium) has been observed with iron chelators of other chemical classes as for example the desferrithiocins (8) and triazoles including ICL670A. In all cases the most serious kidney toxicity was seen with the pharmacologically most effective compounds. This supports the concept that many of the toxic effects seen, particularly kidney toxicity, are a direct result of iron deprivation. Results of studies performed with ICL670A in iron-overloaded animals support this conclusion. Iron deprivation is not expected to occur in patients with iron-overload diseases undergoing chelation therapy.

4.1. Subchronic administration of ICL670A to rats and marmoset monkeys

Four tolerability studies are summarized in Table 4. In these studies ICL670A was given orally to rats and marmoset monkeys. Most of the treatment-related effects in the **four-week oral toxicity in normal rats** were regarded to be a direct result of the pharmacological activity of ICL670A and correlated well with the decreases in liver and kidney iron concentrations. In particular, iron depletion was considered to be responsible for alterations in the tubular epithelium in the kidney. The hematological alterations were consistent with altered erythropoiesis secondary to iron depletion and the myocardial changes appeared to be secondary to anemia. The NOAEL in this study was 10 mg/kg/day.

In order to verify the conclusion that the observed adverse effects in normal rats are related to the pharmacological effect of the iron chelator treatment, an additional **four-week oral toxicity in iron-overloaded and normal rats** was performed. It is concluded that the type of proximal tubular changes seen in kidneys of iron-overloaded rats corresponded to alterations seen in normal animals. However, these were clearly less pronounced in iron-overloaded animals and did not cause premature deaths during the 4-week treatment. This large difference of toxicity in normal and iron-overloaded animals supports the conclusion that the observed adverse effects in normal, non iron-overloaded animals are a direct result of the pharmacological activity of the iron chelator.

The marmoset monkey is one of the smallest non-human primates and is considered to be highly relevant with respect to its iron metabolism and therefore predictive with respect to the pharmacological and toxicological effects of iron chelators. Oral administration of ICL670A in a **four-week oral toxicity study in marmoset monkeys,** resulted in decreased liver and kidney tissue-iron contents. However, due to large interindividual differences a dose dependency could not be established. Even at the highest dose, there was no effect of treatment on copper and zinc levels in liver and kidney. Salient findings in this study were changes in the kidney, liver and gall bladder in occasional animals only and were confined to the high-dose (130 mg/kg/day) male group. The dose related increase in plasma iron seen in all animals receiving ICL670A was probably due to circulating Fe-ICL670A complexes. The NOAEL in this study was considered to be 65 mg/kg/day in males and 130 mg/kg/day in females.

A **two-week oral toxicity study in iron-overloaded marmosets** was performed to investigate further the relationship of the pharmacological effect of the iron chelator, i.e. iron depletion, to the adverse effects observed in normal marmosets. Although the marmosets of the present study were iron-overloaded more than two years previously, their liver iron concentrations were still 5-and 8-fold higher in males and females, respectively, than those of normal marmosets. In iron-overloaded and non-loaded marmosets both the liver and the kidney iron levels were consistently lower in females than in males (as in man). It is concluded from this study that administration of ICL670A at 400 mg/kg/day to iron-overloaded marmosets was well tolerated and did not result in the proximal tubular cell vacuolization noted in a previous study in non overloaded marmosets. The large difference in NOAEL (65 and 130 mg/kg in normal males and females, respectively, and 400 mg/kg in iron-overloaded animals of both sexes) clearly shows that iron-overload protects against proximal tubular cell alterations in these primates.

Table 4. Tolerability of orally administered ICL670A in rat and marmoset monkey.

Study	Doses [mg/kg] (n, male/female)	Observations, adverse effects, tissues/organs affected	NOAEL [mg/kg]
Normal rats, 4 weeks	10 (12/12) 30 (12/12) 100 (12/12)	Mortality at the highest dose, both sexes. Erythrocyte parameters, adrenal and thymus weights decreased at the highest dose. Renal tubular vacuolization (in 75 % of the males, 33 % of the females) and tubular necrosis (in 58 % of the males, 25 % of the females) at the highest dose. Myocardial changes include focal degeneration, inflammation, myocarditis (in 50 % of the males, 8 % of the females) at the highest dose. One male animal was also affected at medium dose. Main target organs: Kidney, gastrointestinal tract, myocardium; effects mostly with highest dose and more pronounced in early death animals. Liver and kidney iron massively decreased at all doses.	10
Normal and iron-overloaded rats, 4 weeks	100 (12/12)	In normal animals serious toxicity, and mortality, no deaths in iron-overloaded animals. Main target organ: Kidney, much more severe effects in normal animals. In iron overloaded rats: Kidney toxicity more severe at interim sacrifice than at the end of the study (indicates regeneration and/or adaptation).	-
Normal marmoset monkey 4 weeks	25 (4/4) 65 (4/4) 130 (4/4)	No deaths. No findings at ophthalmoscopy, hearing examination, ECG. Hematology, bone marrow: No changes. Histopathology: No changes in females, one male of the high dose group presented with moderate vacuolar degeneration of kidney cortical tubules, moderate vacuolization of the intrahepatic bile duct cells and marked acute inflammation of the gall bladder epithelium with moderate fibrosis of the gall bladder wall and moderate vacuolar hyperplasia of the epithelium. Main target organ: Kidney.	65 (male) 130 (female)
Iron-overloaded marmoset monkeys 2 weeks	400 (10/10)	There were no treatment-related effects on body weight, food consumption, hematology, clinical biochemistry or urine analysis and no abnormalities were detected at necropsy. Specifically, there was no indication of renal proximal tubular cell vacuolization.	400

4.2. Chronic Toxicity Studies in normal Rats and Marmoset Monkeys

The purpose of these studies was to establish the toxic effects of ICL670A when administered chronically to rats for up to 26 weeks and to marmosets for up to 39 weeks. Since administration of highly potent iron chelators such as ICL670A rapidly depletes normal animals of iron, some groups of animals received an iron supplemented diet.

Table 5. Overview on chronic toxicity studies with orally administered ICL670

Study type	Species	Doses (mg/kg)	Findings / target organs	NTEL mg/kg
26-week followed by a 4-week recovery period	Rat	0, (no iron supplement) 0, 30, 80, 180 (all iron supplemented feed)	$\geq$ 30 mg/kg: Lenticular aberrations (males); increase in kidney weight. $\geq$ 80 mg/kg: cataracts, haematological effects, and glandular stomach ulceration and erosion (males). 180 mg/kg: Mortality and moribundity (10/30 males, 2/30 females); nephropathy, cytoplasmic vacuolation of adrenal medulla (males); With the exception of the lenticular cataracts, all test-article related effects noted above were at least partially reversible by the end of the 4-week recovery period.	30 (females)
39-week followed by a 4-week recovery period	Marmosets	0, 20, 40, 80 (no iron supplement), 80 (+ iron supplement)	80 mg/kg: Mortality (1/5 males) and moribundity (1/5 males, 1/5 females receiving iron supplement); haematological effects; generalized degeneration and/or acute inflammation of gallbladder, vacuolation of hepatic bile duct cells, nephropathy, atrophy of salivary gland serous cells.	40

In the **26-week oral toxicity study in rats** ICL670-induced mortality/moribundity was attributed to renal toxicity evidenced by slight kidney vacuolation and increase of markers of kidney function.

Clinical signs of ocular opacities were observed at about week 10 in both sexes. By the end of the treatment period the incidence between the sexes was comparable at 180 mg/kg/day. The onset of these changes was somewhat dose related, since ophthalmoscopic findings were observed at the lowest dose level much later in the study (i.e., by 26 weeks). The cataracts did not regress after the 4-week recovery period. Changes of this type have been seen and reported before with iron chelators e.g., Desferal.

Pharmacological changes consisted of decreased levels of iron in the liver and kidneys with concomitant increases in total iron binding capacity and serum iron levels. Though the animals were given an iron-supplemented diet, this could only partly compensate the iron depletion caused by action of ICL670. Following the recovery period, the tissue iron levels improved but did not reach normal levels. Increased splenic erythropoiesis with concomitant reticulocytosis and thrombocytosis were a response to altered iron levels and, in females, anemia. Clinical chemistry, hematology, organ weight macroscopic and microscopic changes were reversed upon cessation of treatment. The no effect level for females was 30 mg/kg/day. A no effect level was not achieved for males.

Toxicokinetics revealed a marked increase in systemic exposure in males upon repeat administration of ICL670. Gender differences in tissue iron content were also noted between control groups, with females clearly exhibiting higher levels of tissue iron with or without iron supplementation. These factors may in part explain the gender differences in toxicity seen in this study, with males being clearly more sensitive to the effects of ICL670.

In the **39-week oral toxicity study in marmoset monkeys** three animals at 80 mg/kg (one female receiving iron supplement) showed clinical signs, bodyweight and food consumption effects, toxic changes in the gall bladder, hepatic bile ducts and kidneys resulting in the premature sacrifice. The histopathological changes (see Table) were only observed in decedent animals. There were no histopathological findings in animals killed on completion of the treatment and the recovery period which could be attributed to the treatment.

There were no clear differences in the magnitude of the effects between animals receiving normal or iron-supplemented diet.

Treatment with ICL670 distinctly decreased liver iron content, by up to 95% for males at 80 mg/kg/day or females receiving 40 or 80 mg/kg/day. Kidney iron content was reduced up to approximately 40%. Zinc and copper levels in either tissue were not affected.

There were no treatment related ophthalmoscopy findings and no effects on electrocardiographs which were attributable to treatment.

5. PHARMACOKINETICS

5.1. Single Dose Kinetics and Bioavailability in Rats, Marmosets and Dogs

ICL670A was rapidly and well absorbed in rats and marmosets following peroral administration of [^{14}C]ICL670 single doses of 10 and 25 mg/kg, respectively. In bile duct-cannulated rats, 75% of the administered dose was absorbed. Plasma concentrations of the free ligand in both species were maximal at 0.5 h postdose and decreased rapidly, reaching levels about 4 times lower at 2 h than at 0.5 h. The radioactivity in plasma of rats consisted mainly of the free ligand, while in marmosets, the plasma concentrations of the ligand were lower than those of ^{14}C concentrations (about 10 times lower at 2 h postdose), indicating the presence of metabolites in the plasma of marmosets. The plasma levels of the iron complex were low, with maximum levels less than 10% of those of the ligand. In both species, the absolute bioavailability compared to an i.v. dose amounted to 30-40% of dose, and the bulk of the dose was recovered with the faeces, mainly in form of metabolites.

The pharmacokinetics of three prototype formulations of ICL670A, two dispersible tablets and a sachet form were investigated in 8 male Beagle dogs for a 10 mg/kg dose. All three oral formulations were rapidly absorbed and maximal plasma concentrations of the free ligand were usually reached within less than 1 hour. After t_{max}, the plasma concentrations declined with a half-life of about 1 h, whereas a minor part of the dose was eliminated more slowly beyond 6 h. The absolute bioavailability was found to be high for all three oral formulations (80 - 100 % compared to i.v. administration) and the plasma concentration profiles were roughly similar (Figure 6).

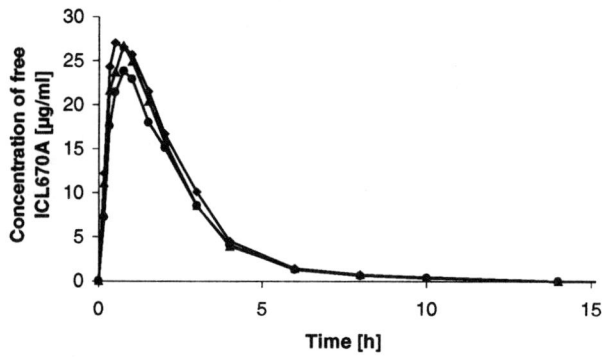

Figure 6. Plasma concentrations of unchanged ICL670A in dogs after oral dosing of three different formulations. n=8. For clarity reasons standard deviations are not shown. Plasma levels at early time points (before tmax) have a larger variation, around tmax the standard deviation is typically 25% of the mean value.

5.2. Multiple Dose Pharmacokinetics in Rats

Concentrations of total ICL670A (free ligand + complex) were determined in plasma of rats during a 2-week and a 4-week oral toxicity study. In both studies ICL670A was given once daily by gavage as a suspension in 0.5% aqueous Klucel. Toxicokinetic analysis revealed rapid absorption (t_{max} between 0.5 and 2 hours). The increase in systemic exposure was overproportional to the dose, suggesting saturation of the elimination process(es). There were no significant gender differences and no evidence of accumulation at dose levels from 10 to 200 mg/kg.

Concentrations of total ICL670A have also been determined in plasma of iron-overloaded and normal rats during a 4-week treatment with 100 mg/kg. Maximal plasma concentrations of total ICL670A were observed at 2 h post dose; these had decreased to one hundredth or less by 24 h. The systemic exposure was lower in iron-overloaded rats than in normal rats, as determined by AUC values. There was no evidence of accumulation in plasma.

5.3. Multiple Dose Pharmacokinetics in Marmosets

Preliminary data are available from the toxicokinetic evaluation of the 2 week studies in marmosets which indicated a rapid absorption (t_{max} 2h) and no marked gender difference in marmosets.

The more detailed investigations during the 4 week toxicity study (administration of ICL670A daily doses as a suspension in 0.45% HPMC/40%PEG 400) revealed rapid absorption at the 25 mg/kg dose level, and a slightly slower absorption at 65 and 130 mg/kg in some animals. On day 1, maximal plasma concentrations of total ICL670A were lower in females than in males whereas no systematic difference between male and female animals was obvious on day 28.

Systemic exposure to total ICL670A was dose proportional to overproportional, particularly at the end of the 4-week study. Since accumulation in plasma was but marginal up to doses of 130 mg/kg, the overproportional systemic exposure suggests saturation of the elimination process(es).

In the two week toxicity study in iron-overloaded marmosets, plasma concentrations of total ICL670A were lower than those found in the previous subchronic toxicity study in non iron-overloaded marmosets at the same dose. There was no marked sex difference in ICL670A plasma concentrations, and accumulation in plasma during treatment, if any, was low.

6. CONCLUSION

For ß-thalassemia patients iron-chelation therapy is life saving. Since its introduction in 1962, Desferal® has remained the only iron chelator approved for general use. However, its modes of application (daily subcutaneous or intravenous infusions) are poorly accepted by patients. Consequently, these patients have been waiting desperately for an alternative which can be taken orally. ICL670A is orally active and is expected to address this clear medical need, as it will be accepted much more widely, will lead to better compliance (and therefore better clinical responses) and will be usable also in areas with a less well developed infrastructure. ICL670A, at this stage of development shows the properties expected of a promising orally active iron chelator:

ICL670A has a high affinity for iron. Orally administered ICL670A is rapidly and well absorbed in rats, marmosets and dogs. The efficient and selective ability of ICL670A to mobilize tissue iron and promote its excretion has been demonstrated in several models: In the bile duct cannulated rat, in a chronic (12 week) study in iron-overloaded rats and in the iron-overloaded marmoset. In all models ICL670A was consistently more efficient than Desferal® and much more efficient than L1.

The "effective oral dose" corresponds to the dose which mobilizes 500 µg of iron per kg body weight, the targeted, daily mobilization rate for iron chelation therapy in ß-thalassaemia patients. In iron overloaded marmosets, the "effective oral dose" was about 22 mg/kg.

Proximal tubular changes seen in kidneys of iron-overloaded rats were clearly less pronounced than in normal animals and did not cause premature deaths during a four-week treatment. The large difference of toxicity in normal and iron-overloaded animals supports the conclusion that the observed adverse effects in normal, non iron-overloaded animals are a direct result of the pharmacological activity of the iron chelator.

Administration of ICL670A at 400 mg/kg/day to iron-overloaded marmosets was well tolerated and did not result in the proximal tubular cell vacuolization noted in a previous study in non overloaded marmoset monkeys. The large difference in NOAEL (65 and 130 mg/kg in normal males and females, respectively, and 400 mg/kg in iron-overloaded animals of both sexes) shows that iron-overload protects against proximal tubular cell alterations in these primates. The therapeutic window increases clearly in iron-overloaded animals.

In chronic toxicity studies in rats nephropathy and alterations of renal function parameters were observed and were responsible for treatment related mortality in a 26-week chronic toxicity study in rats. Iron supplementation of the diet decreased the nephrotoxicity of ICL670A. ICL670A treatment was also associated with gastro-intestinal erosion and ulceration, cytoplasmic vacuolization of the adrenal medulla and the formation of cataracts. Lenticular changes have been seen and reported before with iron chelators e.g. Desferal, in animal species. Neither adrenal, cardiac nor ocular changes were observed in a 39-week chronic toxicity study in marmosets. However, in this species treatment was associated with effects on the gall bladder and intrahepatic bile ducts possibly as a consequence of excretion of ICL670 iron complexes in the bile and subsequent concentration in the gall bladder.

Toxic effects seen in these chronic toxicity studies were in the initial stages all reversible, although the mature cataracts did not regress. The no toxic effect levels in the chronic studies were 30 mg/kg in the rat and 40 mg/kg in the marmoset.

All of our pharmacology studies which were aimed at the quantification of iron excretion in rat and marmoset monkey suggested good oral bioavailability. A bioavailability study in dog confirmed this and showed almost complete absorption of ICL670A after oral administration. In summary, ICL670A emerged from an extensive evaluation process, during which more than 750 compounds from six different chemical classes were examined. ICL670A is an orally active iron chelator which combines high potency, efficiency and tolerability.

7. REFERENCES

1. Brittenham, G.M., Griffith, P.M., Nienhuis, A.W., McLaren, C.E., Young, N.S., Tucker, E.E., Allen, C.J., Farrell, D.E., Harris, J.W. Efficacy of Deferoxamine in Preventing Complications of Iron Overload in Patients with Thalassemia Major. *New Engl. J. Med.* 1994, **331**, 567-573.

2. Gabutti, V., Piga, A. Results of Long-term Iron Chelating Therapy. *Acta Haematol.* 1996, **95**, 26-36.

3. Olivieri, N.F.; Nathan, D.G.; MacMillan, J.H.; Wayne, A.S.; Liu, P.P.; McGee, A.; Martin, M.; Koren, G.; Koren, A.R. Survival in Medically Treated Patients with Homozygous ß-Thalassemia. *New Engl. J. Med.* 1994, **331**, 574-578.

4. Harris, W.R., Raymond, K.N., Weitl. F.L. Ferric Ion Sequestring Agents: The Spectrophotometric and Potentiometric Evaluation of Sulfonated Tricatecholate Ligands. *J. Am. Chem. Soc.* 1981, **103**, 2667-2675.

5. Bergeron, R.J., Wiegand, J., Dionis, J.B., Egli-Karmakka, M., Frei, J., Huxley-Tencer, A., Peter, H.H. Evaluation of Desferrithiocin and its Synthetic Aanalogues as Orally Effective Iron Chelators. *J. Med. Chem.* 1991, **34**, 2072-2078.

6. Pippard, M.J., Groves, M.J., Hider, R.C. Effects of Hydroxypyridinone Chelating Agents on Internal Iron Exchange. *Br. J. Haematol.* 1991, **77** *(Suppl 1)*, 39.

7. Finch, C.A., Ragan, H.A., Dyer, I.A., Cook, J.D. Body Iron Loss in Animals. *Proc. Soc. Exp. Biol. Med.* 1978, **159**, 335-338.

8. Bergeron, R.J., Streiff, R.R., Creary, E.A., Daniels, R.D. Jr., King, W., Luchetta, G., Wiegand, J., Moerker, T., Peter, H.H. A Comparative Study of the Iron-clearing Properties of Desferrithiocin Analogues with Desferrioxamine B in a Cebus Monkey Model. *Blood* 1993, 15, 81(8): 2166-2173.

PYRIDOXAL ISONICOTINOYL HYDRAZONE AND ITS ANALOGUES

Joan L. Buss, Marcelo Hermes-Lima, Prem Ponka*

1. INTRODUCTION

Iron is a precious metal for the organism because of its unsurpassed versatility as a biological catalyst. It is involved in a broad spectrum of essential biological functions such as oxygen transport (hemoglobin), electron transfer (mitochondrial heme and non-heme Fe proteins essential for energy production) and DNA synthesis (ribonucleotide reductase), to name just a few. However, the chemical properties of iron which allow this versatility also lead to the paradoxical situation that acquisition by the organism of an abundant element is exceedingly difficult. At pH 7.4 and physiological oxygen tension, the relatively soluble ferrous ion (Fe^{2+}) is readily oxidized to ferric ion (Fe^{3+}) which is susceptible to hydrolysis, forming virtually insoluble ferric hydroxides. The concentration of aquated Fe^{3+} (pH 7.4) cannot exceed 10^{-17} M. Moreover, unless bound to specific ligands, iron plays a key role in the formation of harmful oxygen radicals which ultimately cause oxidative damage to vital cell structures. Because of this virtual insolubility and potential toxicity, specialized mechanisms and molecules for the acquisition, transport, and storage of iron in a soluble nontoxic form have evolved to meet cellular and organismal iron requirements. In addition, organisms are equipped with sophisticated mechanisms that prevent the expansion of the catalytically active intracellular iron pool, while maintaining sufficient concentrations of the metal for metabolic use.[1-3] However, despite these homeostatic mechanisms, organisms often face the threat of either iron deficiency or iron overload.

Iron metabolism in humans is characterized by limited external exchange and efficient reutilization from internal sources. Because of the limited ability of the body to excrete excess iron, iron overload may develop as a result of prolonged hyperabsorption of dietary iron, administration of excess iron parenterally (*e.g., via* blood transfusions), or a combination of both factors. Patients with iron overload accumulate excessive amounts of

---

* Joan L. Buss and Prem Ponka, Lady Davis Institute for Medical Research, Sir Mortimer B. Davis Jewish General Hospital and Departments of Physiology and Medicine, McGill University, Montreal, Quebec, Canada. Marcelo Hermes-Lima, Oxyradical Research Group, Departamento de Biologia Celular, Universidade de Brasilia, Brasilia, Brazil.

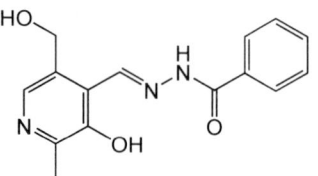

Figure 1. Structures of PIH and some of its analogs. Numbers correspond to the identification scheme used in some studies. [39,41,52]

iron in various organs including the liver, pancreas, and heart and consequently, they suffer from conditions that include cirrhosis, diabetes, and heart dysfunction.[4] Inherited iron overload, including hereditary hemochromatosis, can be treated by phlebotomy, which effectively mobilizes excessive amounts of iron from iron overloaded tissues.

Secondary iron overload develops in transfusion-dependent patients, such as those with thalassemia major, aplastic anemia, and myelodysplastic syndrome. Obviously, such patients cannot be phlebotomized; the only treatment for secondary iron overload is administration of iron chelating agents. Desferrioxamine (DFO) is the only current clinically useful, widely available drug for iron chelation therapy. However, DFO is very costly, may cause toxic side effects, and requires subcutaneous infusion, an inconvenient mode of administration resulting in patient noncompliance.[5,6] There have been considerable efforts to develop new, inexpensive, and orally effective iron chelating drugs with limited success. Although the chelator L1 (deferiprone, 1,2-dimethyl-3-hydroxyprid-4-one) increases urinary iron excretion,[7,8] its use is associated with significant toxicity,[9] which is manifested as severe agranulocytosis, arthritis and gastrointestinal complications.[10-12] Moreover, recent clinical trials revealed that some L1-treated patients actually have an increase in hepatic iron levels, some even developing liver fibrosis.[13,14] A recent review[15] expressed doubt about the usefulness of L1 in the prevention of iron loading in transfused β-thalassemia patients. Because of these concerns regarding L1, the most actively pursued candidate for clinical use, the quest for new iron chelating agents continues.

Pyridoxal isonicotinoyl hydrazone (PIH, Figure 1), produced by the Schiff base condensation of pyridoxal and isonicotinic acid hydrazide (INH) was identified as an effective iron chelator in the late seventies of the last century.[16] PIH and some of its analogues show considerable promise for therapeutic Fe chelation. The iron chelating properties of PIH were discovered serendipitously, during studies of iron metabolism in immature erythroid cells. The first clue that eventually led to the identification of PIH as an iron chelator came from the finding that pyridoxal 5'-phosphate mobilized non-heme ^{59}Fe from mitochondria isolated from reticulocytes incubated with ^{59}Fe-transferrin and INH.[17] Following the observation that pyridoxal together with INH very efficiently mobilized ^{59}Fe from intact reticulocytes loaded with non-heme ^{59}Fe, PIH was synthesized[16] and the iron chelating properties of this compound and some of its analogues were characterized.[18] Subsequently, PIH was found to be effective *in vivo* when given parenterally to mice[18] and orally to rats,[19,20] observations that have been confirmed in numerous laboratories.[21-25] Following the discovery of this compound, several groups investigated the chemistry of PIH and its iron complexes,[26-32] and their biological properties.[33-38] Moreover, numerous PIH analogues were synthesized, some of which have high iron chelation activity both *in vitro* and *in vivo*.[39-43] PIH and some of its analogues were demonstrated to have a strong antioxidant activity related to their chelating properties.[44-49] Furthermore, the PIH analogue, salicylaldehyde isonicotinoyl hydrazone (SIH) has a strong anti-malarial effect.[50,51] In addition, some of the PIH analogues, in particular those derived from 2-hydroxy-1-naphthaldehyde, have strong antiproliferative effects toward cultured cells.[52] Interestingly, erythroid cells[53-56] can efficiently use iron complexed to PIH or some of its analogues, including pyridoxal benzoyl hydrazone (PBH) and SIH, for hemoglobin synthesis (structures are shown in Figure 1). Non-erythroid cells can use FePIH$_2$ and FeSIH$_2$ to support growth.[57-66]

There is an important dilemma that has not always been fully appreciated in proposing iron chelation as a therapeutic strategy for the treatment of iron overload or for the prevention of free-radical-mediated disease processes: it is clear that iron chelation by DFO may be beneficial by virtue of its ability to inhibit the iron catalyzed Haber-Weiss reaction and thus interfere with the production of destructive hydroxyl radicals, however, DFO is also a potent inhibitor of cell proliferation, probably because it sequesters iron required for cellular processes. Pathological conditions whose treatment would benefit most from reduction of iron-catalyzed formation of hydroxyl radicals would also benefit from cell proliferation for tissue repair. An ideal therapeutic chelator to inhibit free-radical-mediated tissue damage would therefore leave iron available for essential cellular metabolic processes. The discovery that PIH complexed with iron supports hemoglobin synthesis[53] and cell proliferation[57-66] makes PIH a candidate antioxidant chelator which has these characteristics.

2. PROPERTIES REQUIRED OF AN EFFECTIVE INTRACELLULAR IRON CHELATOR

In the design of an effective orally available iron chelator, a number of properties must be simultaneously optimized. An effective drug must be well absorbed in the gastrointestinal tract, resulting in high bioavailability, and rapidly distributed to its site of action. It must bind excess iron without affecting the metabolism of other metals, while allowing the uptake of iron required for cellular processes. The iron-chelator complex must be rapidly cleared, to prevent its accumulation in tissues and to minimize the potential reabsorption of its bound iron.

Pyridoxal isonicotinoyl hydrazone (PIH) is a nearly planar[67,68] tridentate ligand, which binds iron *via* its phenolate oxygen, imine nitrogen, and carbonyl oxygen atoms.[68] Analogues synthesized from 2,6-pyridinedicarboxaldehyde and 2-furaldehyde, and therefore lacking the complete iron-binding motif, were inactive at high concentrations in a macrophage model of ^{59}Fe mobilization,[39] indicating the importance of strong affinity for iron in the activity of these compounds. The only biologically active analogues which deviate from the iron-binding motif of PIH are a series of analogs synthesized from 2-pyridylcarboxylic acid,[69] which are also tridentate chelators[70] as effective at ^{59}Fe mobilization as PIH and 2-hydroxy-2-naphthaloyl isonicotinoyl hydrazone (NIH).[69] In aqueous solution, iron-PIH complexes have 1:2 stoichiometry,[16] the structure of which was determined by X-ray crystallography.[67] Its affinity for ferric ion[71] is similar to that of transferrin,[72] though, like many other iron chelators including DFO, it does not remove iron from transferrin at plasma pH.[18] PIH also binds ferrous ion,[71] although the complex is rapidly oxidized to the ferric form in the presence of oxygen.[67]

PIH has four ionizable protons, the *pKa* values of which are 3.0, 4.4, 7.9, and 10.2.[73] At pH 7.4, it exists mainly in its neutrally charged form, with a small fraction in its monoanionic species.[73] Since the fraction of neutral species, which may be expected to penetrate tissues more readily than its charged forms, is highest at pH 6.0,[73] PIH may be absorbed in the small intestine.[74] Its molecular weight of 287 is likely small enough to allow its absorption by the pericellular route[75] as well as the transcellular route.[76]

Synthesis of PIH is achieved by the equilibrium condensation of pyridoxal and INH in ethanol,[77] conditions which favor formation of the hydrazone. Hydrolysis of PIH to pyridoxal and INH, which may be acid- or base-catalyzed, is very slow in buffers of neutral pH.[28] However, at pH 1, hydrolysis occurs with a half-time of approximately 25 min,[28] indicating that effective oral administration of PIH may require measures to ensure that the dose reaches the site of gastrointestinal uptake intact. Such measures may include administering the drug on an empty stomach, or in an enteric-coated formulation. Degradation of PIH, before its absorption may be a significant limiting factor in trials of its chronic efficacy in mobilizing iron in humans,[78] in which PIH was not as effective in mobilizing iron as when PIH was administered to fasting animals.[20,41]

The specificity of PIH for iron may account for its relative non-toxicity *in vivo* and *in vitro*. Its affinity for Ca^{2+} and Mg^{2+} under physiological conditions is negligible,[29] as has also been shown for the interaction of salicylaldehyde benzoyl hydrazone (SBH, Figure 1), an analogue of PIH, with calcium.[79] Although PIH has been shown to form complexes with copper,[80] administration of 250 mg/kg PIH to rats had no effect on the excretion of either Cu^{2+} or Zn^{2+}.[20]

Since PIH is believed to enter cells *via* passive diffusion,[34] its lipophilicity, as well as that of its iron complex, may be a major determinant of its effectiveness. Its ready access to intracellular iron pools may, in part, explain its higher capacity than DFO to mobilize iron in many studies.[18,20,52] The trends in iron mobilization *in vitro* by PIH analogues vary with their lipophilicity,[81,82] rather than their expected affinity for iron, suggesting that access of the chelators to intracellular compartments and release of the iron-chelator complexes, as opposed to the strength of their iron complexes, limit their effectiveness. The lipophilicity of PIH analogues may also influence their bioavailability, since this property affects the rate of absorption through the gastrointestinal epithelium.[83,84] The fecal excretion route of the iron-PIH complex is also likely due to its lipophilicity, as large, lipophilic species tend to be excreted *via* the bile.[85]

3. IRON MOBILIZATION BY PIH AND ITS ANALOGUES

3.1. *In Vitro* Studies

As already mentioned, PIH was discovered using an *in vitro* model in which reticulocytes were labeled with a pulse of ^{59}Fe in plasma (*i.e.*, bound to the plasma iron transport protein, transferrin). In the presence of INH or succinylacetone, which inhibit porphyrin synthesis, ^{59}Fe accumulates in the mitochondria[86,87] in a metabolically available form; if the cells are washed of INH[88] or succinylacetone,[86] this ^{59}Fe is incorporated into heme-containing proteins. In this model, it was found that incubation of labeled cells with INH and pyridoxal or pyridoxal phosphate caused a release of ^{59}Fe into the extracellular medium.[16] The condensation of pyridoxal and INH to form PIH was observed spectrophotometrically.[16] The Fe^{3+} complex of PIH was found, by titration, to have 1:2 stoichiometry.[16] Since incubation of the ^{59}Fe-labeled cells with purified PIH, but not its Fe^{3+} complex, also caused release of ^{59}Fe from the reticulocytes, it was concluded that PIH must bind intracellular iron, and shuttle it through the membrane. While the maximum ^{59}Fe mobilization was achieved by approximately 1 mM PIH, concentrations of PIH as low as 20 µM were effective in mobilizing significant amounts of ^{59}Fe.[18,34]

A study of iron mobilization from reticulocytes by a wide variety of iron chelators provided useful information regarding the properties which are most important for intracellular iron chelation.[35] Many chelators with high affinity for iron, including DFO, ethylenediaminetetraacetic acid (EDTA), and diethylenetriaminepentaacetic acid (DTPA), did not mobilize significant ^{59}Fe in this model, most likely because they do not easily cross cell membranes. A second group of chelators failed to mobilize ^{59}Fe, but was able to inhibit PIH-mediated ^{59}Fe efflux. These compounds, including 2,2'-bipyridine and 1,10-phenanthroline, likely entered intracellular compartments and competed for the ^{59}Fe bound by PIH, but the ^{59}Fe complexes of these chelators were membrane-impermeable, preventing the release of ^{59}Fe into the extracellular environment.[35] These observations have identified three properties necessary for mobilization of intracellular iron; membrane permeability of the chelator, affinity for iron above that of the cellular ligands, and membrane permeability of the iron-chelator complex. It is noteworthy that, within cells, chelators specific for Fe^{2+} can compete for iron bound to PIH, which is presumably in the Fe^{3+} form due to the much stronger affinity of PIH for Fe^{3+} over Fe^{2+}.[71] Apparently, intracellular conditions are such that Fe^{2+}-bipyridine and Fe^{3+}-PIH can exist simultaneously, which does not occur in aqueous solution in the absence of a suitable reducing agent, such as ascorbate.[35]

[14]C-PIH was shown to rapidly enter reticulocytes in a manner affected by neither inhibitors and uncouplers of the electron transport chain nor agents which disturb the cytoskeleton, suggesting a passive mechanism of entry.[34] It is unlikely that diffusion of PIH into cells is rate-limiting for mobilizing intracellular ^{59}Fe, since ^{59}Fe, presumably bound to PIH, accumulates in the cytosol during short incubation periods.[33,34] Furthermore, the kinetics of ^{59}Fe mobilization from reticulocytes were identical to those of [59]FePIH$_2$ efflux.[34] Together, these data indicate that release of FePIH$_2$ is the slowest step in the process, which, in the presence of 1 mM PIH, is complete in approximately one hour.[18,34]

[59]Fe mobilization by PIH was decreased by inhibitors and uncouplers of the electron transport chain,[34] suggesting that the release of the FePIH$_2$ complex is ATP-dependent. Vincristine and vinblastine, which interfere with microtubules, also decreased PIH-mediated ^{59}Fe mobilization[34] These observations were confirmed in a subsequent study in SK-N-MC neuroblastoma cells,[89] which tentatively identified P-glycoprotein as the mechanism by which FePIH$_2$ is transported out of the cell.

From the conditional stability constants of chelators, pM values, which are the negative log of the free iron concentration, can be calculated. In the presence of 1 mM chelator and 1 μM Fe^{3+}, the pM values of PIH[71] and transferrin,[90] respectively, are 27.7 and 25.6, indicating that PIH is a stronger chelator of Fe^{3+} than transferrin. However, PIH analogues do not directly remove iron from transferrin,[18] nor does DFO,[91] which has a pM of 28.6.[92] Therefore, direct competition with transferrin for iron is not likely the mechanism by which PIH acquires iron, either *in vivo* or *in vitro*. PIH analogues do, however, inhibit the uptake of iron from transferrin in hepatocytes[40] and neuroblastoma cells,[52] without affecting the endosomal cycling of ^{125}I-transferrin[40,93] The most likely mechanism by which PIH prevents transferrin-mediated iron uptake is by intercepting iron in the acidified endosome before it is transported through the membrane by DMT1, either before or after its reduction to Fe^{2+}. Supporting this hypothesis is the observation that PIH causes iron release from isolated endosomes.[94] Although interference of PIH analogues with iron uptake has been well documented *in vitro*,[40,52] daily i.p. administration of 200 mg/kg PIH to rats over six days had no effect on the incorporation of a pulse of ^{59}Fe to erythrocytes.[18] Because erythroid cells are the most avid consumers of iron in the body, it is likely that, if delivery of sufficient iron to these cells is possible in the presence of PIH, other tissues also receive an adequate iron supply.

Alternative *in vitro* models of iron mobilization include cultured cells such as neuroblastoma cells,[52] hepatocytes,[39,40] and macrophages.[39] In contrast to reticulocytes, which direct nearly all the iron they take up into heme, these cells store the majority of their iron in ferritin. Although the endosomal pathway of iron uptake from transferrin is common to all these cell types, it may be expected that their intracellular iron handling may differ significantly. Thus, iron mobilization studies have used a variety of cell types with the goals of identifying the most effective analogues, and of understanding the iron pools with which they interact.

A study of ^{59}Fe mobilization in fetal rat hepatocytes demonstrated the superior capacity of PBH and 3-hydroxyisonicotinaldehyde isonicotinoyl hydrazone (IIH) over PIH at inhibiting uptake of ^{59}Fe from transferrin in a manner which did not affect the uptake of transferrin itself,[36,40] indicating that endocytosis was unaffected. The chelators did not remove significant amounts of ^{59}Fe from transferrin directly.[18,36] Likely, they bound ^{59}Fe when it was released from transferrin under the acidic conditions of the endosome, causing its release in a chelator-bound form from cells during exocytosis and "futile cycling" of transferrin. ^{59}Fe mobilization by 100 µM PIH analogues from ^{59}Fe-transferrin-labeled hepatocytes for 24 h decreased in the order IIH > PBH > DFO > PIH > SIH > SBH,[36] indicating that in this cell culture model, the capacity of PIH analogues to inhibit ^{59}Fe uptake from transferrin correlated with mobilization of ^{59}Fe from labeled cells. A kinetic study demonstrated that maximal ^{59}Fe release in this system occurred within 1 h of the addition of PBH,[36] which is similar to that observed for PIH.[35]

The major conclusion from a comprehensive screen of 45 PIH analogues, synthesized by condensation of a series of 15 hydrazides with the aldehydes pyridoxal, salicylaldehyde, and 2-hydroxy-1-naphthaldehyde[95] (structures of selected compounds are shown in Figure 1), in murine peritoneal macrophages, rabbit reticulocytes, and fetal rat hepatocytes was that, regardless of differences in the iron metabolism of these cell types and differences in incubation times and chelator concentrations, a number of analogues were highly effective in all three models,[39] namely PBH, pyridoxal *p*-methoxybenzoyl hydrazone (PMBH), pyridoxal *m*-fluorobenzoyl hydrazone (PFBH), and pyridoxal 2-thiophenecarboxylic acid hydrazone (PTCH),[39,40] the structures of which are shown in Figure 2. In general, analogues synthesized from pyridoxal were more effective than those synthesized from salicylaldehyde or 2-hydroxy-1-naphthaldehyde, an effect which was most pronounced in the hepatocyte model, in which saturated solutions of the chelators were used.[39] These results indicate the solubility of these relatively non-polar analogues as a limiting factor of the effectiveness of these analogues.[39,40]

A further limitation of salicylaldehyde- and 2-hydroxy-1-naphthaldehyde-based analogues is that treatment of ^{59}Fe-labeled hepatocytes with many of these analogues caused accumulation of cytosolic ^{59}Fe, presumably as ^{59}Fe-chelator complexes, indicating a barrier to the release of these complexes from the cells,[39] an observation which was also made ^{59}Fe-labeled SK-N-MB cells treated with NIH.[96] Further evidence that many of these analogues perturbed cellular iron metabolism, although they mobilized little ^{59}Fe, may be found in the observation that, following a 16 h incubation with chelators, the hepatocytes bound significantly more ^{125}I-labeled transferrin.[39] Thus, chelator treatment increased expression of the transferrin receptor, presumably *via* activation of iron regulatory proteins (IRPs) resulting from iron depletion.[97] Incubation of BE-52 cells for 20 h with NIH also caused an increase in expression of the transferrin receptor on the cell surface, indicating that this chelator depletes intracellular iron pools sufficiently to cause upregulation of iron uptake, presumably by an effect on IRP activity.[98] NIH and three novel PIH analogues based on 2-pyridylcarboxaldehyde activated IRPs in SK-N-MC cells, while PIH caused a weaker activation.[69] Interestingly, analogues which were unable to mobilize ^{59}Fe decreased IRP

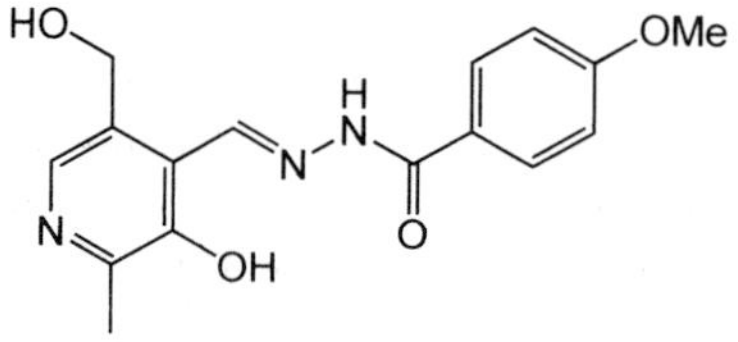

pyridoxal p-methoxybenzoyl hydrazone
(PMBH, 107)

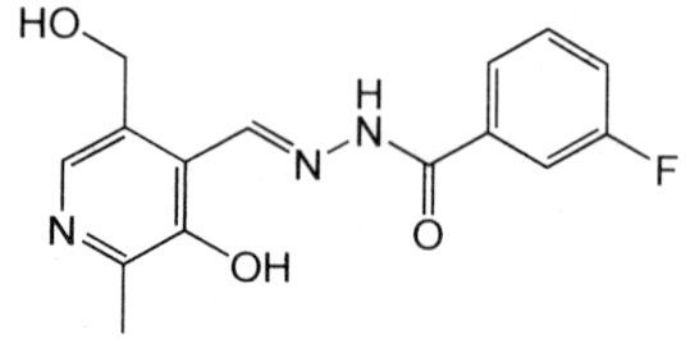

pyridoxal m-fluorobenzoyl hydrazone
(PFBH, 109)

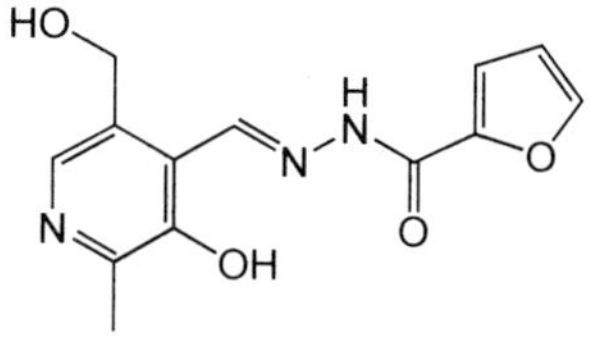

pyridoxal 2-furoyl hydrazone
(PFH, 114)

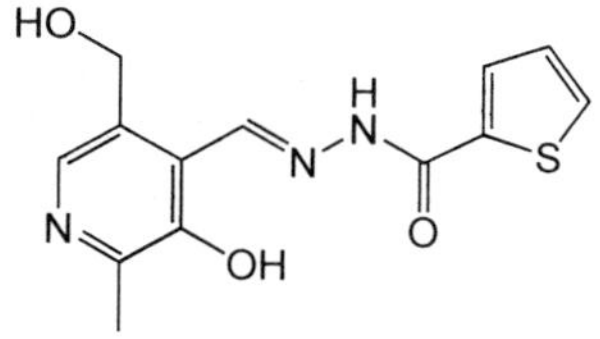

pyridoxal 2-thiophenecarboxyl hydrazone
(PCTH, 115)

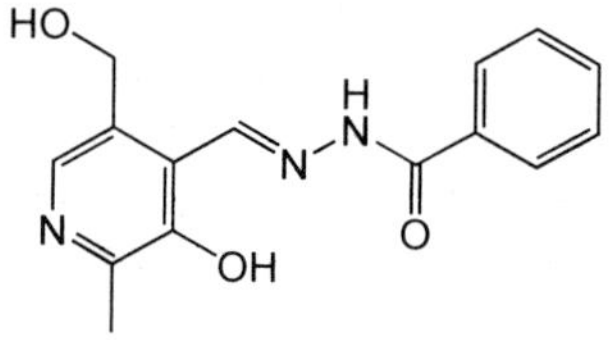

pyridoxal benzoyl hydrazone
(PBH, 101)

Figure 2. Structures of the PIH analogs which are most effective at ^{59}Fe mobilization *in vitro*. Numbers correspond to the identification scheme used in some studies. [39,41,52]

activity, as did ferric ammonium citrate,[69] indicating that they perturb intracellular iron pools without causing iron efflux from the cells. Although NIH increased IRP activity in SK-N-MC and BE-52 cells, it was unable to do so in cell-free extracts,[98] suggesting that the mechanism by which these chelators cause IRP activation is not a direct removal of iron from the Fe-S cluster of IRP1. Accumulation of cytosolic ^{59}Fe-chelator complex indicates the ability of these analogues to enter cells and bind iron, which may be expected to deplete the pool(s) of iron which regulate IRP activity.[1-3,99]

The chelators identified as the most effective at mobilizing cellular ^{59}Fe (PBH, PMBH, PFBH, and PFH[39]), were further investigated in SK-N-MC neuroblastoma cells, in which these analogues were shown to inhibit uptake of ^{59}Fe-transferrin by approximately 90%, in a 24 h incubation, as did PIH.[100] Although both inhibition of ^{59}Fe uptake and ^{59}Fe mobilization by the chelators was similar, 24 h treatment of cells with 100 μM PFBH reduced cell proliferation to 5% of controls, compared to 40% by PIH.[100] For all chelators, cell growth was partially or completely restored by the addition of transferrin to the medium, indicating that the major mechanism of the antiproliferative effects of these compounds involved iron depletion.

^{59}Fe mobilization was, in general, inversely correlated with lipophilicity in a screen of PIH analogues in SK-N-MC cells, which demonstrated that ^{59}Fe mobilization was most efficient by analogues synthesized from pyridoxal, and worst by analogues synthesized from 2-hydroxy-1-naphthaldehyde.[52] This conclusion has been supported by a number of subsequent studies, which demonstrated that ^{59}Fe mobilization from cells is most efficient among analogues which have log octanol/water partition coefficients (log P) predicting both water solubility and membrane permeability.[81,82,101] The optimum range of log P values of PIH analogues includes the compounds most effective at mobilizing intracellular ^{59}Fe, such as PBH, PMBH and PFBH. The explanation of this relationship may lie in the lipophilicity of the iron-chelator complexes, a property linearly related to the lipophilicity of the free chelators.[81] Inability of the complex to be released efficiently from cells may result from a highly polar iron complex, as is the case for DFO,[93] or from a highly lipophilic complex, as was observed for cholylhydroxamic acid.[102] The detection of cytosolic ^{59}Fe complexes of salicylaldehyde- and 2-hydroxy-1-naphthaldehyde-based chelators,[39] which are the most lipophilic of the PIH analogues, is consistent with the hypothesis that highly lipophilic complexes do not easily cross the cell membrane.

3.2. *In vivo* Studies

Daily intraperitoneal administration of 200 mg/kg PIH to mice for six days decreased the accumulation of a pulse of ^{59}Fe, injected as ^{59}Fe-transferrin, in the liver, spleen, and kidney[18] Excretion of this ^{59}Fe was nearly exclusively fecal, and was two- to five-fold higher than in control animals. However, this treatment had no effect on ^{59}Fe levels in the blood, indicating that developing erythrocytes were not iron-deprived in the presence of PIH.

To identify the pool(s) of iron which are available for mobilization by PIH, ^{59}Fe was administered in forms which are taken up by different tissues. Rats were injected with ^{59}Fe-transferrin, which is destined for hepatic parenchymal cells.[103] After 48-72 h, PIH was administered in two doses of 125 mg/kg each, 12 h apart, which resulted in a total ^{59}Fe biliary excretion of approximately 2%, corresponding to a 35-fold increase as compared to control animals[20] (Figure 3). Iron excretion as measured by radioactivity correlated well with total iron excretion, as measured by atomic absorption spectrophotometry.[20] When the same dose of PIH was administered immediately prior to the pulse of ^{59}Fe-transferrin, 7% of the ^{59}Fe was excreted in the bile, and 1.4% in the urine after 24 h.[20] The presence of PIH at the time of ^{59}Fe-transferrin injection probably allowed more efficient interception of ^{59}Fe in the endosomes as it was released from transferrin. This hypothesis is supported by the observation of strong inhibition of ^{59}Fe uptake from transferrin by PIH and some of its analogues.[52]

Administration of two doses of 250 mg/kg PIH, 1 and 13 h after injection of rats with ^{59}Fe-labeled heat-damaged erythrocytes, which are phagocytosed by the reticuloendothelial system, caused excretion of 6.5% of the total ^{59}Fe.[20] The vast majority of the excretion occurred after the first dose; at this time, most of the ^{59}Fe is in a relatively labile form.[104] The

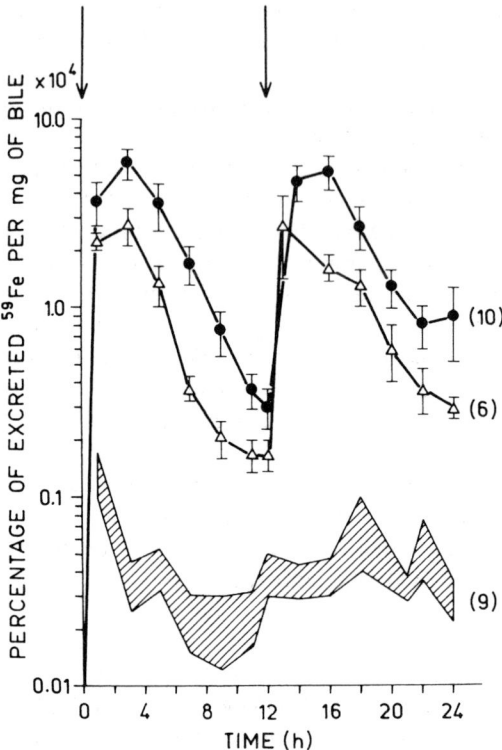

Figure 3. Biliary excretion of ^{59}Fe during 24 h in control (shaded area, indicating 95% confidence limits for means) and in PIH- (circles) or DFO-treated (triangles) rats. ^{59}Fe-transferrin was injected before cannulation of the bile duct. Arrows indicate the time of PIH or DFO injection at 250 mg/kg. Figures in parentheses indicate the number of rats in each group. Reproduced from reference[20] with permission.

second dose of PIH, administered at a time when most of the ^{59}Fe is in reticuloendothelial ferritin,[104] caused a much lower excretion of ^{59}Fe. This suggests that iron in ferritin is relatively inaccessible to PIH *in vivo*, which appears to be the case *in vitro*.[96]

PIH-induced mobilization of ^{59}Fe in rats given ^{59}Fe-labeled, heat-damaged erythrocytes and ^{59}Fe-ferritin yielded similar levels of fecal ^{59}Fe excretion, approximately threefold higher than control animals.[21] Since these forms of iron label reticuloendothelial[21,104] and parenchymal[21] iron, respectively, these results indicate that PIH easily accesses both iron pools.

Regardless of the organismal iron pools labeled, excretion of iron, whether measured by radioisotopic labeling or atomic absorption spectrophotometry, was nearly exclusively fecal.[18,20,41] When both methods of iron determination were used, radioactivity results corresponded with total excreted iron,[20,23] indicating that mobilization of ^{59}Fe-labeled tissue iron stores represents total iron mobilization. Biliary excretion of ^{59}Fe was a major route of elimination in these studies, although excretion through the gastrointestinal wall also contributed to the total fecal iron content.[20] Orally administered ^{59}FePIH$_2$ was not absorbed,[105,106] suggesting that, once the complex reaches the gut, it is not reabsorbed.

Effective iron mobilization by PIH was observed in rats overloaded with dietary iron, in which 25-100 mg/kg PIH increased fecal ^{59}Fe excretion eightfold.[19] A study of normal and hypertransfused rats demonstrated that the redirection of ^{59}Fe from the erythron to the liver, caused by suppressed erythropoiesis, increased the capacity of PIH to mobilize ^{59}Fe from 33% to 66% of the injected ^{59}Fe-transferrin.[21] This effect was not observed in the case of DFO, which bound approximately 85% of the ^{59}Fe in both normal and hypertransfused rats.[21] It is noteworthy that, in iron-overloaded animals, PIH is more effective than in normal animals, which suggests that this chelator preferentially binds iron in pools which are not required for immediate use by tissues.

In a Phase I clinical trial, normal and iron-overloaded patients receiving daily doses of PIH at 30 mg/kg/day for two weeks showed no signs of toxicity.[78] A subsequent randomized, double-blind study examined iron-overloaded patients receiving 30 mg/kg PIH, co-administered with calcium carbonate to reduce the potential of acid-catalyzed hydrolysis in the stomach, three times a day for three weeks. The mean iron excretion, 0.12 mg/kg/day, was sufficient to achieve negative iron balance in non-transfused patients, but insufficient to effectively chelate transfused patients. It may be that absorption of PIH, which was not administered in a soluble form, limited its bioavailability. Furthermore, the dose of PIH in this study was far lower than in the abovementioned animal studies, which, in addition to incomplete drug absorption, may account for the superior performance of PIH in studies with fasting animals. Assessment of the clinical potential of higher doses of PIH, in a formulation designed to maximize its bioavailability, is warranted.

As suggested by the clinical trial, oral administration of PIH may require optimization. A long-term trial in rats, in which PIH was mixed with food pellets, resulted in no ^{59}Fe mobilization as compared to control animals.[24] In contrast, another study of similar design decreased ^{59}Fe levels in the liver and spleen, and caused ^{59}Fe excretion twice that of controls.[107] Since ^{59}FePIH$_2$ was not taken up in the gastrointestinal tract,[105,106] it is unlikely that the reduced activity of PIH in long-term trials is due to increased absorption of dietary iron. It is possible, however, that complexation of iron in the gastrointestinal tract may prevent its absorption, thereby preventing PIH from reaching its sites of action. Indeed, an interaction of PIH which reduces uptake of dietary iron approximately twofold has been demonstrated in humans.[108] To prevent interaction with dietary iron and possible acid-catalyzed hydrolysis, further studies of its bioavailability is warranted.

The kinetics of ^{59}Fe excretion after administration of 250 mg/kg PIH to rats were examined.[20] The level of ^{59}Fe in the bile was maximal almost immediately after PIH was given, and coincided with the appearance of a red-brown substance, presumably FePIH$_2$, in the bile.[20] Biliary ^{59}Fe was cleared with first-order kinetics, the half-time of which was approximately three hours (Figure 3). This rapid clearance of ^{59}FePIH$_2$ is consistent with minimal enterohepatic cycling of the iron mobilized by PIH.[105,106] Both observations suggest that the rate of excretion of FePIH$_2$ minimizes the potential of redistribution and/or reabsorption of the bound iron. In the kinetic study, a second dose of PIH, administered 12 h later, mobilized an equal amount of ^{59}Fe,[20] suggesting that either the pool of iron targeted by PIH is rapidly replenished, or that the first dose was insufficient to deplete it.

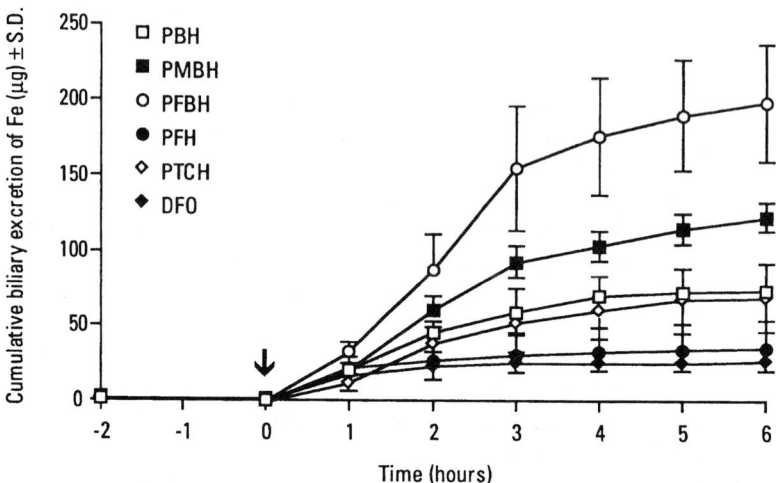

Figure 4. Cumulative excretion of iron in the bile of rats treated with PIH analogs. The bile ducts of normal rats were cannulated and the bile was collected for 2 h, after which the hydrazones were injected i.p. (0.2 mmole/kg). Reproduced from reference[41] with permission.

The superior capacity of the PIH analogue, pyridoxal benzoyl hydrazone (PBH), to mobilize ^{59}Fe was demonstrated in rats.[23] Although administration of i.v. PIH achieved a higher peak concentration of ^{59}Fe in the bile, the half-time of ^{59}Fe mobilization by PBH was longer, resulting in excretion of 58% of the injected ^{59}Fe, as compared to 41% by PIH and 3% in control animals. It is noteworthy that, in this model, the kinetics of ^{59}Fe excretion by PIH were similar to those of DFO[23] (an observation which was also made earlier[20]), which is administered by subcutaneous infusion to overcome difficulties associated with its extremely short plasma half-life.[109] A dose of 50 mg/kg PIH, administered to iron-overloaded rats by i.v. infusion, caused the excretion of 126 μg Fe/h,[106] as compared to 11 μg Fe/h for a single dose of 100 mg/kg administered i.v.,[105] suggesting that the constant presence of PIH dramatically improves its effectiveness.

An *in vivo* study of the PIH analogues which were identified as the most effective in *in vitro* screens[39,40,52] also highlighted the importance of pharmacokinetics in the effectiveness of iron mobilization. The analogues examined (Figure 2), PFBH, PMBH, PBH, PTCH, and PFH, were all more effective than DFO in the rat model of biliary iron excretion, and all but PFH were at least as effective as PIH.[41] In addition to confirming the value of the *in vitro* screens developed to identify effective iron chelators, these results suggest the importance of pharmacokinetics of PIH analogues in their overall effectiveness in mobilizing iron. When the PIH analogues were administered at 0.2 mmole/kg (approximately 60 mg/kg) i.p., their effectiveness in mobilizing iron into the bile decreased in the order PFBH > PMBH > PBH ≈ PTCH ≈ PIH > PFH ≈ DFO (Figure 4), which was also the order of their decreasing half-times of iron mobilization,[41] *i.e.*, the most effective chelators were those which mobilized iron over the longest period of time. Presumably, this is due to a longer plasma half-life of the chelator, but it is possible that the kinetics of iron mobilization are also affected by other factors. It is worth noting that i.p. administration of PFBH mobilized 200 µg iron in six hours, approximately fourfold more than did PIH. When the same doses of the analogues were given orally, their relative effectiveness was altered (PMBH > PFBH > PTCH), but the relationship between the magnitude and the kinetics of iron mobilization was unchanged.[41] Furthermore, oral administration was, overall, less effective than i.p. injections at iron mobilization, an observation reported earlier.[20] It may be that, among the PIH analogues which effectively mobilize iron *in vitro*, the factor limiting effectiveness *in vivo* is pharmacokinetic.

4. IRON COMPLEXES OF PIH AND SIH DELIVER IRON TO CELLS IN A METABOLICALLY AVAILABLE FORM

^{59}FePIH$_2$, at a concentration of 200 µM, is taken up by rabbit reticulocytes at a rate similar to that of ^{59}Fe uptake from a saturating concentration of transferrin.[53] ^{59}Fe acquired from PIH is used for heme biosynthesis.[53] Interestingly, both ^{59}FePBH$_2$[54] and ^{59}FeSIH$_2$[55] provide ^{59}Fe to reticulocytes even more efficiently; significantly more ^{59}Fe-heme is produced, although the percentage of total ^{59}Fe used for heme synthesis is lower, than by incubation of cells with ^{59}Fe-transferrin. A number of studies with non-erythroid cells have demonstrated the ability of FePIH$_2$ to provide iron in a form which supports cell growth.[57-66] It is possible that the ability of PIH to donate iron in a metabolically available form is an important factor in its value *in vivo*.

Two thirds of the ^{59}Fe taken up from ^{59}FePIH$_2$ by reticulocytes was inserted into heme, in comparison to approximately 90% of ^{59}Fe taken up from transferrin, indicating that, although PIH-bound iron delivered to the cell can be used, its intracellular trafficking is not as efficient as the physiological pathway. Furthermore, the level of cytosolic ^{59}Fe, presumably in the form of FePIH$_2$, increases with time,[53] suggesting a saturation of the pathway(s) by which PIH-bound iron reaches its intracellular destinations. In support of this hypothesis, incubation of concanavalin-A-stimulated murine lymphocytes with ^{59}Fe bound to PIH or nitrilotriacetic acid (NTA) showed that, despite the fact that total ^{59}Fe uptake was increased 5- and 25-fold as compared to transferrin, the same total amount of ^{59}Fe was associated with ferritin, transferrin, and other high-molecular-weight species.[110] The remaining ^{59}Fe was found in "insoluble" and "chelatable" forms.[110] Although the fate of iron donated to cells by PIH was not identical to that by transferrin, non-specific iron delivery by FePIH$_2$ was much lower than by FeNTA.[66,110] However, FePIH$_2$ enhanced lymphocyte proliferation, while FeNTA was inhibitory.[110] In CCRF-CEM cells, a lymphoblastoid T-cell line, transcription of protein kinase C-β mRNA is induced by transferrin, while FePIH$_2$ and FeSIH$_2$ had no effect.[111] Hence, it is apparent that cellular handling of these iron donors may be quite different from the physiological pathway of iron acquisition. FePIH$_2$ and FeSIH$_2$, as compared to transferrin, caused a two- and four-fold increase, respectively, in the amount of ^{59}Fe found in ferritin after a 48 h incubation,[111] suggesting that the Fe taken up by the cells during this time exceeded their requirements. Since concentrations of FePIH$_2$ and FeSIH$_2$ in the range of 1-50 µM have been shown to support cell growth in the absence of transferrin,[57-66] it appears that the requirement of cells for iron can be met by these complexes.

5. TOXICITY OF PIH AND ITS ANALOGUES

5.1. *In Vitro* Studies

The *in vitro* toxicity of PIH analogues is complex. It is due in part to depletion of iron which is required for many cellular processes, including DNA synthesis and ATP production. Addition of transferrin to the medium rescues the growth of cells treated with PIH analogues,[100] indicating that an increased supply of iron is sufficient, in this model, to counteract the effects of the chelator. It is noteworthy that transferrin was also able to provide partial protection against the toxicity of PIH analogues, given the strong inhibition of iron uptake *via* transferrin by many PIH analogues.[52] However, analogues synthesized from 2-pyridylcarboxylic acid which mobilized ^{59}Fe from SK-N-MC cells as efficiently as PIH were more toxic than PIH.[69] The toxicity of PIH analogues toward neuroblastoma cells is correlated with neither the efficiency of inhibition of ^{59}Fe uptake from transferrin[52] nor ^{59}Fe mobilization from labeled cells.[52,100] Together, these data suggest a more complex mechanism of toxicity of these chelators than simply the withholding of iron.

Fe chelators are known to inhibit ribonucleotide reductase,[112] the rate-limiting enzyme in the synthesis of DNA, which converts nucleosides into deoxynucleosides. It has been shown that iron chelators act by chelating the labile iron pool from which this enzyme obtains the two Fe^{3+} ions in the active site of the R2 subunit.[113] EPR spectroscopy showed that treatment of CCRF-CEM cells with 0.7 µM NIH for 12 h dramatically reduced the level of active enzyme.[114] A hydroxyurea-resistant cell line developed from CCRF-CEM cells which overexpresses the iron-containing subunit of ribonucleotide reductase was sixfold less sensitive to NIH. Since inhibition of ribonucleotide reductase causes arrest of the cell cycle in the G1/S phase[113,115] which is the point in the cell cycle at which DNA synthesis begins, the effect of iron chelators on the cell cycle is of interest. Consistent with the hypothesis that ribonucleotide reductase is a specific target of iron chelators, many iron chelators cause G1/S arrest.[113,116,117]

In mitogen-stimulated lymphocytes, the antiproliferative effect of a series of PIH analogs was time-dependent; little effect was observed at incubation times less than 8 h, whereas ^{3}H-thymidine incorporation was significantly reduced at 16 h,[118] corresponding to the time at which DNA synthesis is upregulated in response to mitogen stimulation in untreated cells, and was consistent with an effect of the chelators during the late G1 phase of the cell cycle. Treatment of cultured cells with 10 µM NIH for 24 h had variable effects on the cell cycle, depending on the cell type.[96] 10 µM NIH did not cause apoptosis in K562 or IMR32 cells, whereas apoptotic morphology and DNA laddering were detected in SK-N-MC and HL60 cells.[96] Following NIH treatment, K562 and HL60 cells had a lower percentage of cells in the S phase, and a corresponding increase in the G2/M phase, while SK-N-MC and IMR32 cells had a reduced fraction in G2/M phase. Since the effect of NIH on the cell cycle is dependent on the cell type, it is likely that there are several possible cellular targets of NIH, and different cell types may have different sensitivities. DFO, which causes G1/S inhibition in the majority of cell lines tested,[96,113,117] has also been demonstrated to cause G2/M inhibition.[119,120]

Incubation of BE-2, K562, and SK-N-MC cells with NIH enhanced transcription of mRNAs for proteins involved in regulating the cell cycle,[98] including GADD45, which is upregulated in response to DNA damage,[121] MDM-2, which is involved in feedback control of p53 expression,[121,122] and the cdk kinase inhibitor p21, encoded by WAF1. When cells were washed and incubated in fresh medium, WAF1 and GADD45 transcription decreased, although not to baseline levels, but MDM-2 mRNA levels remained elevated, demonstrating that these effects were at least partially reversible.[98] In a more extensive study, it was shown that the large NIH-induced upregulation of WAF1 and GADD45 transcription in SK-N-MC cells resulted in a small or negligible increases in protein levels, indicating the existence of other regulatory factors.[123] Treatment of SK-N-MC cells with NIH for 30 h decreased levels of cyclins A, B1, and D1-3, and increased cyclin E,[123] all of which are involved in progress through the G1 phase of the cell cycle, except B1, which affects progress through the G2 phase. Cdk2 was decreased at NIH concentrations above 1 µM[123] Thus, the NIH-induced G1/S arrest previously observed[96] is likely caused by low and high levels of cyclins D1-D3 and E, respectively.[124] Since cdk2 expression was decreased,[123] formation of the cdk2-E2F complex may be inhibited, causing the observed decrease in hyperphosphorylation of pRB.[124] Another likely consequence of low cdk2 levels is lower cdk2-cyclin A complex formation, which is essential for progression from G1 to S phase. However, whether inhibition of ribonucleotide reductase is the major event which precipitates cell cycle arrest is uncertain, as hydroxyurea, which inhibits ribonucleotide reductase by a mechanism unrelated to iron chelation, had the opposite effects on the levels of cyclins A, B1, and D1-D3. This indicates a significant difference between simple ribonucleotide reductase inhibition and the effects of NIH on cells.[123] Therefore it is unlikely that the only target of NIH is ribonucleotide reductase.

The Fe^{3+} complexes of PIH and SIH are not inert. They have been shown to act as iron donors which are capable of supporting growth in cultured cells[57-66] (see Section 4). Thus, cells are able to remove Fe from the complex, despite the high affinity of PIH analogues for Fe^{3+}.[71] It is also possible that cells acquire iron from PIH complexes by reducing the Fe^{3+} to Fe^{2+}. However, $FePIH_2$ is resistant to reduction in aqueous solutions.[35]

In an effort to determine the structure-activity relationships describing the toxicity of salicylaldehyde benzoyl hydrazone (SBH) analogues, Molt-4 human leukemia cells were used.[125] An inverted parabolic relationship between lipophilicity and toxicity was observed, suggesting that membrane permeability was the property which determined the antiproliferative effects of these compounds for both the free ligands and their Cu^{2+} complexes.[125] It was concluded that chelators and complexes which were either very polar or very lipophilic were unable to cross the cell membrane, thereby limiting their toxicity. In support of this conclusion, a screen of structurally unrelated iron chelators demonstrated that, in general, moderately lipophilic chelators inhibited ^{3}H-thymidine incorporation in K562 and U937 cells, whereas hydrophilic chelators did not.[126]

The toxicity toward SK-N-MC neuroblastoma cells of PIH analogues synthesized from different aldehydes increased in the order pyridoxal < salicylaldehyde < 2-hydroxy-1-naphthaldehyde, which corresponded to their increasing lipophilicity, with PIH among the least toxic.[52] Further study of this trend identified an inverted parabolic relationship between ^{59}Fe mobilization and lipophilicity.[81,82,101,127] Interestingly, the most toxic of the PIH analogues are found among the most lipophilic, which do not effectively mobilize ^{59}Fe. Many of these highly toxic analogues, which are synthesized from salicylaldehyde and 2-hydroxy-1-

naphthaldehyde, caused the accumulation of ^{59}Fe, presumably as Fe-chelator complexes, in the cytosol of hepatocytes,[39] indicating that they readily enter the cell and bind intracellular iron, but are unable to transport it across the cell membrane.

Toxicity of Fe^{3+} complexes of PIH analogues has also been described. Although the growth of U937 cells, in which iron uptake from transferrin was inhibited, was restored at low concentrations of FePIH, higher concentrations caused approximately 20% growth inhibition after 48 h, indicating a balance between beneficial and detrimental effects towards these cells.[61] A similar dependence on $FePIH_2$ concentration of concanavalin-A-stimulated mouse lymphocyte growth under transferrin-free conditions was observed. Cell growth was enhanced up to 50 µM, above which toxicity was observed; at 100 µM, cell growth was undetectable.[65] $FePIH_2$ also supported the growth of WB-F344 hepatic epithelial cells in transferrin-free medium, but concentrations above 10 µM were antiproliferative after 5 days of culture.[64] It seems that delivery of small amounts of iron *via* PIH analogues may be effectively handled by the cell, whereas higher concentrations may donate iron to cells in excess of their capacity to use or store it, thereby causing toxicity.

Generally, the toxicity of a variety of iron chelators toward K562 and U937 cells is diminished by the addition of iron to the incubation medium,[126] an observation which also holds true for PIH analogues.[52] Equimolar amounts of Fe^{3+}, when added within 8 h of a series of PIH analogues, decreased their observed antiproliferative effects toward mitogen-stimulated murine lymph node cells.[118] Likely, Fe^{3+} complexes of the chelators were formed extracellularly, which may be less active, or have lower intrinsic activity. The toxicity of PIH, NIH, and $FePIH_1$ was time- and concentration-dependent in SK-N-MC cells, while no antiproliferative effects of $FeNIH_1$ were observed up to a concentration of 50 µM.[128] One possible reason why toxicity was not observed in this study may be the stoichiometry of the complex, which was 1:1 instead of the 1:2 complex usually used. It is uncertain whether the lower cytotoxicity of iron complexes than of the free chelators is the result of restoration of lost iron to the cells, formation of iron complexes (either extracellularly or intracellularly) which have lower antiproliferative activity than the free chelators, or the relative membrane impermeability of the resulting iron complexes.

In addition to the formation of complexes with iron, PIH analogues have been shown to interact with other metals. The toxicity of gallium, which has significant antitumor activity due to its interference with iron metabolism, is enhanced by non-toxic concentrations of PIH.[52] Ga^{3+} and PIH form a complex analogous to that of Fe^{3+},[52] which may be expected given the similarity of these metal ions in terms of size, charge, and hardness. It is likely that the $GaPIH_2$ complex enters cells readily, as does $FePIH_2$, and that enhanced delivery of gallium to cells is the mechanism by which PIH affects gallium toxicity. Presumably, formation of the $GaPIH_2$ complex enhances the transport of gallium across the cell membrane, thereby increasing intracellular concentrations of this ion, which has been shown to act, at least in part, by interfering with cellular Fe handling.[129-131] Either transferrin or $FePIH_2$ restores the growth of Friend erythroleukemia cells treated with Ga-transferrin.[132]

5.2. In Vivo Studies

Studies of iron mobilization in rats[21,23-25] and humans[78,133] reported no significant adverse effects of PIH administration at doses as high as 200 mg/kg. A study of daily oral administration of 100 mg/kg PIH to rats for 10 weeks reported no toxicity.[24] A study of novel PIH analogs in rats, some of which were nearly as active at mobilizing ^{59}Fe as PIH, demonstrated that single doses of 10 mg[25] and 40 mg were not toxic.[21,25] PIH, at doses of 50-150 mg/kg i.v., caused circulatory collapse and death in neonatal piglets, while it was well tolerated at 10 mg/kg,[45] suggesting that sensitivity to the effects of PIH may be species-specific. The acute oral LD_{50} values for PIH were found to be approximately 5000 mg/kg for both rats and mice,[134] whereas the i.p. LD_{50} values were approximately 1000 mg/kg, well above concentrations which effectively mobilize iron. Subchronic toxicity in rats, including increased serum alkaline phosphatase and aspartate aminotransferase, and hepatocellular vacuolization and degenerated nuclei, was observed at 400 mg/kg/day for 90 days, but not at lower doses.[134] In contrast to the relative non-toxicity of PIH, the acute LD_{50} of i.v. SBH, in contrast, was approximately 75 mg/kg in rats.[23] SBH was therefore administered at 50 mg/kg in this study, and was effective at mobilizing ^{59}Fe administered as ^{59}Fe-ferritin.[23] SBH was not toxic to mice at doses of 400 mg/kg over five days, nor was it active against the growth of P388 lymphocytic leukemia, although its Cu^{2+} complex was toxic at 30 mg/kg.[135]

Given the potential for hydrolysis of PIH analogs,[28,136] either in the acidic environment of the gastrointestinal tract or elsewhere in the body, the toxicity of its hydrolysis products must also be considered. Pyridoxal, a form of vitamin B_6, causes no significant adverse effects below 2000 mg/day in humans,[137] for which the molar equivalent dose of PIH is approximately 40 mg/kg. INH, an antitubercular drug, is usually given at 5-10 mg/kg daily, equivalent to 16 mg/kg PIH. As the extent of hydrolysis of PIH *in vivo* is unknown, it is impossible to determine the exposure of tissues to pyridoxal and INH.

The antimalarial effects of DFO, presumably due to its sequestration of iron from the intracellular parasite, have been shown to depend on its uptake.[138] Since SIH readily enters cells to mobilize iron, its effects against malaria were studied. Doses of 35-65 mg/kg i.p. SIH daily, administered two days after infection with *Plasmodium vinckei petteri*, was well tolerated in mice, and was more effective than DFO in preventing delaying the onset of parasitemia, resulting in lower mortality.[51] PIH, SIH, and 2-hydroxy-1-naphthaloyl *m*-fluorobenzoyl hydrazone were also found active *in vitro* against *Plasmodium falciparum*, and SIH was shown to enhance the antimalarial effect of DFO [50].

6. ANTIOXIDANT ACTIVITY OF PIH AND ITS ANALOGUES

The potential therapeutic value of iron chelators in the treatment of diseases and conditions involving oxidative stress is the subject of considerable attention,[139-141] and the role of iron in ischemia/reperfusion injury,[142,143] chronic inflammatory conditions,[144] cancer,[145] and Parkinson's disease[146] has been clearly documented. The catalytic production of hydroxyl radical by redox cycling of iron in the presence of hydrogen peroxide has been well described.[139-141] Inhibition of redox cycling, a two-step process involving acquisition of an electron by Fe^{3+} to form Fe^{2+}, and transfer of the electron to a substrate to regenerate Fe^{3+}, is thought to be the main mechanism by which iron chelators decrease oxidative damage to lipids, nucleic acids, and proteins. The ready access of PIH analogues to the intracellular environment (see sections 3-5) and their high affinity for iron[71] has prompted many studies of their value as antioxidants.

PIH, SIH, and DFO were equally effective inhibitors of FeEDTA-mediated 2-keto-4-methiobutyric acid degradation and ascorbate oxidation *in vitro*,[147] at concentrations sufficiently low to support the conclusion that they act by inhibiting the catalytic activity of iron, rather than by scavenging the free radicals produced by FeEDTA. EPR spin-trapping studies demonstrated the capacity of PIH to inhibit hydroxyl radical production by Fe^{2+} and hydrogen peroxide in a concentration-dependent manner consistent with the $FePIH_2$ complex having a much poorer capacity to redox cycle than uncomplexed iron.[47] Importantly, PIH was shown to accelerate the oxidation of Fe^{2+} more than 20-fold.[46,47,148] This is likely the result of the higher affinity of PIH for Fe^{3+} over Fe^{2+},[71] which promotes the formation of the $Fe^{3+}(PIH)_2$ complex to such an extent that the Fe^{2+} complex was not detected in solution[35] or isolated in preparation of Fe-PIH for crystallographic study.[67] This preference of PIH for Fe^{3+} enhances the oxidation of $Fe^{2+}(PIH)_2$,[67] and may decrease the rate of its reduction, thereby limiting the overall rate of redox cycling and, therefore, the capacity of $FePIH_2$ to cause oxidative damage.

In vitro studies demonstrated the inhibition by PIH of oxidative damage to biological systems. Fe^{2+}-mediated oxygen consumption by isolated rat liver mitochondria in which respiration was blocked, presumably due to lipid peroxidation, was significantly inhibited by PIH.[148] Mitochondrial swelling and loss of membrane potential were concomitantly decreased, indicating a protective effect of PIH in this model.[148] Micromolar concentrations of PIH also protected plasmid DNA from damage caused by Fe^{2+} and hydrogen peroxide.[46]

Intracellular iron levels were shown to affect the redox status of MEL cells.[149] Transcription of glutathione peroxidase was decreased by treatment with DFO and PIH, and increased by $FePIH_2$ and t-butyl hydroperoxide,[149] indicating regulation of the oxidative stress response by the iron status of the cell. The PIH-mediated decrease in synthesis of an important antioxidant enzyme is likely due to its sequestration of the dynamic iron pools in the cell. It is noteworthy that $FePIH_2$ induced the synthesis of glutathione peroxidase, given the low rate of redox cycling observed in solution studies.[46,47,147,148] It is unknown whether $FePIH_2$ *per se* caused the induced transcription, or whether the iron dissociated from PIH after entering the cells.

The antioxidant activity of SIH was demonstrated in spontaneously beating guinea-pig cardiomyocytes.[49] Challenge with hydrogen peroxide caused increased intracellular calcium levels and contractile irregularities within 30 min.[49] Pre-incubation of the cells with 200 μM SIH fully prevented these events.[49] These results demonstrated the importance of membrane permeability in the prevention of oxidative tissue damage, since DFO, which does not easily cross cell membranes,[150] was only effective at concentrations of 1 mM, and required pre-incubation for 2 h.[49]

The PIH analogue, pyridoxal *m*-fluorobenzoyl hydrazone, prevented lipid peroxidation and hemolysis in phenylhydrazine-challenged, glutathione-depleted erythrocytes *in vitro*.[151] Since the iron complex of the chelator had no effect, it was concluded that iron sequestration was central to the protective mechanism.

Consistent with its ability to cause effective iron excretion *in vivo* (see Section 5.2), PIH was effective at preventing iron-mediated oxidative stress in a neonatal piglet model, in which ventilation of the piglet was interrupted for 5 min.[45] PIH, administered i.v. at a dose of 10 mg/kg, prevented hydroperoxide accumulation and functional impairment of the retina, while the same dose of DFO only prevented the former.[45] This evidence of the potential value of PIH as an antioxidant *in vivo* suggests that its evaluation as a therapeutic agent in the treatment of diseases and conditions involving oxidative stress is warranted.

7. CONCLUSIONS

PIH and its analogues show high iron chelating efficacy, have a high affinity and selectivity for Fe^{3+}, and can be absorbed from the gut. However, to achieve efficient absorption, the ligands should be solubilized before their administration. As compared to DFO, most of the chelators in the PIH class cross cell membranes easily and some, but not all PIH analogues, are efficiently released from cells when complexed to iron. Importantly, PIH and some of its analogues were shown to be strong inhibitors of the production of toxic oxygen free radicals, and this property seems to result from the capacity of these ligands to tightly bind and oxidize iron. Preliminary studies in both animals and humans revealed that PIH exhibits relatively low toxicity. The aforementioned facts make PIH and analogues potentially clinically useful iron chelators. Unfortunately, these promising agents are not proprietary (patentable) and, therefore, their attractiveness to the drug industry is limited. In spite of this drawback, PIH and its available as well as newly synthesized analogues, deserve further vigorous investigation. They are promising candidates in the quest for pharmacological agents to treat iron overload and, possibly, to prevent pathological conditions caused by oxidative stress.

8. ACKNOWLEDGMENTS

The financial assistance of the CIHR (P.P.) and the Cooley's Anemia Foundation (J.L.B.) is gratefully acknowledged. The authors are also grateful to George K. B. Lopes for his assistance in preparing the manuscript, and to Alexander Sheftel and Natasha Szuber for editorial assistance. P. P. is indebted to his numerous collaborators, including Jitka Borová, Miroslav Cikrt, Erica Baker, Des Richardson, Herbert Schulman, Jack Edward, Allen Huang, Sylvain Chemtob, and Magda Horackova for their important contributions in research on PIH and its analogues.

9. REFERENCES

1. Rouault T, Klausner R. Regulation of iron metabolism in eukaryotes. *Curr. Top. Cell Regul.* **35**, 1-19 (1997).
2. Richardson DR, Ponka P. The molecular mechanisms of the metabolism and transport of iron in normal and neoplastic cells. *Biochim. Biophys. Acta.* **1331**, 1-40 (1997).

3. Aisen P, Enns C, Wessling-Resnick M. Chemistry and biology of eukaryotic iron metabolism. *Int. J. Biochem. Cell Biol.* **33**, 940-959 (2001).
4. Hershko C, Link G, Cabantchik I. Pathophysiology of iron overload. *Ann. N.Y. Acad. Sci.* **850**, 191-201 (1998).
5. Porter JB. Practical management of iron overload. *Br. J. Haematol.* **115**, 239-252 (2001).
6. Hershko C, Konijn AM, Link G. Iron chelators for thalassaemia. *Br. J. Haematol.* **101**, 399-406 (1998).
7. Kontoghiorghes GJ, Aldouri MA, Hoffbrand AV, Barr J, Wonke B, Kourouclaris T, Sheppard L. Effective chelation of iron in beta thalassaemia with the oral chelator 1,2-dimethyl-3-hydroxypyrid-4-one. *Br. Med. J.* **295**, 1509-1512 (1987).
8. Olivieri NF, Koren G, St Louis P, Freedman MH, McClelland RA, Templeton DM. Studies of the oral chelator 1,2-dimethyl-3-hydroxypyrid-4-one in thalassemia patients. *Semin. Hematol.* **27**, 101-104 (1990).
9. Berdoukas V, Bentley P, Frost H, Schnebli HP. Toxicity of oral iron chelator L1. *Lancet.* **341**, 1088 (1993).
10. al Refaie FN, Wonke B, Hoffbrand AV. Deferiprone-associated myelotoxicity. *Eur. J. Haematol.* **53**, 298-301 (1994).
11. al Refaie FN, Hoffbrand AV. Oral iron-chelating therapy: the L1 experience. *Bailliere's Clin. Haematol.* **7**, 941-963 (1994).
12. Olivieri NF, Brittenham GM. Iron-chelating therapy and the treatment of thalassemia. *Blood.* **89**, 739-761 (1997).
13. Olivieri NF, Brittenham GM, McLaren CE, Templeton DM, Cameron RG, McClelland RA, Burt A.D, Fleming K.A, Long-term safety and effectiveness of iron-chelation therapy with deferiprone for thalassemia major. *N. Engl. J. Med.* **339**, 417-423 (1998).
14. Richardson DR. The controversial role of deferiprone in the treatment of thalassemia. *J. Lab. Clin. Med.* **137**, 324-329 (2001).
15. Pippard MJ, Weatherall DJ. Oral iron chelation therapy for thalassaemia: an uncertain scene. *Br. J. Haematol.* **111**, 2-5 (2000).
16. Ponka P, Borova J, Neuwirt J, Fuchs O. Mobilization of iron from reticulocytes. Identification of pyridoxal isonicotinoyl hydrazone as a new iron chelating agent. *FEBS Lett.* **97**, 317-321 (1979).
17. Ponka P, Neuwirt J, Borova J, Fuchs O. Control of iron delivery to haemoglobin in erythroid cells. *Ciba Foundation Symposium 51 (new series).* 167-200 (1977).
18. Ponka P, Borova J, Neuwirt J, Fuchs O, Necas E. A study of intracellular iron metabolism using pyridoxal isonicotinoyl hydrazone and other synthetic chelating agents. *Biochim. Biophys. Acta.* **586**, 278-297 (1979).
19. Hoy T, Humphrys J, Jacobs A, Williams A, Ponka P. Effective iron chelation following oral administration of an isoniazid-pyridoxal hydrazone. *Br. J. Haematol.* **43**, 443-449 (1979).
20. Cikrt M, Ponka P, Necas E, Neuwirt J. Biliary iron excretion in rats following pyridoxal isonicotinoyl hydrazone. *Br. J. Haematol.* **45**, 275-283 (1980).
21. Hershko C, Avramovici-Grisaru S, Link G, Gelfand L, Sarel S. Mechanism of *in vivo* iron chelation by pyridoxal isonicotinoyl hydrazone and other imino derivatives of pyridoxal. *J. Lab. Clin. Med.* **98**, 99-108 (1981).
22. Pippard MJ, Johnson DK, Finch CA. A rapid assay for evaluation of iron-chelating agents in rats. *Blood.* **58**, 685-692 (1981).
23. Johnson DK, Pippard MJ, Murphy TB, Rose NJ. An *in vivo* evaluation of iron-chelating drugs derived from pyridoxal and its analogs. *J. Pharm. Exp. Ther.* **221**, 399-403 (1982).
24. Williams A, Hoy T, Pugh A, Jacobs A. Pyridoxal complexes as potential chelating agents for oral therapy in transfusional iron overload. *J. Pharm. Pharmacol.* **34**, 730-732 (1982).
25. Avramovici-Grisaru S, Sarel S, Link G, Hershko C. Syntheses of iron *bis*(pyridoxal isonicotinoylhydrazone)s and the *in vivo* iron-removal properties of some pyridoxal derivatives. *J. Med. Chem.* **26**, 298-302 (1983).
26. Webb J, Vitolo ML. Pyridoxal isonicotinoyl hydrazone (PIH): a promising new iron chelator. *Birth Defects Orig. Art. Ser.* **23**, 63-70 (1988).
27. Vitolo ML, Clare BW, Hefter GT, Webb J. Chemical studies of pyridoxal isonicotinoyl hydrazone relevant to its clinical evaluation. *Birth Defects Orig. Artic. Ser.* **23**, 71-79 (1988).
28. Richardson DR, Vitolo LW, Baker E, Webb J. Pyridoxal isonicotinoyl hydrazone and analogues. Study of their stability in acidic, neutral, and basic aqueous solutions by ultraviolet-visible spectrophotometry. *Biol. Met.* **2**, 69-76 (1989).
29. Richardson DR, Hefter GT, May PM, Webb J, Baker E. Iron chelators of the pyridoxal isonicotinoyl hydrazone class. III. Formation constants with calcium(II), magnesium(II) and zinc(II). *Biol. Met.* **2**, 161-167 (1989).

30. Sarel S, Cohen S, Avramovici-Grisaru S. Iron chelators of the class of pyridoxal acylhydrazone - part 5 - crystal structure and patterns of hydrogen bonding in pyridoxal isonicotinoyl hydrazone (PIH). *Heterocycles.* **47**, 1033-1042 (1998).
31. Souron JP, Quarton M, Robert F, Lyubchova A, Cosse-Barbi A, Doucet JP. Pyridoxal isonicotinoyl hydrazone (PIH), a synthetic ion-chelating agent. *Acta. Cryst. Sect. C.* **51**, 2179-2182 (1995).
32. Colonna P, Cosse-Barbi A, Massat A, Doucet JP. IR studies of iron complexes with pyridoxal isonicotinoyl hydrazone and 3 other similar chelating agents. *Spectroscopy Lett.* **26**, 1065-1072 (1993).
33. Morgan EH. Chelator-mediated iron efflux from reticulocytes. *Biochim. Biophys. Acta.* **733**, 39-50 (1983).
34. Huang AR, Ponka P. A study of the mechanism of action of pyridoxal isonicotinoyl hydrazone at the cellular level using reticulocytes loaded with non-heme ^{59}Fe. *Biochim Biophys. Acta.* **757**, 306-315 (1983).
35. Ponka P, Grady RW, Wilczynska A, Schulman HM. The effect of various chelating agents on the mobilization of iron from reticulocytes in the presence and absence of pyridoxal isonicotinoyl hydrazone. *Biochim. Biophys. Acta.* **802**, 477-489 (1984).
36. Baker E, Vitolo ML, Webb J. Iron chelation by pyridoxal isonicotinoyl hydrazone and analogues in hepatocytes in culture. *Biochem. Pharmacol.* **34**, 3011-3017 (1985).
37. Crowe A, Morgan EH. Effects of chelators on iron uptake and release by the brain in the rat. *Neurochem. Res.* **19**, 71-76 (1994).
38. Hallmann R, Savigni DL, Morgan EH, Baker E. Characterization of iron uptake from transferrin by murine endothelial cells. *Endothelium.* **7**, 135-147 (2000).
39. Ponka P, Richardson D, Baker E, Schulman HM, Edward JT. Effect of pyridoxal isonicotinoyl hydrazone and other hydrazones on iron release from macrophages, reticulocytes and hepatocytes. *Biochim. Biophys. Acta.* **967**, 122-129 (1988).
40. Baker E, Richardson D, Gross S, Ponka P. Evaluation of the iron chelation potential of hydrazones of pyridoxal, salicylaldehyde and 2-hydroxy-1-naphthylaldehyde using the hepatocyte in culture. *Hepatology.* **15**, 492-501 (1992).
41. Blaha K, Cikrt M, Nerudova J, Fornuskova H, Ponka P. Biliary iron excretion in rats following treatment with analogs of pyridoxal isonicotinoyl hydrazone. *Blood.* **91**, 4368-4372 (1998).
42. Richardson DR, Ponka P. Pyridoxal isonicotinoyl hydrazone and its analogs: potential orally effective iron-chelating agents for the treatment of iron overload disease. *J. Lab. Clin. Med.* **131**, 306-315 (1998).
43. Richardson DR, Mouralian C, Ponka P, Becker E. Development of potential iron chelators for the treatment of Friedreich's ataxia: ligands that mobilize mitochondrial iron. *Biochim. Biophys. Acta.* **1536**, 133-140 (2001).
44. Hermes-Lima M, Wang EM, Schulman HM, Storey KB, Ponka P. Deoxyribose degradation catalyzed by Fe(III)-EDTA: kinetic aspects and potential usefulness for submicromolar iron measurements. *Mol. Cell. Biochem.* **137**, 65-73 (1994).
45. Bhattacharya M, Ponka P, Hardy P, Hanna N, Varma DR, Lachapelle P, Chemtob S, Prevention of postasphyxia electroretinal dysfunction with a pyridoxal hydrazone. *Free Rad. Biol. Med.* **22**, 11-16 (1997).
46. Hermes-Lima M, Nagy E, Ponka P, Schulman HM. The iron chelator pyridoxal isonicotinoyl hydrazone (PIH) protects plasmid pUC-18 DNA against *OH-mediated strand breaks. *Free Rad. Biol. Med.* **25**, 875-880 (1998).
47. Hermes-Lima M, Santos NC, Yan J, Andrews M, Schulman HM, Ponka P. EPR spin trapping and 2-deoxyribose degradation studies of the effect of pyridoxal isonicotinoyl hydrazone (PIH) on *OH formation by the Fenton reaction. *Biochim. Biophys. Acta.* **1426**, 475-482 (1999).
48. Hermes-Lima M, Ponka P, Schulman HM. The iron chelator pyridoxal isonicotinoyl hydrazone (PIH) and its analogues prevent damage to 2-deoxyribose mediated by ferric iron plus ascorbate. *Biochim. Biophys. Acta.* **1523**, 154-160 (2000).
49. Horackova M, Ponka P, Byczko Z. The antioxidant effects of a novel iron chelator salicylaldehyde isonicotinoyl hydrazone in the prevention of H(2)O(2) injury in adult cardiomyocytes. *Cardiovasc. Res.* **47**, 529-536 (2000).
50. Tsafack A, Loyevsky M, Ponka P, Cabantchik ZI. Mode of action of iron (III) chelators as antimalarials. IV. Potentiation of desferal action by benzoyl and isonicotinoyl hydrazone derivatives. *J. Lab. Clin. Med.* **127**, 574-582 (1996).
51. Golenser J, Domb A, Teomim D, Tsafack A, Nisim O, Eling W, Cabantchik ZI, The treatment of animal models of malaria with iron chelators by use of a novel polymeric device for slow drug release. *J. Pharm. Exp. Ther.* **281**, 1127-1135 (1997).
52. Richardson DR, Tran EH, Ponka P. The potential of iron chelators of the pyridoxal isonicotinoyl hydrazone class as effective antiproliferative agents. *Blood.* **86**, 4295-4306 (1995).

53. Ponka P, Schulman HM, Wilczynska A. Ferric pyridoxal isonicotinoyl hydrazone can provide iron for heme synthesis in reticulocytes. *Biochim. Biophys. Acta.* **718**, 151-156 (1982).
54. Ponka P, Schulman HM. Regulation of heme synthesis in erythroid cells: hemin inhibits transferrin iron utilization but not protoporphyrin synthesis. *Blood.* **65**, 850-857 (1985).
55. Ponka P, Schulman HM. Acquisition of iron from transferrin regulates reticulocyte heme synthesis. *J. Biol. Chem.* **260**, 14717-14721 (1985).
56. Laskey JD, Ponka P, Schulman HM. Control of heme synthesis during Friend cell differentiation: role of iron and transferrin. *J. Cell. Physiol.* **129**, 185-192 (1986).
57. Landschulz W, Thesleff I, Ekblom P. A lipophilic iron chelator can replace transferrin as a stimulator of cell proliferation and differentiation. *J. Cell. Biol.* **98**, 596-601 (1984).
58. Thesleff I, Partanen AM, Landschulz W, Trowbridge IS, Ekblom P. The role of transferrin receptors and iron delivery in mouse embryonic morphogenesis. *Differentiation.* **30**, 152-158 (1985).
59. Ekblom P, Landschulz W, Andersson LC. A lipophilic iron chelator induces an enhanced proliferation of human erythroleukaemia (HEL) cells. *Scand. J. Haematol.* **36**, 258-262 (1986).
60. Landschulz W, Ekblom P. Iron delivery during proliferation and differentiation of kidney tubules. *J. Biol. Chem.* **260**, 15580-15584 (1985).
61. Forsbeck K, Bjelkenkrantz K, Nilsson K. Role of iron in the proliferation of the established human tumor cell lines U-937 and K-562: effects of suramin and a lipophilic iron chelator (PIH). *Scand. J .Haematol.* **37**, 429-437 (1986).
62. Laskey J, Webb I, Schulman HM, Ponka P. Evidence that transferrin supports cell proliferation by supplying iron for DNA synthesis. *Exp. Cell Res.* **176**, 87-95 (1988).
63. Partanen AM, Thesleff I. Transferrin and tooth morphogenesis: retention of transferrin by mouse embryonic teeth in organ culture. *Differentiation.* **34**, 25-31 (1987).
64. Tsao MS, Sanders GH, Grisham JW. Regulation of growth of cultured hepatic epithelial cells by transferrin. *Exp. Cell Res.* **171**, 52-62. 1987.
65. Brock JH, Stevenson J. Replacement of transferrin in serum-free cultures of mitogen-stimulated mouse lymphocytes by a lipophilic iron chelator. *Immunol. Lett.* **15**, 23-25. 1986.
66. Djeha A, Brock JH. Effect of transferrin, lactoferrin and chelated iron on human T-lymphocytes. *Br. J. Haematol.* **80**, 235-241 (1992).
67. Avramovici-Grisaru S, Sarel S, Cohen S, Bauminger RE. The synthesis, crystal and molecular structure, and oxidation state of iron complex from pyridoxal isonicotinoyl hydrazone and ferrous sulphate. *Israel. J. Chem.* **25**, 288-292 (1985).
68. Murphy TB, Johnson DK, Rose NJ, Aruffo A, Schomaker V. Structural studies of iron(III) complexes of the new iron-binding drug, pyridoxal isonicotinoyl hydrazone. *Inorg. Chim. Acta* **66**, L67-L68. (1982).
69. Becker E, Richardson DR. Development of novel aroylhydrazone ligands for iron chelation therapy: 2-pyridylcarboxaldehyde isonicotinoyl hydrazone analogs. *J. Lab. Clin. Med.* **134**, 510-521 (1999).
70. Richardson DR, Becker E, Bernhardt PV. The biologically active iron chelators 2-pyridylcarboxaldehyde isonicotinoylhydrazone, 2-pyridylcarboxaldehyde benzoylhydrazone monohydrate and 2-furaldehyde isonicotinoylhydrazone. *Acta. Cryst. Sect. C.* **55**, 2102-2105 (1999).
71. Vitolo ML, Hefter GT, Clare BW, Webb J. Iron chelators of the pyridoxal isonicotinoyl hydrazone class Part II. Formation constants with iron(III) and iron(II). *Inorg. Chim. Acta.* **170**, 171-176 (1990).
72. Zak O, Leibman A, Aisen P. Metal-binding properties of a single-sited transferrin fragment. *Biochim. Biophys. Acta.* **742**, 490-495 (1983).
73. Richardson DR, Vitolo ML, Hefter GT, May PM, Clare BW, Webb J. Iron chelators of the pyridoxal isonicotinoyl hydrazone class Part 1. Ionization characteristics of the ligands and their relevance to biological properties. *Inorg. Chim. Acta* **170**, 165-170 (1990).
74. Schmidt RF, Thews J. *Human Physiology* (Springer-Verlag, New York, 1983).
75. Travis S, Menzies IS. Intestinal permeability: functional assessment and significance. *Clin. Sci.* **82**, 471-480 (1992).
76. Maxton DG, Bjarnson I, Reynolds AP, Catt SD, Peters TJ, Menzies IS. ^{51}Cr-labelled EDTA, L-rhamnose and polyethyleneglycol 400 as probe markers for assessment *in vivo* of human intestinal permeability. *Clin. Sci.* **71**, 71-80 (1986).
77. Sah PPT. Nicotinyl and isonicotinyl hydrazones of pyridoxal. *J. Am. Chem. Soc.* **76**, 300 (1954).
78. Brittenham GM. Pyridoxal isonicotinoyl hydrazone: an effective iron-chelator after oral administration. *Semin. Hematol.* **27**, 112-116 (1990).

79. Lyubchova A, Cosse-Barbi A, Doucet JP, El Hage Chahine JM. The interaction of salicylaldehydebenzoylhydrazone with Ca^{2+} and Mg^{2+}. A spectrophotometric study. *J. Chim. Phys.* **94**, 1195-1207. (1997).
80. Singh G, Shastry PSSJ, Lonibala RK, Rao TR. Coordination behaviour of pyridoxalisonicotinoyl hydrazone towards some 3d-metal ions. *Synth. React. Inorg. Met. -Org. Chem.* **22**:1041-1059 (1992).
81. Edward JT, Ponka P, Richardson DR. Partition coefficients of the iron(III) complexes of pyridoxal isonicotinoyl hydrazone and its analogs and the correlation to iron chelation efficacy. *Biometals.* **8**, 209-217 (1995).
82. Edward JT, Chubb FL, Sangster J. Iron chelators of the pyridoxal isonicotinoyl hydrazone class. Relationship of the lipophilicity of the apochelator to its ability to mobilize iron from reticulocytes *in vitro*: reappraisal of reported partition coefficients. *Can. J. Physiol. Pharmacol.* **75**, 1362-1368 (1997).
83. Lea A, Hansch C, Elkins D. Partition coefficients and their uses. *Chem. Rev.* **71**, 525-555 (1971).
84. Hider RC. Potential protection from toxicity by oral iron chelators. *Toxicol. Lett.* **83**, 961-967 (1995).
85. Pappenheimer JR, Karnovsky ML, Maggio JE. Absorption and excretion of undegradable peptides: role of lipid solubility and net charge. *J. Pharm. Exp. Ther.* **280**, 292-300 (1997).
86. Ponka P, Wilczynska A, Schulman HM. Iron utilization in rabbit reticulocytes. A study using succinylacetone as an inhibitor of heme synthesis. *Biochim. Biophys. Acta.* **720**, 96-105 (1982).
87. Borova J, Ponka P, Neuwirt J. Study of intracellular iron distribution in rabbit reticulocytes with normal and inhibited heme synthesis. *Biochim. Biophys. Acta.* **320**, 143-156 (1973).
88. Ponka P, Neuwirt J. The use of reticulocytes with high non-haem iron pool for studies of regulation of haem synthesis. *Br. J. Haematol.* **19**, 593-604 (1970).
89. Richardson DR. Mobilization of iron from neoplastic cells by some iron chelators is an energy-dependent process. *Biochim. Biophys. Acta.* **1320**, 45-57 (1997).
90. Harris WR, Carrano CJ, Raymond KN. Co-ordination chemistry of microbial iron transport compounds: Isolation, characterization, and formation constants of ferric aerobactin. *J. Am. Chem. Soc.* **101**, 2722-2727 (1979).
91. Morgan EH. A study of iron transfer from rabbit transferrin to reticulocytes using synthetic chelating agents. *Biochim. Biophys. Acta.* **244**, 103-116 (1971).
92. Martell AE, Smith RM. *Critical Stability Constants* (New York, 1977).
93. Ponka P, Baker E. The effect of the iron(III) chelator, desferrioxamine, on iron and transferrin uptake by the human malignant melanoma cell. *Cancer Res.* **54**, 685-689 (1994).
94. Bakkeren DL, de Jeu-Jaspars CMH, Kroos MJ, van Eijk HG. Release of iron from endosomes is an early step in the transferrin cycle. *Int. J. Biochem.* **19**, 179-186 (1987).
95. Edward JT, Gauthier M, Chubb FL, Ponka P. Synthesis of new acylhydrazones as iron-chelating compounds. *J. Chem. Eng. Data.* **33**, 538-540 (1988).
96. Richardson DR, Milnes K. The potential of iron chelators of the pyridoxal isonicotinoyl hydrazone class as effective antiproliferative agents II: the mechanism of action of ligands derived from salicylaldehyde benzoyl hydrazone and 2-hydroxy-1-naphthylaldehyde benzoyl hydrazone. *Blood.* **89**, 3025-3038 (1997).
97. Kim S, Ponka P. Effects of interferon-gamma and lipopolysaccharide on macrophage iron metabolism are mediated by nitric oxide-induced degradation of iron regulatory protein 2. *J. Biol. Chem.* **275**, 6220-6226 (2000).
98. Darnell G, Richardson DR. The potential of iron chelators of the pyridoxal isonicotinoyl hydrazone class as effective antiproliferative agents III: The effect of the ligands on molecular targets involved in proliferation. *Blood.* **94**, 781-792 (1999).
99. Eisenstein RS. Iron regulatory proteins and the molecular control of mammalian iron metabolism. *Ann. Rev. Nutr.* **20**, 627-662 (2000).
100. Richardson DR, Ponka P. The iron metabolism of the human neuroblastoma cell: lack of relationship between the efficacy of iron chelation and the inhibition of DNA synthesis. *J. Lab. Clin. Med.* **124**, 660-671 (1994).
101. Ponka P, Richardson DR, Edward JT, Chubb FL. Iron chelators of the pyridoxal isonicotinoyl hydrazone class. Relationship of the lipophilicity of the apochelator to its ability to mobilise iron from reticulocytes *in vitro*. *Can. J. Physiol. Pharmacol.* **72**, 659-666 (1994).
102. Baker E, Page M, Torrance J, Grady R. Effect of desferrioxamine, rhodotorulic acid and cholylhydroxamic acid on transferrin and iron exchange with hepatocytes in culture. *Clin. Physiol. Biochem.* **3**, 277-288 (1985).
103. Hershko C. Determinants of fecal and urinary iron excretion in desferrioxamine-treated rats. *Blood.* **51**, 415-423 (1978).

104. Lipschitz DA, Simon MO, Lynch SR, Dugard J, Bothwell TH, Charlton RW. Some factors affecting the release of iron from reticuloendothelial cells. *Br. J. Haematol.* **21**, 289-303 (1971).
105. Sharma BK, Tavill AS, Louis LN, Wiesen E, Varnes AW. Enteral pyridoxal isonicotinoyl hydrazone (PIH) is an effective chelator in experimental iron overload by promotion of biliary iron excretion. *Hepatology.* **10**, 573 (1989).
106. Sharma BK, Tavill AS, Louis LN, Varnes AW. Predominance of biliary iron chelates in iron-loaded rats *in vivo* during i.v. deferoxamine (DF) or pyridoxal isonicotinoyl hydrazone (PIH). *Hepatology.* **8**, 1240 (1988).
107. Kim BK, Huebers HA, Finch CA. Effectiveness of oral iron chelators assayed in the rat. *Am. J. Hematol.* **24**, 277-284 (1987).
108. Pootrakul P, Yansukon P, Piankijagum A, Muangsub W, Brittenham GM. The interference of pyridoxal isonicotinoyl hydrazone with intestinal iron absorption. *Ann. N. Y. Acad. Sci.* **124**, 582-584 (1990).
109. Summers MR, Jacobs A, Tudway D, Perera P, Ricketts C. Studies in desferrioxamine and ferrioxamine metabolism in normal and iron-loaded subjects. *Br. J. Haematol.* **42**, 547-555 (1979).
110. Djeha A, Brock JH. Uptake and intracellular handling of iron from transferrin and iron chelates by mitogen stimulated mouse lymphocytes. *Biochim. Biophys. Acta.* **1133**, 147-152 (1992).
111. Alcantara O, Obeid L, Hannun Y, Ponka P, Boldt DH. Regulation of protein kinase C (PKC) expression by iron: effect of different iron compounds on PKC-beta and PKC-alpha gene expression and role of the 5'-flanking region of the PKC-beta gene in the response to ferric transferrin. *Blood.* **84**, 3510-3517 (1994).
112. Thelander L, Reichard P. Reduction of ribonucleotides. *Ann. Rev. Biochem.* **48**, 133-158 (1979).
113. Nyholm S, Mann GJ, Johansson AG, Bergeron RJ, Graslund A, Thelander L. Role of ribonucleotide reductase in inhibition of mammalian cell growth by potent iron chelators. *J. Biol. Chem.* **268**, 26200-26205 (1993).
114. Green DA, Antholine WE, Wong SJ, Richardson DR, Chitambar CR. Inhibition of malignant cell growth by 311, a novel iron chelator of the pyridoxal isonicotinoyl hydrazone class: effect on the R2 subunit of ribonucleotide reductase. *Clin. Cancer Res.* **7**, 3574-3579 (2001).
115. Hoffbrand AV, Ganeshaguru K, Hooton JW, Tattersall MH. Effect of iron deficiency and desferrioxamine on DNA synthesis in human cells. *Br. J. Haematol.* **33**, 517-526 (1976).
116. Hoyes KP, Hider RC, Porter JB. Cell cycle synchronization and growth inhibition by 3-hydroxypyridin-4-one iron chelators in leukemia cell lines. *Cancer Res.* **52**, 4591-4599 (1992).
117. Brodie C, Siriwardana G, Lucas J, Schleicher R, Terada N, Szepesi A, Gelfand E, Seligman P. Neuroblastoma sensitivity to growth inhibition by deferrioxamine: evidence for a block in G1 phase of the cell cycle. *Cancer Res.* **53**, 3968-3975 (1993).
118. van Reyk D, Sarel S, Hunt N. Inhibition of *in vitro* lymphoproliferation by three novel iron chelators of the pyridoxal and salicyl aldehyde hydrazone classes. *Biochem. Pharmacol.* **60**, 581-587 (2000).
119. Bomford A, Isaac J, Roberts S, Edwards A, Young S, Williams R. The effect of desferrioxamine on transferrin receptors, the cell cycle and growth rates of human leukaemic cells. *Biochem. J.* **236**, 243-249 (1986).
120. Renton FJ, Jeitner TM. Cell cycle-dependent inhibition of the proliferation of human neural tumor cell lines by iron chelators. *Biochem. Pharmacol.* **51**, 1553-1561 (1996).
121. Levine AJ. p53, the cellular gatekeeper for growth and division. *Cell.* **88**, 323-331 (1997).
122. Momand J, Zambetti GP, Olson DC, George D, Levine AJ. The mdm-2 oncogene product forms a complex with the p53 protein and inhibits p53-mediated transactivation. *Cell.* **69**, 1237-1245 (1992).
123. Gao J, Richardson DR. The potential of iron chelators of the pyridoxal isonicotinoyl hydrazone class as effective antiproliferative agents, IV: The mechanisms involved in inhibiting cell-cycle progression. *Blood.* **98**, 842-850 (2001).
124. Wu X, Bayle JH, Olson D, Levine AJ. The p53-mdm-2 autoregulatory feedback loop. *Genes Dev.* **7**, 1126-1132 (1993).
125. Koh LL, Kon OL, Loh KW, Long YC, Ranford JD, Tan AL, Tjan YY. Complexes of salicylaldehyde acylhydrazones: cytotoxicity, QSAR and crystal structure of the sterically hindered t-butyl dimer. *J. Inorg. Biochem.* **72**, 155-162 (1998).
126. Forsbeck K, Nilsson K, Kontoghiorghes GJ. Variation in iron accumulation, transferrin membrane binding and DNA synthesis in the K-562 and U-937 cell lines induced by chelators and their iron complexes. *Eur. J. Haematol.* **39**, 318-325 (1987).
127. Edward JT. Partition coefficients of the iron (III) complexes of pyridoxal isonicotinoyl hydrazone and its analogs and the correlation to iron chelation efficacy. Correction of some reported partition coefficients. *Biometals.* **11**, 203-205 (1998).

128. Richardson DR, Bernhardt PV. Crystal and molecular structure of 2-hydroxy-1-naphthaldehyde isonicotinoyl hydrazone (NIH) and its iron(III) complex: an iron chelator with anti-tumour activity. *J. Biol. Inorg. Chem.* **4**, 266-273 (1999).
129. Seligman PA, Schleicher RB, Siriwardana G, Domenico J, Gelfand EW. Effects of agents that inhibit cellular iron incorporation on bladder cancer cell proliferation. *Blood.* **82**, 1608-1617 (1993).
130. Chitambar CR, Narasimhan J, Guy J, Sem DS, O'Brien WJ. Inhibition of ribonucleotide reductase by gallium in murine leukemic L1210 cells. *Cancer Res.* **51**, 6199-6201 (1991).
131. Seligman PA, Moran PL, Schleicher RB, Crawford ED. Treatment with gallium nitrate: evidence for interference with iron metabolism *in vivo*. *Am. J. Hematol.* **41**, 232-240 (1992).
132. Chitambar CR, Zivkovic Z. Inhibition of hemoglobin production by transferrin-gallium. *Blood.* **69**, 144-149 (1987).
133. Brittenham GM. Pyridoxal isonicotinoyl hydrazone. Effective iron chelation after oral administration. *Ann. N. Y. Acad. Sci.* **612**, 315-326 (1990).
134. Sookvanichsilp N, Nakornchai S, Weerapradist W. Toxicological study of pyridoxal isonicotinoyl hydrazone: acute and subchronic toxicity. *Drug Chem. Toxicol.* **14**, 395-403 (1991).
135. Mohan M, Kumar A, Kuo YM. Synthesis, characterization and antitumour activity of manganese(II), cobalt(II), nickel(II), copper(II), zinz(II) and platinum(II) complexes of 3- and 5-substituted salicylaldehyde benzoylhydrazones. *Inorg. Chim. Acta.* **136**, 65-74 (1987).
136. Lees-Gayed NJ, Abou-Taleb MA, El-Bitash IA, Iskander MF. Studies on biologically active acylhydrazones. Part 1. Acid-base equilibria and acid hydrolysis of pyridoxal aroylhydrazones and related compounds. *J. Chem. Soc. Perkin 2*, 213-217 (1992).
137. Schaumburg H, Kaplan J, Windebank A, Vick N, Rasmus S, Pleasure D, Brown MJ. Sensory neuropathy from pyridoxine abuse. A new megavitamin syndrome. *N. Engl. J. Med.* **309**, 445-448 (1983).
138. Scott MD, Ranz A, Kuypers FA, Lubin BH, Meshnick SR. Parasite uptake of desferroxamine: a prerequisite for antimalarial activity. *Br. J. Haematol.* **75**, 598-602 (1990).
139. Gutteridge JM, Halliwell B. Iron toxicity and oxygen radicals. *Bailliere's Clin. Haematol.* **2**, 195-256 (1989).
140. Halliwell B, Gutteridge JM. Oxygen toxicity, oxygen radicals, transition metals and disease. *Biochem. J.* **219**, 1-14 (1984).
141. Halliwell B, Gutteridge JM. Role of free radicals and catalytic metal ions in human disease: an overview. *Meth. Enzymol.* **186**, 1-85 (1990).
142. Reddy BR, Kloner RA, Przyklenk K. Early treatment with deferoxamine limits myocardial ischemic/reperfusion injury. *Free Rad. Biol. Med.* **7**, 45-52 (1989).
143. Healing G, Gower J, Fuller B, Green C. Intracellular iron redistribution. An important determinant of reperfusion damage to rabbit kidneys. *Biochem. Pharmacol.* **39**, 1239-1245 (1990).
144. Biemond P, Swaak AJ, van Eijk HG, Koster JF. Superoxide dependent iron release from ferritin in inflammatory diseases. *Free Rad. Biol. Med.* **4**, 185-198 (1988).
145. Toyokuni S. Iron-induced carcinogenesis: the role of redox regulation. *Free Rad. Biol. Med.* **20**, 553-566 (1996).
146. Gassen M, Youdim MB. The potential role of iron chelators in the treatment of Parkinson's disease and related neurological disorders. *Pharmacol. Toxicol.* **80**, 159-166 (1997).
147. Schulman HM, Hermes-Lima M, Wang EM, Ponka P. *In vitro* antioxidant properties of the iron chelator pyridoxal isonicotinoyl hydrazone and some of its analogs. *Redox Rep.* **1**, 373-378 (1995).
148. Santos NC, Castilho RF, Meinicke AR, Hermes-Lima M. The iron chelator pyridoxal isonicotinoyl hydrazone inhibits mitochondrial lipid peroxidation induced by Fe(II)-citrate. *Eur. J. Pharmacol.* **428**, 37-44 (2001).
149. Fuchs O. Effects of intracellular chelatable iron and oxidative stress on transcription of classical cellular glutathione peroxidase gene in murine erythroleukemia cells. *Neoplasma.* **44**, 184-191 (1997).
150. Cable H, Lloyd JB. Cellular uptake and release of two contrasting iron chelators. *J. Pharm. Pharmacol.* **51**, 131-134 (1999).
151. Ferrali M, Signorini C, Ciccoli L, Bambagioni S, Rossi V, Pompella A, Comporti M. Protection of erythrocytes against oxidative damage and autologous immunoglobulin G (IgG) binding by iron chelator fluor-benzoil-pyridoxal hydrazone. *Biochem. Pharmacol.* **59**, 1365-1373 (2000).

THERAPEUTIC POTENTIAL OF IRON CHELATORS IN CANCER THERAPY

Des R. Richardson*

1. GENERAL INTRODUCTION

The development of iron (Fe) chelators for clinical use remains an active research goal. Over the last thirty years desferrioxamine (DFO; Desferal^R; Fig. 1) has been the drug of choice in the treatment of Fe overload disease.[1,2] Despite its considerable success, the problems with this drug remain significant and much research has been invested to obtain alternative ligands (see Chapters 7-9, 13, 14). At present, a number of potential chelators that are orally effective are available for experimental testing (see Chapters 7,13,14). Hence, one can envisage that in the future some of these compounds used alone or in various combinations may provide a better regimen than DFO.

In addition to the use of DFO for the treatment of Fe overload, a wide variety of studies have demonstrated that this ligand has considerable anti-neoplastic activity against cancer, particularly leukemia and neuroblastoma (for review see[3,4]). In addition, investigations with DFO have provided much information about the requirement of Fe for cellular growth and metabolism. The present chapter will summarise the work with DFO as well as the development of other ligands that show much greater anti-proliferative activity. Emphasis will be placed on the molecular targets of these ligands particularly in relation to their effect on the cell cycle. Hence, the study of Fe chelators not only has very important consequences for the development of effective anti-cancer agents, but will also provide information on understanding the role of Fe in cellular metabolism and the cell cycle. Indeed, knowledge on the molecular regulation of cellular Fe uptake and storage have been worked out using DFO as a tool to specifically induce Fe depletion (for reviews see[5-7]).

---

* The Iron Metabolism and Chelation Group, The Heart Research Institute, 145 Missenden Rd, Camperdown, Sydney, New South Wales, 2050 AUSTRALIA.

Iron Chelation Therapy.
Edited by Chaim Hershko, Kluwer Academic/Plenum Publishers, 2002.

It is not generally realised that several anti-cancer drugs can chelate Fe or act on the same biological molecules as those targeted by chelators. For example, the anti-tumor activity of the anthracyclines (eg. doxorubicin) is via a number of mechanisms one of which is chelation of Fe that results in the generation of free radicals that damage DNA.[8] This activity can also generate cardiotoxicity due to the limited protective defenses against free radical damage in the heart (for review see[9]). Another anti-cancer agent, bleomycin, is also a chelator that binds Fe or Cu that then redox cycles to generate cytotoxic radicals.[10] At this point it is important to note that chelators can have variable effects on the redox activity of transition metals. In some cases, such as bleomycin, it can potentiate free radical production, while in the case of ligands such as DFO, this compound binds Fe and prevents radical generation.

Hydroxyurea is used clinically to treat chronic myelogenous leukemia.[11] This drug acts on the enzyme ribonucleotide reductase (RR)[12] that is involved in the conversion of ribonucleotides into deoxyribonucleotides for DNA synthesis. Mammalian RR consists of 2 subunits, a large R1 subunit and a small R2 dimeric subunit.[13] The R1 subunit contains substrate and effector-binding sites while the R2 subunit contains an μ-oxo-bridged Fe atom and a tyrosyl radical.[13] Ribonucleotide reductase is also a target for Fe chelators such as DFO and the α-ketohydroxypyridones.[13,14] However, HU and Fe chelators inhibit this enzyme by different mechanisms. Hydroxyurea acts to directly scavenge the tyrosyl free radical of RR which is involved in catalysis.[12] In contrast, it is thought that DFO and α-ketohydroxypyridone chelators function indirectly to inhibit Fe uptake by the enzyme and prevent formation of the tyrosyl free radical.[13,14]

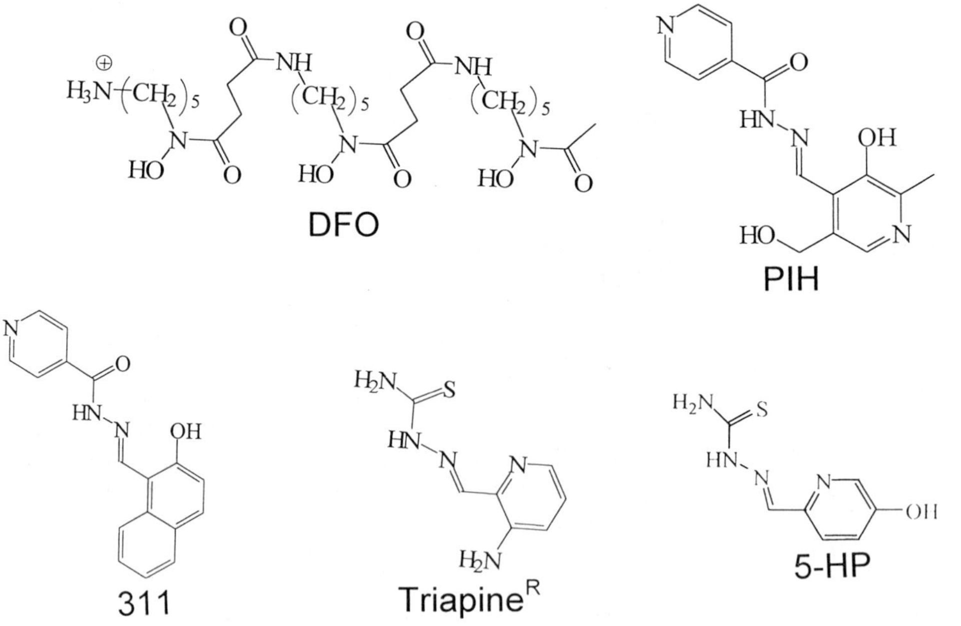

Figure 1. Structural formulae of chelators that show anti-tumour activity: Desferrioxamine (DFO), pyridoxal isonicotinoyl hydrazone (PIH), 2-hydroxy-1-naphthylaldehyde isonicotinoyl hydrazone (311), Triapine[R] (3-aminopyridine-2-carboxaldehyde thiosemicarbazone), 5-hydroxypyridine-2-carboxaldehyde thiosemicarbazone (5-HP).

THERAPEUTIC POTENTIAL OF IRON CHELATORS IN CANCER THERAPY

Unlike HU which does not show pronounced anti-cancer activity, Fe chelators have multiple mechanisms of action and appear to act at a number of sites. For instance, apart from RR, Fe chelators act to directly inhibit Fe uptake from Tf, and to deplete intracellular Fe pools.[15,16] This is a critical effect, as Fe is essential for a wide variety of metabolic processes including energy production. The difference in mechanism is apparent by the fact that cancer cells which show resistance to the action of HU are sensitive to the action of potent chelators.[17] Furthermore, HU has limited potency that is due mainly to its short serum half-life, its low affinity for RR, and the fact that resistance does develop against this agent.[18-20] Very high levels of HU (150 µM) increase the expression of the cell cycle inhibitor p21/*WAF1*, while some chelators of the pyridoxal isonicotinoyl hydrazone (PIH) class (eg. analogue 311; Fig. 1) induce expression at only 2.5 µM, a 60-fold difference.[21] In addition, analogue 311 is far more active at inhibiting RR in CCRF-CEM leukemia cells. For example, 311 (0.7 µM) markedly reduces the RR signal (Fig. 2; Richardson and Chitambar, unpublished) while a similar effect is observed at a HU concentration of 250 µM, a 350-fold difference.[22] Another important difference between Fe chelators and HU is that these ligands inhibit cells in the G_1 phase of the cell cycle, while HU halts progression in S.[23]

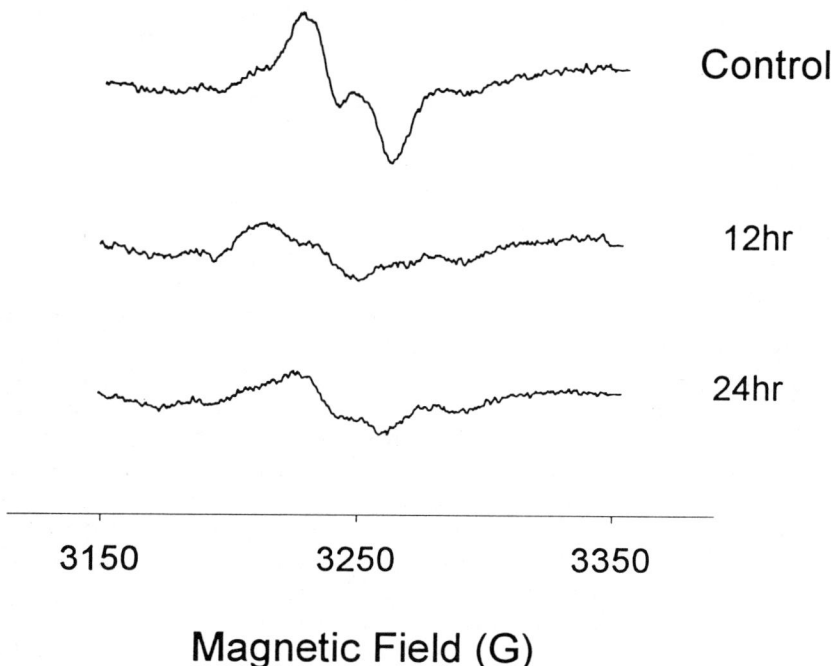

Figure 2. Very low concentrations (0.7 µm) of the PIH analogue 2-hydroxy-1-naphthylaldehyde isonicotinoyl hydrazone (311) markedly diminish the intensity of the tyrosyl radical of ribonucleotide reductase in CCRF-CEM leukemia cells. From Richardson, D.R. and Chitambar, C.R. (submitted).

Iron is thus a legitimate target for anti-cancer agents, and obviously new methods of treating cancer are essential for tumors that remain refractory or have become resistant to standard chemotherapy. Iron chelators could be classed in a similar category as the antimetabolite, methotrexate, which inhibits folate metabolism that is required for DNA synthesis.[24] Hence, chelation of Fe from cells will not only damage neoplastic cells but also inhibit the proliferation of rapidly growing normal cells. However, it has already been shown that the therapeutic index of chelators is appreciable. For instance, DFO is capable of a potent cytotoxic effect on NB cells, while having little effect on normal or other neoplastic cell types.[25,26] In fact, it has been reported that while DFO was cytotoxic to NB cells, normal bone marrow cells were 10-fold less sensitive.[26] More importantly, a single 5-day course of DFO given as an 8 h i.v. infusion at 150 mg/kg/day, resulted within 2 days in 7 of the 9 patients having more than a 50 % decrease in bone marrow infiltration of tumor cells.[27] In addition, in 1 patient, a 48 % decrease of tumor size was observed. In contrast to most cytotoxic drug treatments, there were no significant side effects.[27]

In order to understand the effect of Fe chelators at inhibiting cellular proliferation, their effects on Fe metabolism are essential to assess. In the following section the mechanisms of Fe uptake and release are examined and the possible sites of interaction of Fe chelators described.

2. CELLULAR IRON METABOLISM AND THE TARGETS OF IRON CHELATORS

The sensitivity of neoplastic cells to chelators is thought to be due to the high Fe requirement of rapidly growing cells. Iron is essential component of the active sites for crucial enzymes that are involved in a variety of metabolic processes including energy production and DNA synthesis. These proteins include the cytochromes, Fe-S proteins, RR, lipoxygenases, and cyclooxygenases. Of these, RR has been the most studied and is an important target of Fe chelation, but as described above, it is not the only target. Iron is transported to cells via the Fe-transport protein transferrin (Tf). Some chelators such as 1,2-dimethyl-3-hydroxypyridone (also known as L1, deferiprone, DMHP, or CP20) can remove Fe from Tf,[28] while others act to bind Fe after it has been released from Tf within the cell (eg. PIH and its analogues).[29] Transferrin binds to two receptors on the cell surface known as transferrin receptor 1 (TfR1)[30] and TfR2.[31,32] Up until recently only the TfR1 was clearly identified to be involved in Fe uptake from Tf.[30] However, the cloning of the TfR2 in 1999 has suggested an alternative Fe uptake process,[31] although its significance remains unclear.

For many years prior to the discovery of the TfR2, a second process of Fe uptake from Tf had been reported to be significant in some normal and neoplastic cell types[33-36] but not others.[37] However, this latter mechanism was thought to be via a pinocytotic mechanism.[33-36,38] The role of the TfR2 in Fe uptake and cellular proliferation is questioned by previous studies demonstrating that MoAbs specific for the TfR1 can inhibit tumor cell growth either alone[39] or in combination with chelators.[40,41] In addition, in a mouse TfR1 knockout model, TfR2 was not able to compensate for the loss of TfR1, resulting in death of the animals after day 12.5 due to anaemia and neurological abnormalities.[42] Recent studies have shown that the affinity of the TfR2 for Tf is less than that found for the TfR1.[32] Yet, CHO cells expressing transfected TfR2 grow into larger tumors than those expressing TfR1.[32] These puzzling results may suggest that the molecule may have another role apart from Fe uptake, and further studies on its role in proliferation in normal and neoplastic cells are essential.

Iron uptake from Tf via the TfR1 occurs by receptor-mediated endocytosis.[30] The process of Fe uptake by the TfR1 is regulated by intracellular Fe levels via the well described iron-regulatory protein (IRP) and iron-responsive element (IRE) regulatory system (for reviews see[5,6]). Obviously, this regulatory mechanism is a well known target of Fe chelators via their ability to chelate the intracellular Fe pool, rather than a direct effect on the [4Fe-4S] cluster in this protein.[21] The expression of the TfR1 can also be controlled at the transcriptional level by oxygen levels and the state of cellular proliferation and differentiation.[43-46] In contrast, TfR2 is not regulated by Fe,[32] and its expression appears to be linked to the cell cycle.[32] Neoplastic cells express high levels of the TfR1[30,47] and take up Fe at a greater rate than their normal counterparts.[35,38,48] This has been suggested to be the reason for their higher sensitivity to chelators. The high Fe requirement is related to the use of Fe for metabolic needs such as DNA synthesis and energy production. The higher uptake of Tf-bound Fe by cancer cells is reflected by the fact that tumors can be radiolocalised[49] and treated with gallium salts[50,51] which bind to Tf and is delivered to cells via its interaction with the TfR1.[52-54]

Iron is released from Tf by a decrease in endosomal pH. The Fe is then transported across the endosomal membrane by the metal ion transporter Nramp2[55,56] and subsequently enters the intracellular Fe pool (for review see[6]). The identity of this latter compartment remains a mystery, but it appears to be a labile form of Fe that is targeted by chelators. The importance of this pool is also clear from the fact that it plays a major role in regulating the RNA-binding activity of IRP1 (for reviews see[5,6]).

Iron from the pool can stored in the protein ferritin.[57] Iron in ferritin is rather insoluble and is not effectively bound by chelators. Ferritin can also be secreted from tumor cells, although the pathophysiological significance of this phenomenon remains unclear (for review see[4]).

Apart from Fe uptake from Tf, Fe can be released from cells via the ferroxidase activity of ceruloplasmin.[58-60] Recently, a putative transport molecule involved in Fe efflux has been cloned and is known as ferroportin 1.[61] The release of Fe from cells is an obvious target for chelators and may be highly significant for more hydrophilic compounds such as DFO.

3. CHELATORS THAT SHOW PRONOUNCED ANTI-PROLIFERATIVE ACTIVITY

3.1 Desferrioxamine

Desferrioxamine (Fig. 1) has been shown in a wide variety of *in vitro*, *in vivo*, and clinical trials to demonstrate anti-cancer activity.[27,62-65] Most of the clinical studies have either been in patients with neuroblastoma or leukemia, tumor cells that respond well to DFO treatment *in vitro*. Reviews on various aspects of this activity are available,[3,4,66,67] and an exhaustive description will not be attempted here. In some clinical trials DFO has been combined with a variety of cytotoxic agents, and using these regimens some impressive results have been obtained. For example, Donfrancesco and colleagues[65] have reported that the combined use of DFO with cytotoxic agents (etoposide, cytoxan, carboplatin, thio-TEPA) in 57 patients suffering from neuroblastoma resulted in 24 complete responses, 5 very good responses, 21 partial responses, 3 minor responses, and 4 with progressive disease. Since NB is one of the most aggressive malignant solid tumors of childhood,[68] these results must be considered impressive. However, in some *in vivo* studies using mice bearing neuroblastoma xenografts[69] and clinical trials,[70] DFO has shown little activity and this may be due to the fact that this chelator suffers from poor membrane permeability and has a short half-life in the serum. Indeed, in order to induce apoptosis in neoplastic cells in culture, very high concentrations of DFO (2.5 mM) were required.[16] These problems combined with the known disadvantages of DFO therapy,[1,2] namely, the requirement for long s.c. infusions (12-24 h/day, 5-6 days/week) and high cost, certainly indicate that further studies on more effective chelators are essential.

3.2 Thiosemicarbazones

A wide variety of thiosemicarbazones and their metal complexes have been prepared and their anti-tumour activity has been known for some time.[71-74] Structurally these ligands are similar to those of the aroylhydrazones described below (Fig. 1). However, in contrast, none have proven to demonstrate high Fe chelation efficacy and low anti-proliferative effects. Thus, these compounds are generally more cytotoxic than some aroylhydrazones (eg. PIH). Indeed, it has been suggested that some members of the α-heterocyclic thiosemicarbazones are among the most effective inhibitors of RR yet identified.[72-74]

Of the wide variety of thiosemicarbazones that have been produced over the last 30-40 years, only two have reached clinical trials. The first of these compounds, 5-hydroxypyridine-2-carboxaldehyde thiosemicarbazone (Fig. 1) was rapidly metabolised due to the production of a O-glucuronide conjugate and these studies were discontinued.[75,76] More recently, Vion Pharmaceuticals Ltd (California, USA) have initiated a phase I clinical trial with 3-aminopyridine-2-carboxaldehyde thiosemicarbazone (3-APT or Triapine[R]; Fig. 1) that has shown highly promising results in studies *in vitro* in cell culture and *in vivo* in animals.[17] Studies on the properties of important thiosemicarbazones have recently been reviewed,[67] and further investigation of this ligand and its analogues (see below) are vital in terms of developing anti-cancer agents.

3.3 Aroylhydrazones

In 1954 Sah[77] announced the synthesis of pyridoxal isonicotinoyl hydrazone (PIH; Fig. 1) and at the same time described that this compound had distinct anti-tumour activity against leukemia and mammary carcinoma in mice. Twenty-five years then elapsed before Ponka and his colleagues demonstrated that this ligand showed pronounced Fe chelation activity *in vitro* and *in vivo* in animals.[78,79] Subsequent studies by others *in vitro* in cell culture models and *in vivo* in animals confirmed these initial observations.[29,80-83] In 1990, a phase I clinical trial demonstrated high Fe chelation efficacy in humans despite the low dose of PIH used and the fact that the compound was given in a non-soluble form that had limited bioavailability.[84] Indeed, PIH resulted in Fe excretion that would effectively maintain Fe balance in non-transfused thalassemia intermedia patients.[84]

Apart from PIH, a broad range of PIH analogues were synthesized belonging to 3 groups, namely the pyridoxal benzoyl hydrazones (100 series), salicylaldehyde benzoyl hydrazones (200 series), and the 2-hydroxy-1-naphthylaldehyde benzoyl hydrazones (300 series). Systematic substitutions in these groups enabled a useful examination of the structure-activity relationships of these chelators. Surprisingly, despite the same Fe-binding site, marked differences in activity were found.[15] The 100 series of chelators were found to be highly effective chelators but poor anti-proliferative agents. Hence, these compounds had properties suitable for the treatment of Fe-overload disease that requires long-term therapy.[15] In contrast, the 300 series of ligands had high Fe chelation efficacy and marked anti-proliferative effects, being far more active than DFO. These latter chelators obviously had potential as agents to treat cancer.[15] The anti-proliferative activity of the 300 series of compounds has been reviewed,[66,67] and will only be discussed here in terms of their effects on molecules involved in cell cycle progression.

4. THE EFFECT OF IRON CHELATORS ON CELL CYCLE PROGRESSION

It is surprizing that very little is understood concerning the effects of Fe on the cell cycle. However, it is well known that Fe chelators can induce arrest of the cell cycle at G_1/S.[23,85,86] For many years the major target was thought to be RR. However, it has become clear that Fe chelators probably act at a number of sites within the cycle and induce the expression of molecules that play major roles in controlling cell cycle progression. It is important to assess how chelators affect the cell cycle in terms of understanding the structural features of the ligand that determine if they are useful for the treatment of Fe overload or neoplasia. Below I will summarize the mechanisms involved in cell cycle control and cell death and what is known concerning the effects of Fe chelators on these processes. Due to the importance of p53 in the G_1/S transition, the effects of chelators on this molecule and its downstream targets will be the primary focus.

4.1 The Effect of Iron Chelators on the Transcription Factor p53 and its Downstream Targets

The tumour suppressor gene, p53, functions as a transcription factor that plays a key role in cell cycle control via transactivation of p53-dependent genes eg. p21/*WAF1* (wild-type p53 activating fragment 1), *GADD-45* (growth arrest and DNA damage gene), *mdm-2* (murine double minute gene), and *Bax* [87] (Fig. 3). The *p53* gene is inactivated more frequently in human cancer than any other[88] and acts as a "tumor suppressor" by inhibiting the cell cycle and allowing DNA repair. In 50% of human cancers *p53* is inactivated, while in the other 50% p53-independent regulatory pathways (see below) are probably lost.[87]

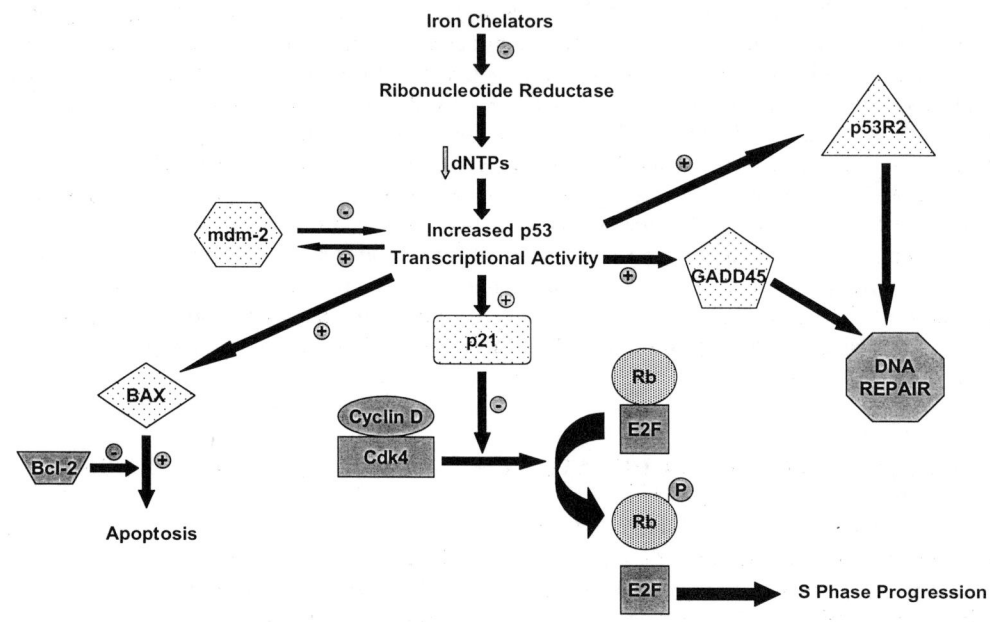

Figure 3. The possible effect of iron chelators on the p53 transcription factor and some of its target genes. It is known that p53 can act as a metabolic sensor of nucleotide levels. Indirect inhibition of ribonucleotide reductase by the ability chelators to deplete intracellular Fe pools may increase p53 activity resulting in transactivation of target genes such as *WAF1* and *GADD45* that result in proteins which inhibit cell cycle progression. See text for details.

Of the p53 target genes listed above, p21/*WAF1* is well known to induce cell cycle arrest due to its ability to act as a universal inhibitor of cyclin-dependent kinases (cdks;[89] Fig. 3). Inhibition of cdk activity prevents phosphorylation of the retinoblastoma protein (Rb) that binds the transcription factor E2F which plays an important role in the transcription of genes necessary for G_1/S transition.[87] Only after phosphorylation of Rb does it release E2F that is essential for cell cycle progression (Fig. 3). Expression of *GADD45* is induced upon DNA-damage and can arrest the cell cycle and is involved in DNA nucleotide excision repair.[90,91] The human homologue of the *mdm-2* gene acts like a feedback control of mechanism of p53 activity.[90] Transactivation of *Bax* results in apoptosis, while the anti-apoptotic protein Bcl-2 can prevent this process (Fig. 3).[87] More recently a RR gene (*p53R2*) encoding a homologue of the R2 subunit of this enzyme has been identified that is regulated by p53 transcriptional activity[92] (Fig. 3). Expression of *p53R2* in p53-deficient cells restores G_2/M arrest and cell survival after DNA damage. Inhibition of *p53R2* expression in cells that have an intact p53-dependent DNA damage checkpoint reduced RR activity activity, DNA repair, and cell survival after exposure to genotoxins.[92] The R2 subunit contains Fe, and obviously the effect of Fe chelators on the activity of this enzyme is important to understand.

In pioneering work, Terada and associates[93] used T cells to examine the effects of DFO on Rb and p53. These workers found that this chelator completely prevented Rb phosphorylation, but the expression of p53 was unaffected.[93] In contrast, in Raji and ML-1 cells, Fe-deprivation induced by DFO was shown to increase the expression of p53 protein but not mRNA, suggesting a post-transcriptional mechanism.[94] Both an increase in the expression of wild-type p53 protein (in ML-1 cells) and mutant p53 protein (Raji cells) was observed. An increase in the expression of *WAF1* was also seen in ML-1 cells which possess wild-type p53, but not in Raji cells that have mutant p53.[94]

In a cell-free system p53 appears to be subject to redox regulation, as oxidants or chelators disrupt wild-type p53 conformation and decrease its DNA-binding activity in cell lysates.[95-97] In marked contrast, using whole cell systems it has been shown that the Fe chelator 1,10-phenanthroline can induce p53 transactivation activity as well as sequence-specific DNA-binding in a dose-dependent manner.[98] In addition, this ligand induced the expression of *WAF1* and *mdm-2*, but not *Bax*, *GADD45*, or the proliferating cell nuclear antigen (PCNA).[98] The increase in p53 activity induced by 1,10-phenanthroline was not due to elevated p53 mRNA or protein levels.[98] These results are different from our previous studies[21] which showed that 311 or DFO increased the expression of *GADD45* and *WAF1* but not *mdm-2* in cells that do not possess wild-type p53.[99,100] These results suggest that the increased expression of these genes was via a p53-independent pathway. In good correlation with the anti-proliferative effects of these chelators,[15,16] only very low levels of 311 (eg. 2.5-5 µM) were required to induce *WAF1* and *GADD45* expression (Fig. 4). In contrast to 311, much higher concentrations of DFO (150 µM) were necessary to induce similar levels of expression (Fig. 4).[21] Recently, we have also found that DFO and 311 increase the expression of *GADD45* and *WAF1* in a range of cell types which possess wild-type p53[101] (Fig. 4). Collectively, the results suggest that there may be two pathways that respond to Fe chelation, namely p53-dependent and p53-independent pathways (see Section 4.2), and this requires additional investigation.

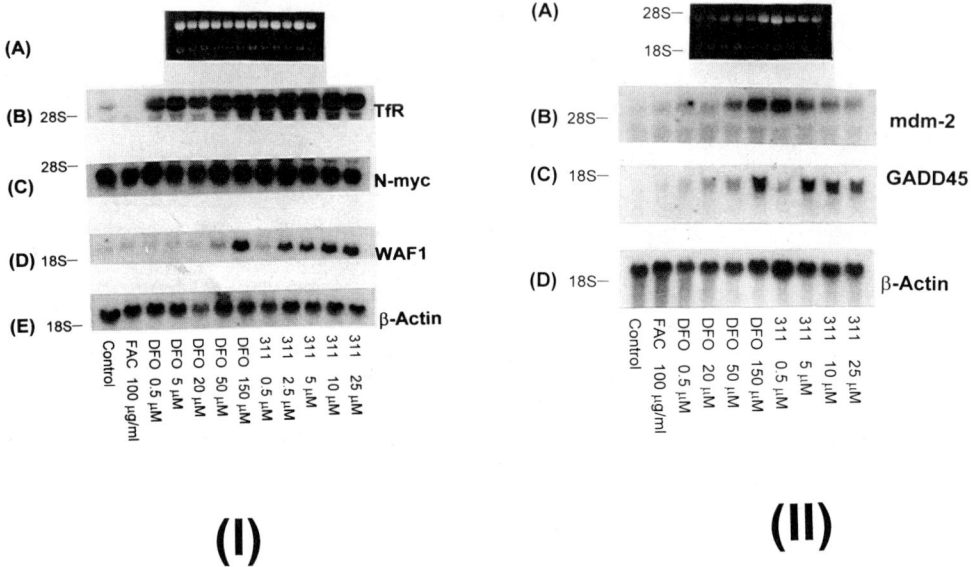

Figure 4. The effect of the concentration of DFO or 311 on mRNA levels of the transferrin receptor (TfR), N-myc, WAF1, β-actin, mdm-2, and Gadd-45 in BE-2 neuroblastoma cells. **(I-A)** Ethidium bromide staining of the agarose gel; **(I-B)** TfR; **(I-C)** N-myc; **(I-D)** WAF1; **(I-E)** β-actin; **(II-A)** Ethidium bromide staining of the agarose gel; **(II-B)** mdm-2 **(II-C)** GADD45; **(II-D)** β-actin. The result illustrated is a typical experiment from 3 experiments performed. Taken from reference 21.

The mechanism whereby Fe chelation induces the expression of genes that are transactivated by p53 remains unknown. However, it is of interest that p53 can act as a metabolic sensor.[102] When nucleotide pools are disturbed by a variety of inhibitors this results in a reversible G_0/G_1 cell cycle arrest associated with induction of *p53* and *WAF1*.[102] Since DFO and 311 inhibit ^{3}H-thymidine incorporation and chelate Fe from cells,[15,16,21] it is likely that they inhibit RR which reduces ribonucleotides to deoxyribonucleotides (dNTP). Hence, the decrease in dNTP levels could play a role in increasing *WAF1* and *GADD45* expression in cells with wild-type p53. However, it is unknown whether a similar response occurs in cells which lack functional p53 (eg. SK-N-MC NB cells and K562 cells;[99,100]), and thus explain our results.[21] Our previous studies have demonstrated that the RR inhibitor, HU, increases *WAF1* and *GADD45* expression in cells which lack functional p53, supporting a regulatory role of dNTP levels.[21]

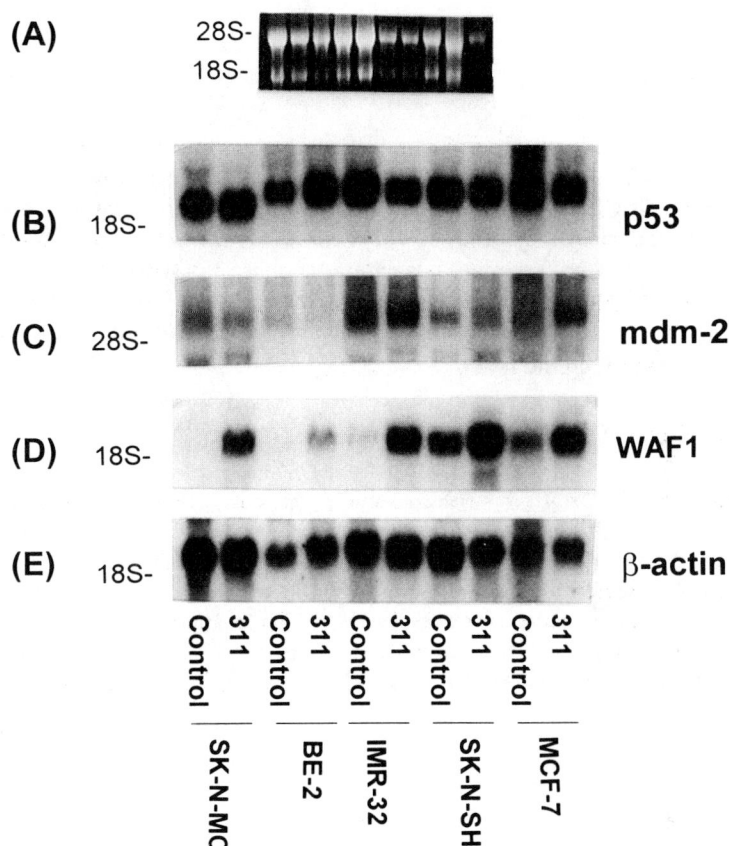

Figure 5. The effect of the PIH analogue 2-hydroxy-1-naphthylaldehyde isonicotinoyl hydrazone (311) on the mRNA levels of p53, mdm-2, WAF1 and β-actin in SK-N-MC neuroepithelioma cells, BE-2 neuroblastoma cells, IMR-32 neuroblastoma cells, SK-N-SH neuroblastoma cells and MCF-7 breast carcinoma cells. (A) Ethidium bromide staining of the agarose gel; (B) p53; (C) mdm-2; (D) WAF1; (E) β-actin. The result illustrated is a typical experiment from 3 experiments performed. From Richardson, D.R., and Gao, J. (submitted).

1.2 The Effect of Iron Chelators on the p53-Independent Pathway

The process responsible for the p53-independent pathway of increasing WAF1 and GADD45 expression after exposure to chelators remains unknown.[21] However, it is of interest that the p53 homologues, p63 and p73, may play important roles in this process (for reviews see[103,104]). Both p63 and p73 bind to the p53-response elements in both the *WAF1* and *GADD45* promoters.[104,105] Obviously, the ability of Fe chelators to inhibit cancer cells by the p53-independent pathway is of great interest, as p53 is the most frequently mutated gene in human cancer.[103]

5. SUMMARY

The success of DFO at markedly inhibiting the growth of aggressive tumors such as neuroblastoma and leukemia justifies interest in the development of chelators as anti-neoplastic agents. This is emphasized by the fact that DFO has suboptimal properties, namely poor membrane permeability and a very short serum half-life. More recently, the thiosemicarbazone chelator, Triapine, has entered a phase I clinical trial again confirming the potential of these compounds. Further studies examining the effects of chelators on neoplastic cells will not only be valuable in terms of identifing novel anti-cancer agents, but will also provide new information on the role of Fe in cell cycle control.

6. ACKNOWLDEGEMENTS

I gratefully thank Dr. Len Kritharides for critically reviewing this chapter prior to submission. I also kindly acknowledge excellent typesetting and secretarial help from Ms. Kaylene Thomas. The authors research work presented in this review was supported by grants from the Medical Research Council of Canada, National Cancer Institute of Canada, National Health and Medical Research Council of Australia, The Kathleen Cuningham Foundation for Breast Cancer Research, and the Australian Research Council Large and Small Grants Scheme. I also kindly thank the Heart Research Institute for financial support.

References

1. Olivieri, N.F., and Brittenham, G.M., 1997, Iron chelating therapy and the treatment of thalassemia, *Blood* 89:739.

2. Richardson, D.R., and Ponka, P., 1998, The development of iron chelators to treat iron overload disease and their use as experimental tools to study intracellular iron metabolism, *Am. J. Hematol.* 58:299.

3. Donfrancesco, A., Deb, G., De Sio, L., Cozza, R., and Castellano, A., 1996, Role of desferrioxamine in tumor therapy, *Acta Haematol.* 95:66.

4. Richardson, D.R., 1997, Iron chelators as effective anti-proliferative agents, *Can. J. Physiol. Pharmacol.* 75:1164.

5. Hentze, M.W., and Kuhn, L.C., 1996, Molecular control of vertebrate iron metabolism: mRNA-based regulatory circuits operated by iron, nitric oxide, and oxidative stress, *Proc. Natl. Acad. Sci. USA*, 93:8175.

6. Richardson, D.R., and Ponka, P., 1997, The molecular mechanisms of the metabolism and transport of iron in normal and neoplastic cells, *Biochim. Biophys. Acta* 1331:1.

7. Ponka, P., Beaumont, C., and Richardson, D.R., 1998, Function and regulation of transferrin and ferritin, *Semin Hematol.* 35:35.

8. Myers, C., Gianni, L., Zweier, J., Muindi, J., Sinha, K., and Eliot T., 1986, Role of iron in adriamycin biochemistry, *Fed. Proc.* 45:2792.

9. Kwok, J., and Richardson, D.R., 2000, The cardioprotective effect of the iron chelator dexrazoxane (ICRF-187) on anthracycline-mediated cardiotoxicity, *Redox Report* 6: (in press)

10. Burger, R.M., Peisach, J., and Horwitz, S.B. 1981, Activated bleomycin: a transient complex of drug, iron and oxygen that degrades DNA, *J. Biol. Chem.* 256:11636.

11. Salmon, S.E., Part VII, 1980, Chemotherapeutic Agents-Cancer Chemotherapy. In: Meyers, F.H., Jawetz, E., Goldfien, E., Goldfien, A., eds. Review of Medical Pharmacology.Los Altos: Lange Medical Publications, 477.

12. Nyholm, S., Thelander, L., and Graslund, A., 1993, Reduction and loss of the iron center in the reaction of the small subunit of mouse ribonucleotide reductase with hydroxyurea, *Biochemistry* 32:11569.

13. Nyholm, S., Mann, G.J., Johansson, A.G., Bergeron, R.J., Graslund, A., and Thelander, L., 1993, Role of ribonucleotide reductase in inhibition of mammalian cell growth by potent iron chelators, *J. Biol. Chem.* 268: 26200.

14. Cooper, C.E., Lynagh, G.R., Hoyes, K.P., Hider, R.C., Cammack, R., and Porter, J.B. 1996, The relationship of intracellular iron chelation to the inhibition and regeneration of human ribonucleotide reductase. *J. Biol. Chem.* 271: 20291.

15. Richardson, D.R., Tran, E., and Ponka, P., 1995, The potential of iron chelators of the pyridoxal isonicotinoyl hydrazone class as effective antiproliferative agents, *Blood* 86:4295.

16. Richardson, D.R., and Milnes, K., 1997, The potential of iron chelators of the pyridoxal isonicotinoyl hydrazone class as effective anti-proliferative agents II: The mechanism of action of ligands derived from salicylaldehyde benzoyl hydrazone and 2-hydroxy-1-naphthylaldehyde benzoyl hydrazone, *Blood* 89:3025.

17. Finch, R.A., Liu, M-C., Grill, S.P., Rose, W.C., Loomis, R., Vasquez, K.M., Cheng, Y-C., and Sartorelli, A.C., 2000, Triapine (3-aminopyridine-2-carboxaldehyde-thiosemicarbazone): A potent inhibitor of ribonucleotide reductase activity with broad spectrum antitumor activity, *Biochem. Pharmacol.* 59:983.

18. Beckloff, G.L., Lerner, H.J., Frost, D., Russo-Alesi, F.M., and Gitomer, S., 1965, Hydroxyurea (NSC-32065) in biological fluids: Dose-concentration relationship, *Cancer Chemother. Rep.* 48:57.

19. Gwilt, P.R., and Tracewell, W.G., 1998, Pharmacokinetics and pharmacodynamics of hydroxyurea, *Clin. Pharmacokinet.* 34:347.

20. Moore, E.C., and Hurlbert, R.B., 1989, The inhibition of ribonucleoside diphosphate reductase by hydroxyurea, guanazole and pyrazoloimidazole (IMPY). In: *Inhibitors of Ribonucleoside Diphosphate Reductase Activity*, Cory, J.G., Cory, A.H., eds., Pergamon Press, Oxford, pp. 165-201.

21. Darnell, G., and Richardson, D.R., 1999, The potential of analogues of the pyridoxal isonicotinoyl hydrazone class as effective anti-proliferative agents III: The effect of the ligands on molecular targets involved in proliferation, *Blood* 94:781.

22. Chitambar, C.R., and Wereley, J., 1995, Effect of hydroxyurea on cellular iron metabolism in human leukemic CCRF-CEM cells: Changes in iron uptake and the regulation of transferrin receptor and ferritin gene expression following inhibition of DNA synthesis, *Cancer Res.* 55:4361.

23. Brodie, C., Siriwardana, G., Lucas, J., Schleicher, R., Terada, N., Szepesi, A., Gelfand, E., and Seligman, P., 1993, Neuroblastoma sensitivity to growth inhibition by deferoxamine: Evidence for a block in the G_1 phase of the cell cycle, *Cancer Res.* 53: 3968.

24. Rang, H.P., Dale, M.M., and Ritter, M.M., 1995, *Pharmacology*, 3rd Edn., Churchill Livingston Publishers.

25. Blatt, J., and Stitely, S., 1987, Antineuroblastoma activity of desferrioxamine in human cell lines, *Cancer Res.* 47:1749.

26. Becton, D.L., and Bryles, P., 1988, Deferoxamine inhibition of human neuroblastoma viability and proliferation, *Cancer Res.* 48:7189.

27. Donfrancesco, A., Deb, G., Dominici, C., Pileggi, D., Castello, M.A., and Helson, L., 1990, Effects of a single course of deferoxamine in neuroblastoma patients, *Cancer Res.* 50:4929.

28. Kontoghiorghes, G.J., and Evans, R.W. 1985, Site specificity of iron removal from transferrin by α-ketohydroxypyridone chelators, *FEBS Lett.* 189: 141.

29. Baker, E., Richardson, D.R., Gross, S., and Ponka, P., 1992, Evaluation of the iron chelation potential of pyridoxal, salicylaldehyde and 2-hydroxy-1-naphthylaldehyde using the hepatocyte in culture, *Hepatology,* 15:492.

30. Morgan, E.H., 1981, Transferrin: biochemistry, physiology and clinical significance, *Mol. Aspects Med.* 4:1.

31. Kawabata, H., Yang, R., Hirama, T., Vuong, P.T., Kawano, S., Gombart, A.F., and Koeffler, H.P., 1999, Molecular cloning of transferrin receptor 2: A new member of the transferrin receptor-like family, *J. Biol. Chem.* 274:20826.

32. Kawabata, H., Germain, R.S., Vuong, P.T., Nakamaki, T., Said, J.W., and Koeffler, H.P., 2000, Transferrin receptor 2-α supports cell growth both in iron-chelated cultured cells and in vivo, *J. Biol. Chem.* 275:16618.

33. Page, M.A., Baker, E., and Morgan, E.H., 1984, Transferrin and iron uptake by rat hepatocytes in culture, *Am. J. Physiol.* 246:G26.

34. Trinder, D., Morgan, E.H., and Baker, E., 1986, The mechanisms of iron uptake by rat fetal hepatocytes, *Hepatology* 6:852.

35. Richardson, D.R., and Baker, E., 1990, The uptake of iron and transferrin by the human melanoma cell, *Biochim Biophys. Acta* 1053:1.

36. Richardson, D.R., and Baker, E., 1994, Two saturable mechanisms of iron uptake from transferrin in human melanoma cells: The effect of transferrin concentration, chelators and metabolic probes on transferrin and iron uptake, *J. Cell Physiol.* 161:160.

37. Richardson, D.R., and Ponka, P., 1994, The iron metabolism of the human neuroblastoma cell. Lack of relationship between the efficacy of iron chelation and the inhibition of DNA synthesis, *J. Lab. Clin. Med.* 124: 660.

38. Trinder, D., Zak, O., and Aisen, P., 1996, Transferrin receptor-independent uptake of diferric transferrin by human hepatoma cells with antisense inhibition of receptor expression, *Hepatology* 23:1512.

39. Trowbridge, I.S., and Lopez, F., 1982, Monoclonal antibody to transferrin receptor blocks transferrin binding and inhibits tumor cell growth in vitro, *Proc. Natl. Acad. Sci. USA* 79:1175.

40. Kemp, J.D., Smith, K.M., Kanner, L.J., Gomez, F., Thorson, J.A., and Naumann, P.W., 1990, Synergistic inhibition of lymphoid tumor growth in vitro by combined treatment with the iron chelator deferoxamine and an immunoglobulin G monoclonal antibody against the transferrin receptor, *Blood* 76:991.

41. Kemp, J.D., Thorson, J.A., Stewart, B.C., and Naumann, P.W., 1992, Inhibition of hematopoietic tumor growth by combined treatment with deferoxamine and an IgG monoclonal antibody against the transferrin receptor: evidence for a threshold model of iron deprivation toxicity, *Cancer Res.* 52: 4144.

42. Levy, J.E., Jin, O., Fujiwara, Y., Kuo, F., and Andrews, N.C., 1999, Transferrin receptor is necessary for the development of erythrocytes and the nervous system, *Nature Genet.* 21:396.

43. Chan, L-N.L., and Gerhardt, E.M., 1992, Transferrin receptor gene is hyperexpressed and transcriptionally regulated in differentiating erythroid cells, *J. Biol. Chem.* 267:8254.

44. Bianchi, L., Tacchini, L., and Cairo, G., 1999, HIF-1-mediated activation of transferrin receptor gene transcription by iron chelation, *Nucleic Acids Res.* 27:4223.

45. Lok, C.N., and Ponka, P., 1999, Identification of a hypoxia response element in the transferrin receptor gene. *J. Biol.Chem.* 274:24147.

46. Tacchini, L., Bianchi, L., Bernelli-Zazzera, A., and Cairo, G., 1999, Transferrin receptor induction by hypoxia. HIF-1 mediated transcriptional activation and cell specific post-transcriptional regulation, *J. Biol. Chem.* 274:24142.

47. Larrick, J.W., and Cresswell, P., 1979, Modulation of cell surface iron transferrin receptors by cellular density and the state of activation, *J. Supramol. Struct.* 11:579.

48. Richardson, D.R., and Baker, E., 1992, Two mechanisms of iron uptake from transferrin by melanoma cells. The effect of desferrioxamine and ferric ammonium citrate, *J. Biol. Chem.* 267:13972.

49. Chan, S.M., Hoffer, P.B., Maric, N., and Duray, P., 1987, Inhibition of gallium-67 uptake in melanoma by an anti-human transferrin receptor monoclonal antibody, *J. Nucl. Med.* 28:1303.

50. Crawford, E.D., Saiers, J.H., Baker, L.H., Costanzi, J.H., and Bukowski, R.M., 1991, Gallium nitrate in advanced bladder carcinoma: southwest oncology group study, *Urology,* 38:355.

51. Chitambar, C.R., Zahir, S.A., Ritch, P.S., and Anderson, T., 1997, Evaluation of continuous-infusion gallium nitrate and hydroxyurea in combination for the treatment of refractory non-Hodgkin's lymphoma, *Am. J. Clin. Oncol.* 20:173.

52. Chitamber, C.R. and Seligman, P.A., 1986, Effects of different transferrin forms on transferrin receptor expression, iron uptake, and cellular proliferation of human leukemic HL60 cells: mechanisms responsible for the specific cytotoxicity of transferrin-gallium, *J. Clin. Invest.* 78: 1538.

53. Chitambar, C.R., and Zivkovic, Z., 1987, Uptake of gallium-67 by human leukemic cells: Demonstration of transferrin receptor-dependent and transferrin receptor-independent mechanisms, *Cancer Res.* 47:3929.

54. Lovejoy, D.B., and Richardson, D.R., 2000, Complexes of gallium(III) and other metal ions and their potential in the treatment of neoplasia, *Expert Opin. Invest. Drugs* 9:1257.

55. Gunshin, H., MacKenzie, B., Berger, U.V., Gunshin, Y., Romero, M.F., Boron, W.F., Nussberger, S., Gollan, J.L., and Hediger, M.A., 1997, Cloning and characterization of a mammalian proton-coupled metal-ion transporter, *Nature* 388:482.

56. Fleming, M.D., Trenor, C.C., Su, M.A., Foernzler, D., Beier, D.R., Dietrich, W.F., and Andrews, N.C., 1997, Microcytic anemia mice have a mutation in *Nramp2*, a candidate iron transporter gene, *Nat. Genet.* 16:383.

57. Harrison, P.M., and Arosio, P., 1996, The ferritins: molecular properties, iron storage function and cellular regulation, *Biochim. Biophys. Acta* 1275:161.

58. Osaki, S., Johnson, D.A., and Frieden, E., 1971, The mobilization of iron from the perfused mammalian liver by a serum copper enzyme, ferroxidase I, *J. Biol. Chem.* 246:3018.

59. Young, S.P., Fahmy, M., and Golding, S., 1997, Ceruloplasmin, transferrin and apotransferrin facilitate iron release from human liver cells, *FEBS Lett.* 411:93.

60. Richardson, D.R., 1999, The role of ceruloplasmin and ascorbate in cellular iron release, *J. Lab. Clin. Med.* 134:454.

61. Donovan, A., Brownlie, A., Zhou, Y., Shepard, J., Pratt, S.J., Moynihan, J., Paw, B.H., Drejer, A., Barut, B., Zapata, A., Law, T.C., Brugnara, C., Lux, S.E., Pinkus, G.S., Pinkus, J.L., Kingsley, P.D., Palis, J., Fleming, M.D., Andrews, N.C., and Zon, L.I., 2000, Positional cloning of zebrafish *ferroportin 1* identifies a conserved vertebrate iron exporter, *Nature* 403:776.

62. Estrov, Z., Tawa, A., Wang, X-H., Dube, I.D., Sulh, H., Cohen, A., Gelfand, E.W., and Freedman, M.H., 1987, In vivo and in vitro effects of desferrioxamine in neonatal acute leukemia, *Blood* 69:757.

63. Dezza, L., Cazzola, M., Danova, M., Carlo-Stella, C., Bergamaschi, G., Brugnatelli, S., Invernizzi, R., Mazzini, G., Riccardi, A., and Ascari, E., 1989, Effects of desferrioxamine on normal and leukemic human hematopoietic cell growth: in vitro and in vivo studies, *Leukemia* 3:104.

64. Donfrancesco, A., Deb, G., Dominici, C., Angioni, A., Caniglia, M., De Sio, L., Fidani, P., Amici, A., and Helson, L., 1992, Deferoxamine, cyclophosphamide, etoposide, carboplatin, and thiotepa (D-CECat): A new

cytoreductive chelation-chemotherapy regimen in patients with advanced neuroblastoma, *Am. J. Clin. Oncol.* 15:319.

65. Donfrancesco, A., De Bernardi, B., Carli, M., Mancini, A., Nigro, M., De Sio, L., Casale, F., Bagnulo, S., Helson, L., and Deb, G., 1995, Deferoxamine (D) followed by cytoxan (C), etoposide (E), carboplatin (Ca), thio-TEPA (T), induction regimen in advanced neuroblastoma, *Eur. J. Cancer* 31A: 612.

66. Richardson, D.R., 1998, Analogues of pyridoxal isonicotinoyl hydrazone (PIH) as potential iron chelators for the treatment of neoplasia, *Leukemia & Lymphoma* 31:47.

67. Richardson, D.R., 2001, The use of iron chelators as therapeutic agents for the treatment of cancer, *Crit. Rev. Oncol. Hematol.* (in press).

68. Voute, P.A., 1984, Neuroblastoma, In: *Clinical Pediatric Oncology*, Mosby Publ. Co., 1984:559.

69. Selig, R.A., White, L., Gramacho, C., Sterlinglevis, K., Fraser, I.W., and Naidoo, D., 1998, Failure of iron chelators to reduce tumor growth in human neuroblastoma xenografts, *Cancer Res.* 58:473.

70. Blatt, J., 1994, Deferoxamine in children with recurrent neuroblastoma, *Anticancer Res.* 14:2109.

71. Sartorelli, A.C., Agrawal, K.C., and Moore, E.C., 1971, Mechanism of inhibition of ribonucleoside diphosphate reductase by α-(N)-heterocyclic aldehyde thiosemicarbazones, *Biochem. Pharmacol.* 20:3119.

72. French, F.A., Blanz, E.J. JR., Schaddix, S.C., and Brockman, R.W., 1974, α-(N)-Formylheteroaromatic thiosemicarbazones. Inhibition of tumor derived ribonucleoside diphosphate reductase and correlation with in vivo anti-tumor activity, *J. Med. Chem.* 17:172.

73. Liu, M-C., Lin, T-S., and Sartorelli, A.C., 1995, Chemical and biological properties of cytotoxic α-(N)-heterocyclic carboxaldehyde thiosemicarbazones. In: *Progress in Medicinal Chemistry*, Volume 32. Ellis, G.P., Luscombe, D.K. (Eds.), Elsevier Science B.V. 1-35.

74. Agrawal, K.C., and Sartorelli, A.C., 1978, The chemistry and biological activity of α-(N)-heterocyclic carboxaldehyde thiosemicarbazones, *Prog. Med. Chem.* 15:321.

75. DeConti, R.C., Toftness, B.R., Agrawal, K.C., Tomchick, R., Mead, J.A.R., Bertino, J.R., Sartorelli, A.C., and Creasey, W.A., 1972, Clinical and pharmacological studies with 5-hydroxy-2-formylpyridine thiosemicarbazone, *Cancer Res.* 32:1455.

76. Krakoff, I.H., Etcubanas, E., Tan, C., Mayer, K., Bethune, V., and Burchenai, J.H., 1974, Clinical trial of 5-hydroxypicolinaldehyde thiosemicarbazone (5-HP: NSC-107392), with special reference to its Fe chelating properties, *Cancer Chemother. Rep.* 53:207.

77. Sah, P., 1954, Nicotinoyl and isonicotinoyl hydrazones of pyridoxal, *J. Am. Chem. Soc.*, 76:300.

78. Ponka, P., Borova, J., Neuwirt, J., and Fuchs, O., 1979, Mobilization of iron from reticulocytes. Identification of pyridoxal isonicotinoyl hydrazone as a new iron chelating agent, *FEBS Lett.* 97:317.

79. Ponka, P., Borova, J, Neuwirt, J., Fuchs, O., and Necas, E., 1979, A study of intracellular iron metabolism using pyridoxal isonicotinoyl hydrazone and other synthetic chelating agents, *Biochim. Biophys. Acta* 586:278.

80. Cikrt, M., Ponka, P., Necas, E., and Neuwirt, J., 1980, Biliary iron excretion in rats following pyridoxal isonicotinoyl hydrazone, *Br. J. Haematol.* 45:275.

81. Hoy, T., Humphreys, J., Jacobs, A., Williams, A., and Ponka, P., 1979, Effective iron chelation following oral adminstration of an isoniazid pyridoxal hydrazone, *Br. J. Haematol.* 43:443.

82. Baker, E., Vitolo, M.L., and Webb, J.M., 1985, Iron chelation by pyridoxal isonicotinoyl hydrazone and analogues in hepatocytes in culture, *Biochem. Pharmacol.* 34:3011.

83. Richardson, D.R., and Ponka, P., 1998, Orally effective iron chelators for the treatment of iron overload disease: The case for a further look at pyridoxal isonicotinoyl hydrazone (PIH) and its analogs, *J. Lab. Clin. Med.* 132:351.

84. Brittenham, G.M., 1990, Pyridoxal isonicotinoyl hydrazone: an effective chelator after oral administration, *Semin. Hematol.* 27:112.

85. Hoyes, K.P., Hider, R.C., and Porter, J.B., 1992, Cell cycle synchronization and growth inhibition by 3-hydroxypyridin-4-one iron chelators in leukemic cell lines, *Cancer Res.* 52: 4591.

86. Lucas, J.J., Terada, N., Szepesi, A., and Gelfand, E.W., 1992, Regulation of synthesis of p34cdc2 and its homologues and their relationship to p110Rb phosphorylation during cell cycle progression of normal human T cells, *J. Immunol.* 148:1804.

87. Agarwal, M.L., Taylor WR, Chernov MV, Chernova OB, and Stark, GR., 1998, The p53 network, *J. Biol. Chem.* 273:1.

88. Greenblatt, M.S., Bennett, W.P., Hollstein, and M., Harris, C.C., 1994, Mutations in the p53 suppressor gene: clues to cancer etiology and molecular pathogenesis, *Cancer Res.* 54:4855.

89. Xiong, Y., Hannon, G.J., Zhang, H., Casso, D., Kobayashi, R., and Beach, D., 1993, p21 is a universal inhibitor of cyclin kinases, *Nature* 366:701.

90. Levine, A.J., 1997, p53, the cellular gatekeeper for growth and division, *Cell* 88:323.

91. Katan, M.B., Zhan, Q., El-Deiry, W.S., Carrier. F., Jacks, T., Walsh, W.V., Plunkett, B.S., Vogelstein, B., and Fornace, A.J. Jr, 1992, A mammalian cell cycle check point pathway utilizing p53 and gadd45 is defective in ataxia telangiectasia, *Cell* 71:587.

92. Tanaka, H., Arakawa, H., Yamaguchi, T., Shiraishi, K., Fukuda, S., Matsui, K., Takei, Y., and Nakamura, Y., 2000, A ribonucleotide reductase gene involved in a p53-dependent cell-cycle checkpoint for DNA damage, *Nature* 404:42.

93. Terada, N., Lucas, J.J., and Gelfand, E.W., 1991, Differential regulation of the tumor suppressor molecules, retinoblastoma susceptibility gene product (Rb) and p53, during cell cycle progression of normal human T cells, *J. Immunol.* 147: 698.

94. Fukuchi, K., Tomoyasu, S., Watanabe, H., Kaetsu, S., Tsuruoka, N., and Gomi, K., 1995, Iron deprivation results in an increase in p53 expression, *Biol. Chem. Hoppe Seyler,* 376:627.

95. Russo, T., Zambrano N., Esposito, F., Ammendola, R., Cimino, F., Fiscella, M., Jackman, J., O Connor, P.M., Anderson, C.W., and Appella, E.,1995, A p53-independent pathway for activation of WAF1/CIP1 expression following oxidative stress, *J. Biol. Chem.* 270:29386.

96. Hainaut, P., and Milner, J., 1993, A structural role for metal ions in "wild-type" conformation of the tumor suppressor protein p53, *Cancer Res.* 53:1739.

97. Hainaut, P., Butcher, S., and Milner, J., 1995, Temperature sensitivity for conformation is an intrinsic property of wild-type p53, *Br. J. Cancer*, 71: 227.

98. Sun, Y., Bian, J., Wang, Y., and Jacobs, C., 1997, Activation of p53 transcriptional activity by 1,10-phenanthroline, a metal chelator and redox sensitive compound, *Oncogene* 14:385.

99. Davidoff, A.M., Pence, J.C., Shorter, N.A., Ingelhart, J.D., and Marks, J.R., 1992, Expression of p53 in human neuroblastoma- and neuroepithelioma-derived cell lines, *Oncogene* 7:127.

100. Bi, S., Hughes, T., Bungey, J., Chase, A., de Fabritiis, P., and Goldman, J.A., 1992, p53 in chronic myeloid leukemia cell lines, *Leukemia* 6:839

101. Gao, J., Lovejoy, D., and Richardson, D.R., 1999, Effect of iron chelators with potent anti-proliferative activity on the expression of molecules involved in cell cycle progression and proliferation, *Redox Report* 4:311.

102. Linke, S.P., Clarkin, K.C., Di Leonardo, A., Tsou, A., and Wahl, G.M., 1996, A reversible, p53-dependent Go/G1 cell cycle arrest induced by ribonucleotide depletion in the absence of detectable DNA damage, *Genes and Develop.* 10:934.

103. Levrero, M., De Laurenzi, V., Costanzo, A., Gong, J., Melino, G., and Wang, J.Y., 1999, Structure, function and regulation of p63 and p73, *Cell Death Differ.* 6:1146.

104. Lohrum, M.A., and Vousden, K.H., 2000, Regulation and function of the p53-related proteins: same family, different rules, *Trends Cell Biol.*10:197.

105. De Laurenzi, V., Costanzo, A., Barcaroli, D., Terrinoni, A., Falco, M., Annicchiarico-Petruzzelli, M., Levrero, M., and Melino, O., 1998, Two new p73 splice variants, gamma and delta, with different transcriptional activity, *J. Exp. Med.* 188:1763.

ANTIMALARIAL EFFECT OF IRON CHELATORS

Victor R. Gordeuk and Mark Loyevsky

1. INTRODUCTION

Malaria is a major cause of morbidity and mortality in tropical areas of the world. Despite work on controlling the mosquito vector and on developing an effective vaccine, anti-malarial chemotherapy is at present the fundamental approach to controlloing this infection. Because of widespread development of resistance on the part of the parasite to the presently available anti-malarial drugs, it is important to develop new agents with new mechanisms of action. Iron is an essential nutrient for the malaria parasite, and this chapter reviews the information available on the anti-malarial potential of iron chelating agents.

2. IRON-RELATED PATHWAYS OF *P. FALCIPARUM*

2.1. Plasmodial Metabolic Pathways that are Dependent on Iron: Many enzymes of the erythrocytic malaria parasite are dependent on iron and Table 1 presents a partial listing. These enzymes are involved in DNA synthesis[1,2,3], de novo synthesis of heme[4], and normal mitochondrial function and electron transport[5,6]. The withholding of iron by iron chelators could possibly cause malfunction of certain of these enzymes for which iron is indispensable. Two of the plasmodial enzymes that are dependent on iron will be considered in more detail, namely ribonucleotide reductase and delta-aminolivulinate synthase.

Ribonucleotide reductase is an iron-containing enzyme of the malaria trophozoite that is essential for DNA synthesis[1,2]. This enzyme catalyzes the reduction of ribonucleoside diphosphates to deoxyribonucleoside diphosphates, the precursors of DNA[7]. *In vitro*, iron chelation reversibly inhibits ribonucleotide reductase[7,8,9] and produces a potent inhibition of DNA synthesis in various cellular systems[8]. Exposure to desferrioxamine leads to decreased levels of mRNA for the small B2 subunit of ribonucleotide reductase in *P. falciparum* parasites cultured in erythrocytes [10].

Table 1. Some enzymes of the erythrocytic trophozoite that are dependent on iron.

Enzyme	Function	Reference
Ribonucleotide reductase	Synthesis of DNA	1,2
Dihydroorotate dehydrogenase	Synthesis of pyrimidine	3
Delta-aminolevulinate synthase	Synthesis of heme	4
Cytochrome oxidase and reductase, Cytochrome b	Transport of electrons in the mitochondria	5,6

Delta-aminoleulinate synthase, the first enzyme in the biosynthetic pathway of heme, appears to be synthesized by the malaria parasite[4]. It is of interest that *P. falciparum* and other plasmodial species synthesize heme *de novo*, despite the fact that the parasite is located within the red blood cell, a virtual "red sea" of host hemoglobin. To synthesize heme, it appears that the parasite produces the first enzyme in the pathway, delta-aminolevulinate synthase. It also appears that other enzymes in the heme synthetic pathway, such as delta-aminolevulinate dehydrase, coproporphyrinogen oxidase and ferrochelatase, are of host origin and are transported into the parasite from the host red blood cell compartment[4]. In mammalian erythroid cells, iron chelators cause a down-regulation in delta-aminolevulinate synthase synthesis[11], resulting in a reduced ability to synthesize heme. Iron chelators might exert a similar effect in the erythrocytic trophozoite, which could prevent cytochrome synthesis. The *de novo* synthesis of heme by the malarial parasite might therefore represent a novel target for antimalarial therapy[12,4].

2.2 Acquisition of Iron by the Intraerythrocytic Parasite: The intraerythrocytic parasite lies within a parasitophorous vacuole. The trophozoite obtains nutrients by ingesting host cell cytoplasm including hemoglobin by means of a cytostome[13] and may possibly take up molecules from the outer medium directly through a parasitophorous duct[14,15]. Host hemoglobin is degraded in the food vacuole of the parasite, and heme liberated in this process is polymerized to form hemozoin[16]. How the intraerythrocytic phase parasite acquires iron has not yet been determined, and several possible sources have been postulated.

Iron bound to host transferrin in the plasma has been proposed as one potential source of iron for the erythrocytic parasite[17], but transferrin receptors do not appear to be expressed on mature parasitized erythrocytes[18,19], and the bulk of the evidence indicates that host transferrin iron is not taken up by parasitized red cells[20,21]. Iron contained in host erythrocyte ferritin is another theoreitical source of iron for the erythrocytic plasmodium. Although the mature erythrocyte cannot synthesize ferritin, it does contain residual ferritin that was produced during the earlier erythroblast phase[22]. The possibility that host erythrocyte ferritin serves as a source of iron for the parasite has not been investigated.

Iron derived from host hemoglobin is a third possibility as a source of iron for the erythrocytic parasite. Heme released during the proteolysis of host hemoglobin in the food vacuole of the parasite contains a substantial amount of iron, which, if liberated from heme, might be available for the parasite's metabolic needs[23]. Although firm evidence that the parasite utilizes iron derived from host heme is lacking at present, the possibility that a small amount of heme in the food vacuole is degraded in a controlled manner to release iron for the metabolic processes of the parasite represents one of the plausible sources of iron for the intraerythrocytic parasite.

Iron contained in a labile pool of iron in the cytoplasm of the erythrocyte is a fourth possible source of iron for the erythrocytic parasite[20]. In support of this possible source, gel filtration and ultrafiltration studies on hemolysates of rat red cells parasitized with *P. berghei* revealed a labile pool of iron that is chelatable by preincubation of the intact cells with the iron chelator, desferrioxamine[20]. Further evidence in support of this hypothesis was obtained recently by monitoring the concentration of labile iron in parasitized and non-parasitized erythrocytes with the fluorescent iron-sensing probe, calcein. Labile iron pools were lower in parasitized than non-parasitized erythrocytes, suggesting that labile iron of the host red cell may be either utilized or stored during plasmodial growth[24]. On the other hand, two studies found that when iron-chelating agents are introduced into the cytoplasm of erythrocytes but not into the parasite compartment within the parasitophorous vacuole, no plasmodial growth inhibition occurs[15,25]. It is possible that both host labile iron and another source of iron, such as host hemoglobin iron, are used by the parasite, and that the abrogation of only one source will not prevent parasite growth.

3. ANTI-MALARIAL ACTIVITY OF IRON CHELATORS *IN VITRO*

3.1 Classes of Iron Chelators that Have Antimalarial Activity *in vitro*: As shown in Table 2, several classes of iron-chelating compounds suppress the growth of *P. falciparum* in erythrocytes *in vitro*. A number of these compounds are naturally occurring siderophores, molecules produced by microorganisms to acquire iron from the environment. The antimalarial iron chelators can be placed into two major categories depending on the predominant mechanism of inhibition of parasite growth. One mechanism is the withholding of iron from plasmodial metabolic pathways by iron III chelators such as desferrioxamine and deferiprone[26,27,28,29,30]. The other is the formation of complexes with iron that are toxic to the parasite by iron II chelators such as bipyridyl[31,32,30]. For both of these categories, an interaction with iron is the focus of the antimalarial activity.

Table 2. Classes of iron chelators that inhibit growth of P. falciparum in erythrocytes.

Class of compound and selected specific agents	ID_{50}*	Reference
A. AGENTS THAT INHIBIT PARASITE GROWTH BY WITHHOLDING IRON		
Hydroxamate siderophores and derivatives		
Desferrioxamine	4-35 µM	26
Desferrithiocin**	25 µM	77
Desferricrocin	30-40 µM	77
Reversed siderophores	0.3-70 µM	34
Catecholamide and catecholate siderophores		
Vibriobactin	2-5 µM	78
Parabactin	2-3 µM	78
Gamma amino butyric acid (GABA)	4-5 µM	78
Alpha-ketohydroxypyridinones		
Deferiprone	15-45 µM	27
Dihydroxycoumarins		
Daphnetin (ash tree bark extract)	25-40 µM	79
Polyanionic amines		
HBED	5 µM	50
Acylhydrazones		
Pyridoxal isonicotinoyl hydrazone (PIH)	30 µM	29
Salicylaldehyde isonicotinoyl hydrazone (SIH)	18-30 µM	45
Aminothiols		
Ethane-1,2-bis(N-1-amino-3-ethylbutyl-3-thiol) (BAT)	6-9 µM	30
Aminophenols		
Aminophenol II	0.5-0.7µM	24
Bis-cyclic imides		
Dexrazoxane***	32-36 µM	47
B. AGENTS THAT INHIBIT PARASITE GROWTH BY FORMING TOXIC COMPLEXES WITH IRON		
2,2'-Bipyridyl	12-14 µM	80
8-Hydroxyquinoline	8.3 nM	31

* Concentrations of iron chelator that produce 50% growth inhibition after 48-72 hours of culture.
** Effective against both erythrocytic and hepatic stages of the parasite.
*** Effective against hepatic stage but not erythrocytic stage of the parasite.

3.2 Physical Characteristics of Antimalarial Iron Chelators: The iron withheld by iron chelators in the process of inhibiting the growth of intraerythrocytic malaria parasites most likely resides within the parasitic compartment of the infected red blood cell[20,15,25,24]. One would thus predict and it has been observed that effective antimalarial iron chelators have the ability to cross lipid membranes well, have a high affinity for iron (II) or iron (III), and selectively bind iron as compared to other trace metals[33,34,35,36,30]. The iron atom has six coordination sites, and hexadentate chelators would be expected to form the most stable complexes with the metal. Pentadentate and quadridentate chelators may leave one or two coordination sites of the iron atom unbound and potentially available to participate in toxic reactions that could damage host tissues. Tridentate and bidentate chelators could fully occupy the coordination sites of iron by forming 2:1 or 3:1 complexes with the metal, but, especially at low chelator concentrations, partial dissociation from iron might occur and expose coordination sites to participate in toxic reactions.

3.3 Desferrioxamine: Desferrioxamine is a naturally occurring trihydroxamic acid derived from cultures of *Streptomyces pilosus* that is widely available for clinical use as an iron chelator. The agent is remarkably safe and nontoxic, but to be effective it must be administered parenterally as a slow infusion. As shown in Table 2, when given as a single agent desferrioxamine suppresses the growth of *P. falciparum* in parasitized erythrocytes *in vitro* in concentrations achieved and tolerated in the blood of patients. The antimalarial activity of deseferrioxamine appears to be related to its ability to enter the erythrocytic trophozoite and to chelate a pool of parasite-associated iron. It has been suggested that desferrioxamine may enter the parasite directly through a parasitophorous duct that invaginates from the red cell membrane and communicates with the parasitophorous vacuole[14], thus bypassing the host red cell cytoplasm[15].

Experiments with synchronized *in vitro* cultures of *P. falciparum* showed that desferrioxamine has a cytocidal effect on late trophozoites and early schizonts, and that the critical duration of exposure may be as short as six hours at this stage of parasite development[37]. Ultrastructural lesions included the breakdown of the nuclear envelope into small membranous fragments and progressive vacuolization of the nucleoplasm[38]. Other organelles, including food vacuoles and mitochondria, were not visually affected, though the most recent biochemical data indicate that the levels of mRNAs encoding cytochrome c oxidase and cytochrome b are affected[10]. Desferrioxamine has an additive inhibitory effect on the *in vitro* growth of *P. falciparum* when it is combined with classical antimalarials[39], although in a single report it failed to enhance the activity of chloroquine[40]. It has been reported that zinc-desferrioxamine complexes have greater anti-parasitic activity than desferrioxamine alone on *P. falciparum* cultured in erythrocytes[41]. A potential explanation for this observation is that Zn-desferrioxamine may penetrate the cell membranes and exchange bound zinc for ferric ions because the affinity of desferrioxamine for iron is greater than its affinity for zinc. In summary, the available studies indicate that the action of desferrioxamine on the intraerythrocytic parasite is both stage-specific and cytocidal. In contrast, in mammalian cells this compound displays only a cytostatic inhibitory effect, which is reversed upon the removal of the drug from the suspension[42,43]. The differential effect of desferrioxamine on malaria-infected erythrocytes and mammalian cells provides the basis for the selective action of desferrioxamine as an antimalarial.

3.4 Deferiprone: Deferiprone is a neutral bidentate ligand with a high specificity for ferric iron. The stability constant for the iron complex (log K_a = 37) is six orders of magnitude higher than that of desferrioxamine. Unlike desferrioxamine, deferiprone has some effecacy to treat iron overload when administered orally. *In vitro*, deferiprone exhibits a dose-related suppression of *P. falciparum* growth[27,33,44]. The agent inhibits the growth of *P. falciparum* by more than 50% at concentrations ranging from 5 to 100 µM, when exposure to the chelator is continuous.

3.5 Combinations of Iron Chelators: The combination of slowly permeating desferrioxamine with more liphophilic and rapidly permeating iron chelators such as the reversed siderophore, RSFilem2, or the acylhydrazone, SIH, has led to a potentiation of the antimalarial effect *in vitro* of these chelators as single agents[42,45]. The more rapidly permeating agents seem to have the ability to inhibit parasite growth both at the ring and trophozoite stages, while desferrioxmaine is active at the trophozoite stage.

3.6 Inhibition of the Hepatic Phase of Malarial Growth by Iron Chelators: The studies summarized in Table 2 and demonstrating a suppressive effect on the growth of *P. falciparum* by iron chelators were performed on the erythrocytic phase of the life cycle of the parasite. It was also reported that desferrioxamine and desferrithiocin[46] and dexrazoxane[47] are able to inhibit the growth of the exoerythrocytic hepatic phase of *P. falciparum* or *P. yoelii* in culture systems employing human or mouse hepatocytes. These studies suggest that iron chelation represents a potential antimalarial strategy with effectiveness against both the erythrocytic and hepatic phases of the parasite. Dexrazoxane is an iron-chelating prodrug that must undergo intracellular hydrolysis to bind iron[48]. As a single agent, dexrazoxane inhibits synchronized cultures of *P. falciparum* in human erythrocytes only at supra-pharmacologic concentrations (>200 µM). In contrast, pharmacologic concentrations of dexrazoxane (50-200 µM) as a single agent inhibit the progression of *P. yoelii* from sporozoites to schizonts in cultured mouse hepatocytes by 45% to 69%[47].

4. ANTI-MALARIAL ACTIVITY OF IRON CHELATORS IN LABORATORY ANIMALS

4.1. Desferrioxamine: In the only animal study investigating iron chelation therapy to suppress parasitemia with *P. falciparum*, desferrioxamine was active against the erythrocytic phase of the parasite in *Aotus* monkeys[49]. Similar observations were made with *P. berghei* and *P. vinckei* infections in rodents[20,50,28,51]. These animal studies demonstrated an antimalarial effect of desferrioxamine at doses that overlap with acceptable doses in humans (up to 100 mg/kg per day). Continuous subcutaneous infusions or divided doses of desferrioxamine were more effective than single daily doses in reducing parasitemia and mortality. Desferrioxamine administered subcutaneously in liposomes has also been effective to combat *P. vinckei* infection in mice[51].

4.2. Other Iron Chelators as Single Agents: Although *in vitro* studies suggest that the oral administration of hydroxypyridinone (deferiprone) in safe doses might result in a clinically detectable antimalarial effect, the single reported animal study of deferiprone proved to be negative. Deferiprone in three divided doses of 300 mg/kg per day for 13 days did not suppress *P. berghei* infection in 6 female Wistar rats [52]. The lack of effectiveness of deferiprone in this animal model, as compared to the studies in vitro, was attributed to intermittent attainment of suppressive plasma concentrations with the subcutaneous or oral mode of administration and to the relatively low lipophilicity of deferiprone, which would limit entry into the red cell under these circumstances. The highly lipophilic reversed siderophore, $RSFileu_{m2}$, was delivered in fractionated coconut oil (miglyol 840) via subcutaneous injections to mice infected with *P. vinckei petteri*, and some antimalarial activity was observed (36).

4.3. Combinations of Iron Chelators: Desferrioxamine and SIH were administered as a combination to Swiss mice infected with *P. vinckei petteri* or *P. berghei*. The drugs were delivered by several routes: single intraperitoneal injection, multiple intraperitoneal injections or subcutaneous insertion of a drug-containing polymeric device designed for slow, continuous drug release over seven days. As single agents administered in doses of 125 to 500 mg/kg/day in these manners, all three agents led to delays and reductions in peak parasitemias and to reduced mortality. The combination of DFO and SIH led to greater speed of drug action and greater inhibition of parasitemia than either agent alone. The antimalarial action of this combination was greatest when the drugs were slowly released into the circulation by means of a biodegradable polymer that was implanted subcutaneously[53].

5. IRON CHELATION THERAPY FOR HUMAN MALARIA

A total of eight clinical studies of the use of iron chleators in the treatment of malaria in humans have been published. These include two studies of desferrioxamine as a single agent in adults with asymptomatic *P. falciparum* parasitemia[54,55], one study of deferiprone as a single agent in adults with asymptomatic parasitemia[56], a study of desferrioxamine as a single agent in adults with uncomplicated clinical malaria[57], a study of desferrioxamine in addition to chloroquine for symptomatic malaria[58], a study of desferrioxamine in addition to artesunate in symptomatic malaria[59] and two studies of desferrioxamine in combination with standard quinine therapy for cerebral malaria in Zambian children[60,61]. The results of these studies are evaluated in this section.

5.1. Iron Chelators in Adults with Asymptomatic *P. falciparum* Infection.

5.1a. Desferrioxamine as a single agent: Desferrioxamine (100 mg/kg per day by continuous 72 hour subcutaneous infusions) was administered to 65 adult subjects in Zambia with asymptomatic infection with *P. falciparum* in the course of two randomized, double blind, placebo-controlled, crossover trials[54,55]. In both of these studies, desferrioxamine as a single agent significantly enhanced the clearance of parasitemia. For the purpose of this review, we have prepared a new analysis of the combined data sets of these two studies.

Table 3. Characteristics of Zambinan adults with asymptomatic parasitemia who participated in randomized controlled trials of desferrioxamine therapy. Combined results of two studies.

	Desferrioxamine (N = 31)	Placebo (N = 35)	P
Age in years (mean ± SD)	22 ± 6	23 ± 10	0.6
Female sex (no. and %)	17 (55)	24 (62)	0.3
Initial parasite count in no./µL (median and range)	813 (17-8,268)	406 (0-18,943)	0.07

As shown in Table 3, the subjects in the combined data set who were randomized to receive desferrioxamine or placebo initially were comparable in terms of age, sex and initial parasitemia. Figure 1a shows the time to 90% parasite clearance for 31 subjects originally given desferrioxamine and 35 subjects originally given placebo. The median 90% parasite clearance time was 60 hours for desferrioxamine but more than 72 hours for placebo. In fact, only one-fifth of the subjects receiving placebo had 90% parasite clearance by the end of the 72-hour treatment period. As assessed by the log rank test, parasite clearance was significantly faster with desferrioxamine than with placebo ($P < 0.001$). In Cox proportional hazards modeling, the administration of desferrioxamine was associated with a 4.2 fold increase in the rate of parasite clearance after adjustment for initial parasitemia (95 confidence interval of 1.4 to 12.5). Figure 1b shows time to 90% parasite clearance for 17 subjects who originally received placebo by continuous subcutaneous infusion for 72 hours and then were crossed over to receive desferrioxamine for 72 hours. Again, desferrioxamine was associated with a median parasite clearance time of 60 hours.

Serum concentrations of desferrioxamine + ferrioxamine (the iron complex of desferrioxamine) were measured in 26 subjects with asymptomatic parasitemia receiving desferrioxamine, 100 mg/kg per day by continuous subcutaneous infusion for 72 hours[55]. Mean ± SEM steady state concentrations were 6.9 ± 0.6 µM at 36 hours and 7.7 ± 0.7

µM at 72 hours. These levels are at the lower end of the range of values reported for the ID_{50} for desferrioxamine against *P. falciparum* as determined *in vitro* (Table 2). While results obtained with low levels of parasitemia in partially immune adults cannot necessarily be extrapolated to patients with severe infection, these findings suggested that iron chelation might be a potential chemotherapeutic strategy for human infection with *P. falciparum*.

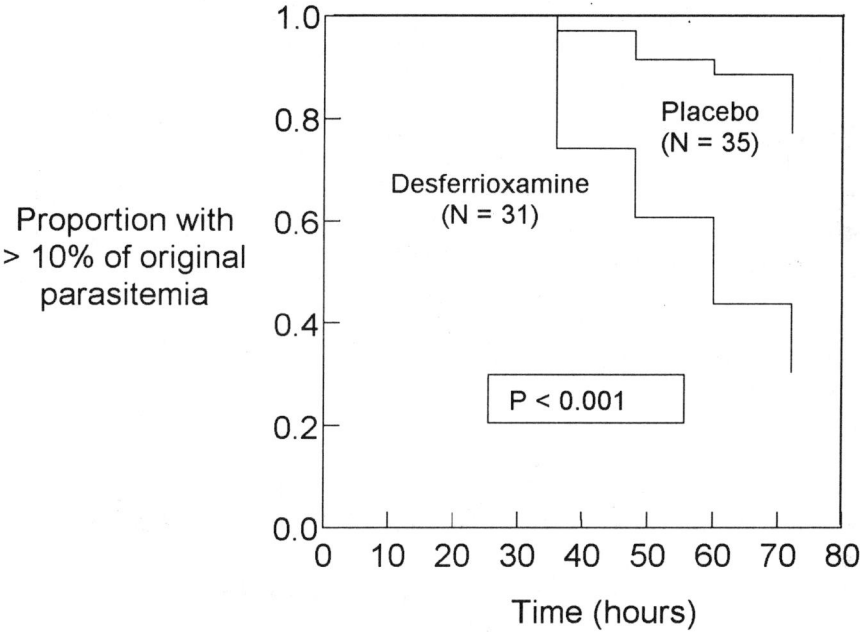

Figure 1a. Combined results of two studies. Tme to 90% parasite clearance in Zambian adults with asymptomatic *P. falciparum* parasitemia given either placebo or desferrioxamine, 100 mg/kg per day, by continuous infusion for three days.

5.2. Deferiprone as a single agent: A prospective, double blind, placebo-controlled crossover trial of deferiprone administered orally was conducted in 25 adult Zambians with asymptomatic *P. falciparum* parasitemia[56]. Deferiprone was administered daily for three or four days in divided doses of 75 or 100 mg/kg body weight. No reduction in asexual intraerythrocytic parasites was observed during or after deferiprone treatment. The mean peak plasma concentration of deferiprone (108.2 ± 24.9 µmol/L) achieved was within the range demonstrated to inhibit the growth of *P. falciparum in vitro*. However, the times to reach peak plasma levels and to clear the drug from the plasma were short, and plasma levels of deferiprone were only in the range of a modest antimalarial effect for much of the time between the oral doses used in these studies.

Because of the risk of neutropenia and other adverse effects with higher doses or prolonged use of the chelator, additional trials of deferiprone as an antimalarial would not seem to be justified.

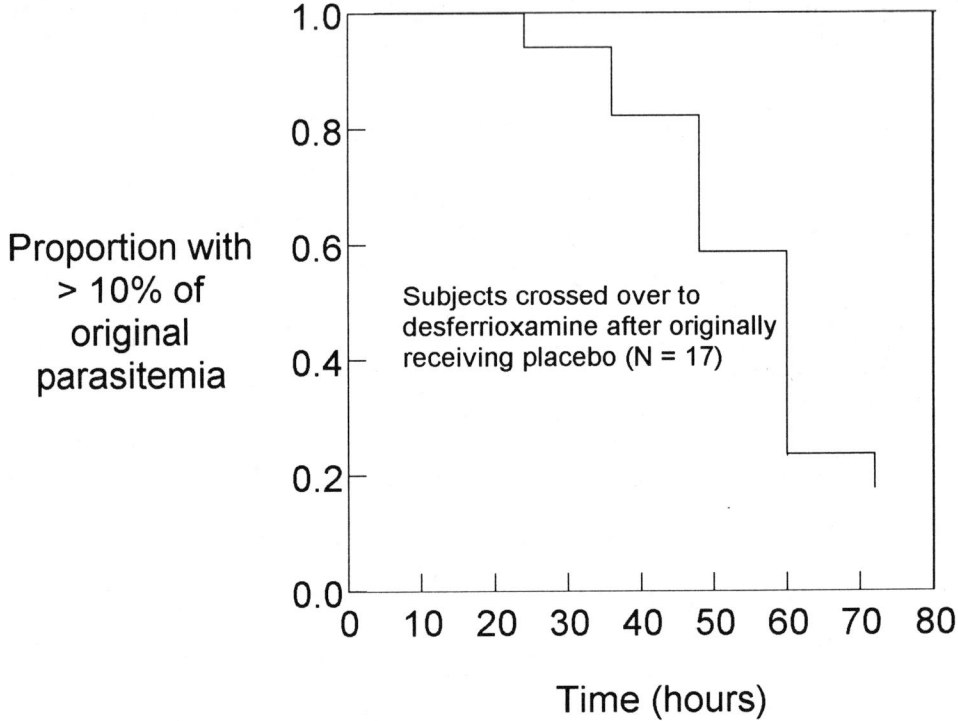

Figure 1b. Combined results of two studies. Time to 90% parasite clearance for 17 subjects who originally received placebo by continuous subcutaneous infusion for 72 hours and then were crossed over to receive desferrioxamine for 72 hours.

5.3. Desferrioxamine in Symptomatic, Uncomplicated Falciparum and Vivax Malaria.

5.3a. Desferrioxamine as a single agent: In Thailand, 14 adult males with *P. falciparum* infection and 14 adult males with *P. vivax* infection were given desferrioxamine, 100 mg/kg per day, as a continuous intravenous infusion for three consecutive days. Desferrioxamine as a single agent reduced parasitemia to zero within 57 hours for the falciparum group and 106 hours for the vivax group. Desferrioxamine was in general well tolerated, but about one-third of the subjects experienced transient visual blurring. Recrudescence was observed in all subjects, occurring on the average 10 days after start of therapy in the falciparum group and 15 days in the vivax group. This study demonstrated that iron chelation with desferrioxamine is effective as a single agent in both uncomplicated falciparum and vivax malaria, and that this therapy can clear moderate degrees of parasitemia. It also showed that the dose and duration of iron chelation therapy employed in this study failed to achieve a radical cure[57].

5.3b. Desferrioxamine in combination with other agents: Desferrioxamine was given along with chloroquine to six subjects with falicpaurm malaria in Cameroon, and no adverse effects were seen[58]. Although the combination of desferrioxamine and artemisinin derivatives might be expected to be antagonistic, no evidence of adverse interaction was found in a cohort of Thai patients with either uncomplicated or severe malaria who received both desferrioxamine and artesunate[59].

5.4. Effect of Desferrioxamine on Recovery from Coma and Mortality in Children with Cerebral Malaria: While the pathophysiology of cerebral malaria is incompletely understood, obstruction of the cerebral microvasculature by *P. falciparum*-infected erythrocytes leading to ischemia and microhemorrhage probably contributes to the development of this conditon[62,63,64]. Under these circumstances, free hemoglobin can act as a biologic Fenton reagent to provide iron for electron transfer and the generation of the hydroxyl radical, which then could contribute to central nervous system toxicity[65]. Desferrioxamine inhibits peroxidant damage to lung tissue in mice[66], to the myocardium in rabbits[67] and to the central nervous system in cats[68]. It therefore seems possible that iron chelation with desferrioxamine might protect against damage to the central nervous system by inhibiting iron-induced peroxidant damage to cells and subcellular structures of the brain. Desferrioxamine might also be beneficial in cerebral malaria (i) by enhancing parasite clearance through withholding iron from a vital metabolic pathway of the parasite, and/or (ii) by enhancing Th-1 cell mediated immunity[69].

To determine if iron chelation added to standard quinine therapy can speed recovery of full consiousness in children with cerebral malaria, a prospective, randomized, double-blind trial of desferrioxamine or placebo added to standard quinine treatment was conducted in 83 Zambian children at a single rural institution. Each child received quinine, 10 mg/kg, every eight hours for five days and a single dose of sulfadoxine/pyrimethamine, 25/1.2 mg/kg. No loading dose of quinine was given. In addition either DFO, 100mg/kg per day, or placebo were given as a 72-hour intravenous infusion. The addition of DFO to the conventional therapy shortened the time to clearance of parasitemia and the time to recovery of full consciousness in children with deep coma, each by about 2-fold[60]. Further analysis of the results of this study suggested that iron chelation with DFO appeared to be associated with faster recovery from coma specifically in children with high transferrin saturations of > 43%[70].

To examine the effect of iron chelation on mortality in cerebral malaria, 352 children were enrolled into a second clinical trial of desferrioxamine in addition to standard quinine therapy at two centers in Zambia, one rural and one urban[61]. The study design was the same as the first study except that a loading dose of quinine was given (20 mg/kg). Overall mortality was 18.3% (32/175) in the desferrioxamine plus quinine group and 10.7% (19/177) in the placebo plus quinine group (p=0.074). Parasite clearance was not different between the two treatment arms, but, among survivors, there was a trend to faster recovery from coma in the desferrioxamine group. This study did not provide evidence for a beneficial effect on mortality in children with cerebral malaria when desferrioxamine was added to quinine in a regimen that included a loading dose of quinine; rather there was a trend to higher mortality with desferrioxamine. The lack of a positive effect of desferrioxamine on parasite clearance in this study, in contrast to the earlier work[60], might be attributable to the impact of a loading dose of quinine used in the present study but not in the previous one. Our data indicate that a loading dose of quinine has a substantial benefit in the treatment of cerebral malaria[71], and a relatively delayed beneficial effect of desferrioxamine in cerebral malaria may have been masked by the loading dose of quinine.

For the purpose of this review, we have prepared an analysis of the combined data from the two published trials of desferrioxamine in addition to quinine for cerebral malaria in Zambian children just described[54,56]. The characteristics of the study participants are summarized in Table 4 according to the type of experimental treatment. Recovery from coma was examined by Cox proportional hazards models that adjusted for site of the study, loading dose of quinine, age, sex, history of seizures before presentation, duration of coma before presentation, depth of coma, glucose and hemoglobin concentrations, and initial parasitemia. Figure 2a shows that, among all 435 children involved in the study, there was a trend to faster recovery of consciousness with desferrioxamine plus quinine as compared to placebo plus quinine, but the trend was not statistically significant (P = 0.15). The addition of desferrioxamine to standard quinine therapy was associated with an estimated 1.2 fold increase in the adjusted rate of recovery of full consciousness (95% confidence interval of 0.9 to 1.5). Figure 2b shows that, among 192 children with transferrin saturations >43%, recovery of consciousness was significantly faster with desferrioxamine plus quinine as compared to placebo plus quinine (P = 0.027). The addition of desferrioxamine to standard quinine therapy was associated with a 1.5 fold increase in the adjusted rate of recovery of full consiousness (95% confidence interval of 1.0 to 2.2).

Despite the trend to faster recovery of consciousness with the addition of desferrioxamine to quinine in this combined analysis of data from two studies of cerebral malaria in Zambian children, the use of desferrioxamine was not associated with a reduction in mortality. Among all children, 39 (18%) of 217 receiving desferrioxamine plus quinine died compared to 28 (13%) of 218 receiving placebo plus quinine (P = 0.15). In a logistic regression model that adjusted for site of the study, loading dose of quinine, age, sex, depth of coma, glucose concentration and size of the liver, the addition of desferrioxamine to standard quinine therapy was associated with a

Table 4. Characteristics of Zambian children with cerebral malaria who participated in randomized placebo-controlled trials of desferrioxamine therapy in addition to standard quinine treatment. Combined results for two studies.

	Desferrioxamine (N = 217)	Placebo (N = 218)	P
Age in months (mean ± SD)	35 ± 16	38 ± 16	0.065
Female sex (no. and %)	94 (43)	111 (51)	0.1
Studied at rural hospital (no. and %)	159 (73)	159 (73)	1.0
Loading dose of quinine given (no. and %)	175 (81)	177 (81)	0.9
Duration of fever before presentation in hours (median and range)	49 (0-504)	51 (0-326)	0.8
Duration of coma before presentation in hours (median and range)	10 (1-107)	9 (1-100)	0.6
History of chloroquine before presentation (no. and %)	165 (76)	153 (71)	0.2
History of traditional medicine before presentation (no. and %)	99 (46)	86 (39)	0.2
History of seizures before presentation (no. and %)s	192 (88)	198 (91)	0.4
Initial coma score (no. and %)			1.0
0	2 (1)	2 (1)	
1	18 (8)	18 (8)	
2	124 (57)	132 (61)	
3	57 (26)	51 (23)	
4	16 (7)	15 (7)	
Spleen size in cm below right costal margin (median and range)	2 (0-8)	1 (0-8)	0.2
Liver size in cm below right costal margin (median and range)	2 (0-10)	2 (0-6)	0.07
Initial parasite count in no./µ/L (median and range)	49,800 (22-1,345,025)	45,753 (20-1,145,844)	0.07
Hemoglobin in g/dL (mean ± SD)	6.8 ± 2.3	6.9 ± 2.2	0.4
White Blood Cells in no/µ/L (mean ± SD)	11,292 ± 6,830	10,795 ± 5,276	0.4
Glucose in mg/dL (mean ± SD)	86 ± 50	88 ± 53	0.7
Transferrin saturation (mean ± SD)	60 ± 46*	55 ± 41**	0.2

*N = 177
**N = 163

1.3-fold increase in the odds of death (95% confidence interval of 0.8 to 2.3; P = 0.3). Also, the addition of desferrioxamine to standard quinine therapy did not have a significant effect on parasite clearance (P = 0.6) in a Cox proportional hazards model that adjusted for study site, loading dose of quinine, age, sex, duration of coma before presentation, size of the liver, glucose concentration and degree of parasitemia. Similarly, the addition of desferrioxamine to standard quinine therapy did not have a significant effect on fever clearance (P = 0.7), in a model that adjusted for study site, loading dose of quinine, age, sex, and the sizes of the liver and spleen. In these models, the loading dose of quinine was associated with an estimated 1.5 fold increase in the rate of recovery from coma (P = 0.015), a 1.9-fold increase in the rate of parasite clearance (P < 0.001), a 2.5-fold increase in the rate of fever clearance (P < 0.005), and a two-fold reduction in the odds of death (P = 0.09).

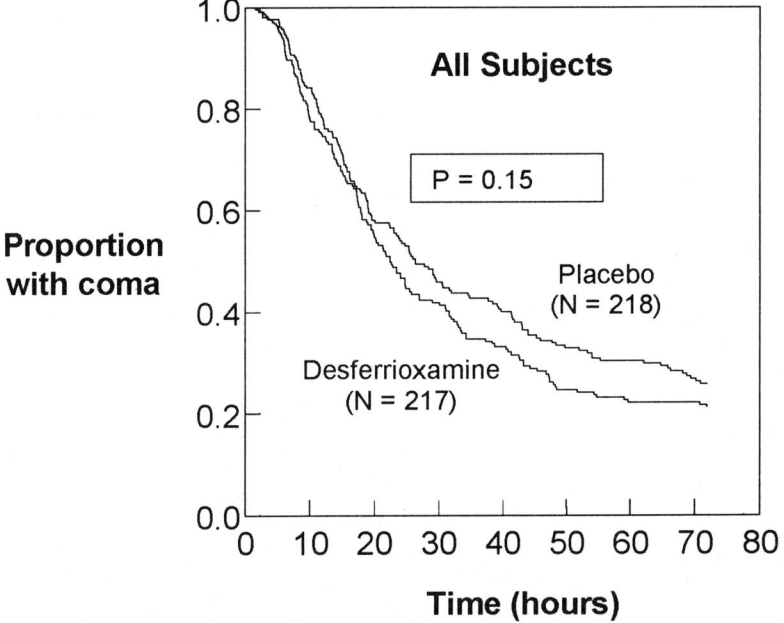

Figure 2a. Combined results of two studies of cerebral malaria. Among all 435 children involved in the studies, there was a trend to faster recovery of consciousness with desferrioxamine plus quinine as compared to placebo plus quinine, but the trend was not statistically significant (P = 0.15).

5.5. Iron Chelators and the Immune Response in the Setting of Malaria: Two different substudies[72,69] of Zambian children who were enrolled in placebo-controlled trials of DFO in addition to quinine for cerebral malaria[54,56] suggested a possible effect of iron chelation on Th-1 mediated immune function in the setting of malaria. In one substudy, serum levels of neopterin, an indirect marker of Th-1 cell mediated immune function[73,74], did not change significantly in children receiving desferrioxamine plus quinine but did decline significantly in children receiving placebo plus quinine[69]. In the same substudy, serum concentrations of NO_2^-/NO_3^-, the stable end products of NO

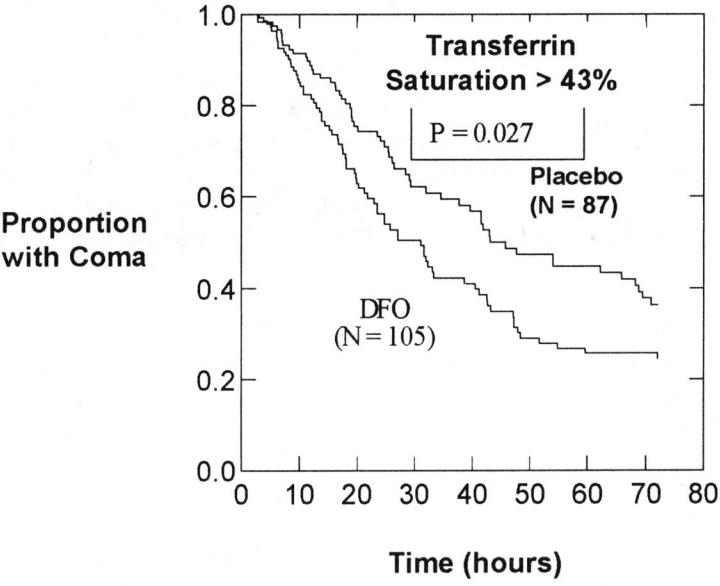

Figure 2b. Combined results of two studies of cerebral malaria. Among 192 children with transferrin saturations >43%, recovery of consciousness was significantly faster with desferrioxamine plus quinine as compared to placebo plus quinine.

degradation, increased significantly with desferrioxamine plus quinine but not placebo plus quinine. These observations made in patients are compatible with *in vitro* results demonstrating that desferrioxamine enhances neopterin formation by positively modulating IFN-γ activity[75]. In another substudy, serum concentrations of IL-4, a Th-2 related cytokine, increased with placebo plus quinine but not desferrioxamine plus quinine[72]. Taken together, these studies raise the possibility that iron chelation therapy in severe malaria may strengthen Th-1 cell mediated immune function by enhancing IFN-γ activity, as reflected by neopterin and NO formation, while reducing the production of Th-2 mediated cytokines such as IL-4. The directing of the immune response towards a Th-1 effector mechanism should be beneficial in the early phase of parasitic infections.

The potential for desferrioxamine to influence the immune response in the setting of malaria was examined further *in vitro* in a system in which macrophages were cocultured with malaria parasite-infected human erythrocytes[75]. Cytokine stimulation of murine macrophages resulted in increased nitric oxide (NO) formation by the macrophages and decreased survival of plasmodia within cocultured human erythrocytes. The addition of desferrioxamine to the system before cytokine stimulation enhanced both the NO formation and parasite killing associated with cytokine stimulation. At the same time, desferrioxamine had no effect on the presence of the inhibitor of NO formation, L-N6-(1-iminoethyl)-lysine. Moreover, peroxynitrite, which is formed after the chemical reaction of NO with superoxide, appeared to be the principal effector molecule for macrophage-mediated cytotoxicity toward *P. falciparum*, and interferon-gamma was a major regulatory cytokine for this process. Thus, this laboratory study *in vitro* is consistent with the hypotheses that desferrioxamine may contribute to parasite clearance *in vivo* by enhanced generation of NO on the part of macrophages in addition to or instead of limitation of iron availability to the parasite.

6. CONCLUSIONS

The discussion in this chapter is based on early and fragmentary knowledge of the iron metabolism of *P. falciparum* and the clinical role of iron chelators as antimalarials. Nevertheless, some conclusions can be drawn:

1. A variety of iron chelators have antimalarial activity *in vitro* and in animals with experimental plasmodial infections.

2. The antimalarial mechanism of action of iron III chelators ***in vitro*** appears to be related to the withholding of iron from critical metabolic pathways of the parasite. Enhancement of macrophage-mediated cytotoxicity toward the parasite may also contribute to the antimalarial activity.

3. Desferrioxamine, the only iron chelator that is widely available for use in humans, has clinical antimalarial activity. However, this agent has slow onset of activity and cannot be relied upon to effect a radical cure. When given together with a highly effective standard regimen such as quinine with a loading dose, any additional antimalarial effect of desferrioxamine is difficult to discern. Moreover, the addition of desferrioxamine to standard quinine therapy does not appear to improve the outcome of severe malaria in terms of survival.

4. Deferiprone, the only other iron chelator that is available for human use in some countries, does not appear to have clinical antimalarial activity at acceptable doses for administration to humans.

Given this mixed picture at the present time, we would like to end this chapter by expressing our view that the strategy of iron chelation as a chemotherapeutic approach to malaria deserves further consideration and development. In this regard, it will be critical to develop iron chelators designed for the treatment of malaria rather than iron overload. Such antimalarial iron chelating compounds should ideally be effective orally, be safe for a short course of antimalarial therapy, and be specifically targeted to bind parasite-associated iron, promote the host immune response against malaria, and provide anti-oxidant protection to host tissues. Advances are also needed in understanding the iron metabolism of the malaria parasite, specifically the determination of the source of the iron that is essential for the growth of the intraerythrocytic parasite and the identification of the metabolic pathways with which iron chelation interferes. Finally, it is also important to further investigate the potential influence of iron chelators in enhancing host immunity.

REFERENCES

1. Chakrabarti D, Schuster SM, Chakrabarti R. Cloning and characterization of subunit genes of ribonucleotide reductase, a cell-cycle-regulated enzyme, from *Plasmodium falciparum*. *Proc Natl Acad Sci U S A* 1993 Dec 15:90(24):12020-4

2. Rubin H, Salem HS, Li LS, Yang FD, Mama S, Wang ZM, Fisher A, Hamann CS, Cooperman BS. Cloning sequence determination, and regulation of the ribonucleotide reductase subunits from Plasmodium falciparum: a target for antimalarial therapy. *Proc Natl Acad Sci USA* 1993; **90**: 9280-9284.

3. Krungkrai J, Cerami A, Henderson GB. Purification and characterization of dihydroorotate dehydrogenase from the rodent malaria parasite *Plasmodium berghei*. *Biochemistry* 1991; **30**: 1934-9.

4. Bonday ZQ, Taketani S, Gupta PD, Padmanaban G. Heme biosynthesis by the malarial parasite. Import of delta-aminolevulinate dehydrase from the host red cell. *J Biol Chem* 1997; **272**: 21839-21846.

5. Krungkrai J, Krungkrai SR, Suraveratum N, Prapunwattana P. Mitochondrial ubiquinol-cytochrome c reductase and cytochrome c oxidase: chemotherapeutic targets in malarial parasites. *Biochem Mol Biol Int* 1997; **42**: 1007-14.

6. Petmitr S, Krungkrai J. Mitochondrial cytochrome b gene in two developmental stages of human malarial parasite Plasmodium falciparum. *Southeast Asian J Trop Med Public Health* 1995; **Dec:26(4)**:600-5

7. Reichard, P, Ehrenberg, A. Ribonucleotide reductase - a radical enzyme. *Science* 1983; **221**: 514-519.

8. Cavanaugh PF, Porter CW, Tukalo D, Frankfurt DS, Pavelic ZP, Bergeron RJ. Characterization of L1210 cell growth inhibition by the bacterial iron chelators parabactin and compound II. *Cancer Res* 1985; **45**: 4754-4759.

9. Nyholm S, Mann GJ, Johansson AG, Bergeron RJ, Graslund A, Thelander L. Role of ribonucleotide reductase in inhibition of mammalian cell growth by potent iron chelators. *J Biol Chem* 1993; **268**: 26200-26205.

10. Moormann AM, Hossler PA, Meshnick SR. Deferoxamine effects on *Plasmodium falciparum* gene expression. *Mol Biochem Parasitol* 1999; **98**: 279-283.

11. Fuchs O, Ponka P. The role of iron supply in the regulation of 5-aminolevulinate synthase mRNA levels in murine erythroleukemia cells. *Neoplasma* 1996; **43**: 31-36.

12. Surolia N, Padmanaban G. *De novo* biosynthesis of heme offers a new chemotherapeutic target in the human malarial parasite. *Biochem Biophys Res Commun* 1992; **187**: 744-750.

13. Aikawa M, Wernsdorfer WH, McGregor I, Livingstone C, Edinburgh. Fine structure of malaria parasites in the various stages of development, *In*: Principles and Practice of Malariology 1988; 97-130.

14. Pouvelle B, Spiegel R, Hsiao L, Howard RJ, Morris RL, Thomas AP, Taraschi TF. Direct access to serum macromolecules by intraerythrocytic malaria parasites. *Nature* 1991; **353**: 73-75.

15. Loyevsky M, Lytton SD, Mester B, Libman J, Shanzer A, Cabantchik ZI. The antimalarial action of Desferal involves a direct access route to erythrocytic (*Plasmodium falciparum*) parasites. *J Clin Invest* 1993; **91**: 218-224.

16. Slater AF, Cerami A. Inhibition by chloroquine of a novel heme polymerase enzyme activity in malaria trophozoites. *Nature* 1992; **355**: 167-169.

17. Pollack S, Flemming J. *P. falciparum* takes up iron from transferrin. *Brit J Haematol* 1984; **58**: 289-293.

18. Pollack S, Schnelle V. Inability to detect transferrin receptors on *P. falciparum* parasitized red cells. *Brit J Haematol* 1988; **68**: 125-129.

19. Sanchez-Lopez R, Halder K. A transferrin-independent iron uptake activity in Plasmodium falciparum-infected and uninfected erythrocytes. *Mol Biochem Pharmacol* 1992; **55**: 9-20.

20. Hershko C, Peto TE. Deferoxamine inhibition of malaria is independent of host iron status. *J Exp Med* 1998; **168**:

375-387.

21. Peto TE, Thompson J L. A reappraisal of the effects of iron and desferrioxamine on the growth of *Plasmodium falciparum* 'in vitro': the unimportance of serum iron. *Brit J Haematol* 1986; **63**: 273-280.

22. Cazzola M, Arioso P, Barosi G, Bergamaschi G, Dezza L, Ascari E. Ferritin in the red cells of normal subjects and patients with iron deficiency and iron overload. *Brit. J. Haematol* 1983; **53**: 659-665.

23. Gabay, T, Ginsburg, H. Haemoglobin denaturation and iron release in acidified red blood cell lysate - a possible source of iron for intraerythrocytic malaria parasites. *Exp. Parasitol* 1993; **77**: 261-272.

24. Loyevsky M, John C, Dickens B, Hu V, Gordeuk VR. Chelation of iron within the erythrocytic *Plasmodium falciparum* parasite by iron chelators. *Mol Biochem Parasitol* 1999; **101**:43-59.

25. Scott MD, Ranz A, Kuypers FA, Lubin BH, Meshnick SR. Parasite uptake of desferrioxamine: a prerequisite for antimalarial activity. *Brit J Haematol* 1990; **75**: 598-603.

26. Raventos-Suarez C, Pollack S, Nagel RL. *Plasmodium falciparum*: inhibition of in vitro growth by desferrioxamine. *Am J Trop Med Hyg* 1982; **31**: 919-922.

27. Heppner DG, Hallaway PE, Kontoghiorghes GJ, Eaton JW. Antimalarial properties of orally active iron chelators. *Blood* 1998; **72**: 358-361.

28. Fritsch G, Sawatzki G, Treumer J, Jung A, Spira DT. *Plasmodium falciparum*: inhibition *in vitro* with lactoferrin, desferrithiocin and desferricrocin. *Exp Parasitol* 1987; **63**:1-9.

29. Clarke CJ, Eaton JM. Hydrophobic iron chelators as new antimalarial drugs. *Clin Res* 1990; **38**: 300A.

30. Loyevsky M, John C, Zaloujnyi I, Gordeuk V. Aminothiol multidentate chelators as antimalarials. *Biochem Pharmacol* 1997; **54**: 451-458.

31. Scheibel LW, Adler A. Anti-malarial activity of selected aromatic chelators. *Mol Pharmacol* 1980; **18**: 320-325.

32. Scheibel LW, Stanton GG. Anti-malarial activity of selected aromatic chelators IV. Cation uptake of Plasmodium falciparum in the presence of oxines and siderochromes. *Mol Pharmacol* 1986; **30**: 364-369.

33. Hershko C, Theanacho EN, Spira DT, Peter HH, Dobbin P, Hider RC. The effect of N-alkyl modification on the antimalarial activity of 3-hydroxyppyridin-4-one oral iron chelators. *Blood* 1991; **77**: 637-643.

34. Shanzer A, Libman J, Lytton SD, Glickstein H, Cabantchik ZI. Reversed siderophores act as anti-malarial agents. *Proc Natl Acad Sci USA* 1991; **88**: 6585-6589.

35. Lytton SD, Loyevsky M, Mester B, Libman J, Landau I, Shanzer A, Cabantchik ZI. In vivo antimalarial action of a lipophilic iron (III) chelator: suppression of Plasmodium vinckei infection by reversed siderophore. *Am J Hematol* 1993; **43**: 217-220.

36. Cabantchik ZI, Glickstein H, Golenser J, Loyevsky M, Tsafack A. Iron chelators: mode of action as antimalarials. *Acta Haematol.* 1996; **95(1)**: 70-7

37. Whitehead S, Peto TE. Stage-dependent effect of deferoxamine on growth of Plasmodium falciparum in vitro. *Blood* 1990; **76**: 1250-1255.

38. Atkinson CT, Bayne MT, Gordeuk VR, Brittenham GM, Aikawa M. Stage-specific ultrastructural effects of desferrioxamine on *Plasmodium falciparum*. *Am J Trop Med Hyg* 1991; **45**:593-601.

39. van Zyl RL, Havlik I, Hempelman E, MacPhail AP, McNamara L. Malaria pigment and extracellular iron: possible target for iron chelating agents. *Biochem Pharmacol* 1993; **45**: 1431-1436.

40. Basco LK, Le Bras J. In vitro activity of chloroquine and quinine in combination with desferrioxamine against *Plasmodium falciparum*. *Am J Haematol* 1993; **42**: 389-391.

41. Chevion M, Chuang L, Golenser J. Effects of zinc-desferrioxamine on Plasmodium falciparum in culture. *Antimicrob Agents Chemother* 1995; **39**: 1902-1905.

42. Lytton SD, Mester B, Libman J, Shanzer A, Cabantchik, ZI. Mode of action of iron (III) chelators as antimalarials: II. Evidence for differential effects on parasite iron-dependent nucleic acid synthesis. *Blood* 1994; **84**: 910-915.

43. Glickstein H, Breuer B, Loyevsky M, Konijn A, Libman J, Shanzer A, Cabantchik ZI. Differential cytotoxicity of iron chelators on malaria-infected cells versus mammalian cells. *Blood* 1996; **87**: 4871-4878.

44. Pattanapanyasat K, Thaithong S, Kyle DE, Udomsangpetch R, Yongvanitchit K, Hider RC Webster HK. Flow cytometric assessment of hydroxypyridone iron chelators on *in vitro* growth of drug-resistant malaria. *Cytometry* 1997; **27**: 84-91.

45. Tsafack A, Loyevsky M, Ponka P, Cabantchik ZI. Mode of action of iron (III) chelators as anti-malarials. IV. Potentiation of desferal action by benzoyl and isonicotinoyl hydrazone derivatives. *J Lab Clin Med* 1996; **127**: 575-582.

46. Stahel E, Mazier D, Guillouzo A, Miltgen P, Landau I, Mellouk S, Beaudoin RL, Langlois P, Gentilini M. Iron chelators: in vitro inhibitory effect on the liver stage of rodent and human malaria. *Am J Trop Med Hyg* 1988; **39**: 236-240.

47. Loyevsky M, Sacci JB Jr, Boehme P, Weglicki W, John C, Gordeuk VR. *Plasmodium falciparum and Plasmodium yoelii:* effect of the iron chelation prodrug dexrazoxane on *in vitro* cultures. *Exp Parasitol* 1999; **91**: 105-114.

48. Hasinoff BB, Reiders FX, Clark V. The enzymatic hydrolysis-activation of the adriamycin cardioprotective agent (+)-1,2-bis(3,5-dioxopiperazinyl-1-yl)propane. *Drug Metabolism and Disposition* 1991; **19**: 74-80.

49. Pollack S, Rossan RN, Davidson DE, Escajadillo A. Desferrioxamine suppresses *Plasmodium falciparum* in Aotus Monkeys. *Proc Soc Experiment Biol Med* 1987; **184**: 162-164.

50. Yinnon AM, Theanacho EN, Grady RW, Spira DT, Hershko C. Antimalarial effect of HBED and other phenolic and catecholic iron chelators. *Blood* 1989; **74**: 2166-2171.

51. Postma NS, Hermsen CC, Zuidema J, Eling WM. Plasmodium vinckei: optimization of desferrioxamine B delivery in the treatment of murine malaria. *Exp Parasitol* 1998; **89**: 323-330.

52. Hershko C, Gordeuk VR, Thuma PE, Theanacho EN, Spira DT, Hider RC, Peto TE, Brittenham GM. The antimalarial effect of iron chelators in animal models and in humans with mild falciparum malaria. *J Inorg Biochem* 1992; **47**: 267-277.

53. Golenser J, Tsafack A, Amichai Y, Libman J, Shanzer A, Cabantchik ZI. Antimalarial action of hydroxamate-based iron chelators and potentiation of desferrioxamine action by reversed siderophores. *Antimicrob Agents Chemotherap* 1995; **39**: 61-65.

54. Gordeuk VR, Thuma PE, Brittenham GM, Zulu S, Simwanza G, Mhangu A, Flesch G, Parry D. Iron chelation with deferoxamine B in adults with asymptomatic *Plasmodium falciparum* parasitemia. *Blood* 1992a; **79**: 308-312.

55. Gordeuk VR, Thuma PE, Brittenham GM, Biemba G, Zulu S, Simwanza G, Kalense P, M'hango A, Parry D, Poltera AA, Aikawa M. Iron chelation as a chemotherapeutic strategy for falciparum malaria. *Am J Trop Med Hyg* 1993; **48**: 193-197.

56. Thuma PE, Olivieri NF, Mabeza GF, Biemba G, Parry D, Zulu S, Fassos FF, McClelland RA, Koren G, Brittenham GM, Gordeuk VR. Assessment of the effect of the oral iron chelator deferiprone on mild *Plasmodium falciparum* parasitemia in humans. *Am J Trop Med Hyg* 1998a; **58**:358-364.

57. Bunnag D, Poltera AA, Viravan C, Looaresuwan S, Harinasuta T, Schundlery C. Plasmodicidal effect of desferrioxamine in human vivax or falciparum malaria from Thailand. *Acta Trop* 1992; **52**: 59-67.

58. Traore O, Carnevale P, Kaptue-Noche L, M'Bede J, Desfontaine M, Elion J, Labie D, Nagel RL. Preliminary report on the use of desferrioxamine in the treatment of *Plasmodium falciparum* malaria. *Am J Hematol* 1991; **37**: 206-208.

59. Looaresuwan S, Wilairatana P, Vannaphan S, Gordeuk VR, Taylor TE, Meshnick SR, Brittenham GM. Co-

administration of desferrioxamine B with artesunate in malaria: an assessment of safety and tolerance. *Ann Trop Med Parsaitol* 1996; **90**: 551-554.

60. Gordeuk VR, Thuma PE, Brittenham GM, McLaren C, Parry D, Backenstose AR, Biemba G, Msiska R, Holmes L, McKinley E, Vargas L, Gilkeson RC, Poltera AA. Effect of iron chelation therapy on recovery from deep coma in children with cerebral malaria. *New Engl J Med* 1992b; **327**: 1473-1477.

61. Thuma PE, Mabeza GF, Biemba G, Bhat GJ, McLaren C, Moyo VM, Zulu S, Khumalo H, Mabeza P, M'hango A, Parry D, Poltera AA, Brittenham GM, Gordeuk VR. Effect of iron chelation therapy on mortality in Zambian children with cerebral malaria. *Trans Roy Soc Trop Med Hyg* 1998b; **92**: 214-218.

62. MacPherson GG, Warrell MJ, White NJ, Looareesuwan S, Warrell DA. Human cerebral malaria: a quantitative ultrastructural analysis of parasitized erythrocyte sequestration. *Am J Pathol* 1985; **119**: 385-401.

63. Oo MM, Aikawa M, Than T, Aye TM, Myint PT, Igarashi I, Schoene WC. Human cerebral malaria: a pathological study. *J Neuropathol Exp Neurol* 1987; **46**: 223-231.

64. Aikawa M, Iseki M, Barnwell JW, Taylor D, Oo MM, Howard RJ. The pathology of human cerebral malaria. *Am J Trop Med and Hyg* 1990; **43 (Suppl)**: 30-37.

65. Sadrzadeh SM, Graf E, Panter SS, Hallaway PE, Eaton JW. Hemoglobin: a biologic Fenton reagent. *J Biol Chem* 1984; **259**: 14354-14356.

66. Ward PA, Till GO, Kunkel R, Beauchamp C. Evidence for role of hydroxyl radical in complement and neutrophil-dependent injury. *J Clin Invest* 1983; **72**: 369-371.

67. Ambrosio G, Zweier JL, Jacobus WE, Weisfeldt ML, Flaherty JT. Improvement of post-ischemic myocardial function and metabolism induced by administration of deferoxamine at time of reflow: role of iron in the pathogenesis of reperfusion injury. *Circulation* 1987; **76**: 906-915.

68. Sadrzadeh SM, Anderson DK, Panter SS, Hallaway PE, Eaton JW. Hemoglobin potentiates central nervous system damage. *J Clin Invest* 1987; **79**: 662-664.

69. Weiss G, Thuma P, Mabeza G, Werner ER, Herold M, Gordeuk VR. Modulatory potential of iron chelation therapy on nitric oxide formation in cerebral malaria. *J Infect Dis* 1997; **175**:226-230.

70. Gordeuk VR, Thuma PE, McLaren CE, Biemba G, Zulu S, Poltera AA, Askin JE, Brittenham GM. Transferrin saturation and recovery from coma in cerebral malaria. *Blood* 1995; **85**: 3297-3301.

71. van der Torn M, Thuma PE, Mabeza GF, Biemba G, Moyo VM, McLaren CE, Brittenham GM, Gordeuk VR. Loading dose of quinine in African children with cerebral malaria. *Trans Roy Soc Trop Med Hyg* 1998; **92**:325-331.

72. Thuma PE, Weiss G, Herold M, Gordeuk V. Serum neopterin, interleukin-4 and interleukin-6 concentrations in cerebral malaria patients and the effect of iron chelation therapy. *Am J Trop Med Hyg* 1996; **54**: 164-168.

73. Huber C, Batchelor JR, Fuchs D, Hausen A, Lang A, Niederwieser D, Reibnegger G, Swetly P, Troppmair J, Wachter H. Immune response associated production of neopterin. *J Exp Med* 1984; **160**: 310-316.

74. Fuchs D, Hausen A, Reibnegger G, Werner ER, Deitrich MP, Wachter H. Neopterin as a marker of activated cell-mediated immunity: Application on HIV infection. *Immunol. Today* 1988; **9**: 150-155.

75. Weiss G, Fuchs D, Hausen A, Reibnegger G, Werner ER, Werner-Felmayer G, Wachter H. Iron modulates interferon-gamma effects in human myelomonocytic cell line THP-1. *Exp Hematol* 1992; **20**: 605-10.

76. Fritsche G, Larcher C, Schennach H, Weiss G. Regulatory Interactions between Iron and Nitric Oxide Metabolism for Immune Defense against Plasmodium falciparum Infection. *J Infect Dis* 2001; **183**:1388-94

77. Bienzle U, Fritsch KG, Hoth G, Rozdzinski E, Kohler K, Kalinowski M, Kremsner P, Rosenkaimer F, Feldmeier H. Inhibition of Plasmodium vinckei-malaria in mice by recombinant murine interferon-gamma. *Acta Trop.* 1988; **Sep;45(3)**:289-90.

78. Scheibel LW, Rodriguez S. Anti-malarial activity of selected aromatic chelators. V. localization of 59Fe in

Plasmodium falciparum in the presence of oxines. *Prog Clin Biol Res* 1989; **313**: 119-149.

79. Yang Y, Ranz A, Pan HZ, Zhang ZN, Lin XB, Meshnick SR. Daphnetin: a novel anti-malarial agent with in vitro and in vivo activity. *Am J Trop Med Hyg* 1992; **46**: 15-20.

80. Jairam KT, Havlik I, Monteagudo FSE. Possible mechanism of action of desferrioxamine and 2,2'-bipyridyl on inhibiting the *in vitro* growth of *Plasmodium falciparum* (3 strain). *Biochem Pharmacol* 1991; **8**: 1633-1634.

INDEX

(S)-α-methylcysteine, 175
[4Fe-4S] iron-sulfur cluster, 23
1,10-phenanthroline, 209, 238, 248
1,2 diethyl-3- hydroxypyridin-4-one, 134
1,2-dimethyl-3-hydroxypyridin-4-one (deferiprone), 144
1-hydroxypyridin-2-one, 144
1-methyl-3-hydroxypyridin-2-one, 144
2,2′-bipyridyl, 141, 209
2-methyl-3-hydroxypyran-4-one (maltol), 144
2-hydroxy-1-naphthaldehyde, 207, 211, 213
3-hydroxy isonicotinaldehyde isonicotinoyl hydrazone (IIH), 211
3-hydroxypyridin-4-ones, 143, 146, 148, 150, 159, 161, 163, 165–166
4-hydroxynonenal, 83

ABC7 transporter protein, 10
Aceruloplasminemia, 9
Acetohydroxamic acid, 144
Aconitase, 27
Action potential, 83
Activating transcription factor, 22
Activator protein (AP)-1, 21
Activator protein-1 (c-Fos/c-Jun), 22
Acute iron poisoning, 91
Adrenal dysfunction, 103
Adrenal steatosis, 132
Agranulocytosis, 132–133, 136
Alanine-transaminase (ALT), 134
Alas2, 11
Alkoxyl (RO·), 20
Aluminium(III), 141
Aminocarboxylates, 143, 152
Aminolevulinic acid synthase (*ALAS2*), 10
Anaemia of chronic disease (ACD), 26
Anthracyclines, 27, 231
Antimalarial therapy, 251, 265, 267
Aplasia, 132
Apotransferrin, 45
Area under the curve [AUC], 94
Arthropathy, 132, 133, 207

Arylaldehydeisonicotynoylhydrazones, 64
Ascorbate, 42, 46–47, 59, 62, 83, 87, 92, 96–97, 106, 114, 122, 139
ATM1, 10
Autoxidation, 141, 161

Basolateral iron transfer, 3
Bax, 236, 238
Beagle dogs, 200
Belgrade (*b*) rats, 4
Beta-2 microglobulin (*B2m*), 6
Bidentate, 142–143, 146–148, 150, 153–154, 156–157, 171, 189
Biliary excretion, 79
Bilirubin, 21
Bioavailability, 150, 152–153, 162, 164, 176, 186, 200, 202
Bivalent metals, 141
Blackfan Diamond anaemia, 133
Bleomycin, 231, 242
Blood-brain barrier, 150–151, 157, 159, 163
Bone marrow transplantation, 46
Branched carbohydrate chains (glycans), 45
Bruch's membrane, 110

C282Y mutation, 7
Ca^{2+}/Na^+ exchanger, 22
Ca^{2+}-ATPase, 22
Cadaverine, 170
Calcein, 58, 71
Calcineurin, 22
cAMP responsive element binding protein, 22
Carbonyl iron, 29
Cardiac disease, 100, 130
Cardiolipin, 31–32, 81, 88
Cardiomyocytes, 80
Cardioprotection, 77
Cardiopulmonary bypass surgery, 46
Cardiotoxicity, 27
Catalase, 21
Cataracts, 110
Catecholates, 141, 143, 153

Catechol-*O*-methyl transferase, 5
CCl_4, 37
CCRF-CEM leukemia cells, 217–218, 232, 243
CD4 T cells, 17, 136
CD8 T cells, 136
Cebus apella monkeys, 156, 174, 189, 184
Ceruloplasmin, 3
CGP65015, 189
Chelatable pool, 77, 78
Chloroquine, 256, 260, 262, 267–268
Cholylhydroxamic acid, 213
Cirrhosis, 6–7, 28–29, 33, 35–37, 43, 107, 134
c-Jun N terminal kinase, 22
C_{max}, 94
Coagulation defect, 50
Coma, 260–263, 270
Complex IV, 31
Compliance, 93
Concanavalin-A, 217
Congenital atransferrinemia, 15, 49
Cycle inhibitor p21/*WAF1*, 232
Cyclooxygenases, 233
Cytochrome b, 251
Cytochrome c oxidase (complex IV), 81
Cytochrome oxidase and reductase, 31, 251
Cytochrome P-450, 30

Daphnetin, 253, 271
DCT1, 2
DDTA, 189
Deferiprone, 56, 76, 90, 113, 116, 118, 124, 127–134, 136–140, 206, 233, 253, 255–256, 258–259, 269
Deferoxamine (DFO), 32, 51, 56–59, 61, 63–65, 68–74, 77–79, 83–97, 99–116, 118–127, 130–132, 134, 136, 137–140, 146, 148, 151–152, 155, 170, 172–174, 188–189, 191, 206–211, 213–216, 218, 230–239, 241, 256, 260, 263
Delta-aminolivulinate synthase, 250–251
Desazadesferrithiocin, 182
Desferrioxamine-chelatable iron (DCI), 32, 47, 69–70, 72–74, 136
Desferrithiocin, 127, 154, 164, 171, 173–174, 181, 183–184, 195, 203
Dexrazoxane, 242, 255, 269
Diabetes, 6, 9–10, 16, 28, 84, 102, 104, 109, 114, 116, 121, 133
Dialkylhydroxypyridinones, 156
Dibasic cations, 141, 143, 147
Diethylenetriaminepentaacetic acid (DTPA), 57, 59, 61–62, 79, 126, 127, 143, 152, 162–163, 171–172, 183, 209
Dihydroorotate dehydrogenase, 251
Dihydroxybenzoic acid, 127
DMT1, 2, 4–6, 8, 14, 48
Doxorubicin, 27, 231

Endotoxins, 37
Enterobactin, 153, 170, 183
Enterocytes, 3
Enterohepatic cycling, 215
Epidermal growth factor receptor, 22
Erythron, 190
Ethylenediaminetetraacetic acid (EDTA), 46, 57–58, 143, 209
Excitation-contraction coupling, 83
Excitatory depolarization, 83
Extracellular matrix (ECM), 33

FDM, 57
Ferrioxamine B, 91
Ferritin 12, 15, 17, 23–24, 29–30, 38–40, 45–49, 58, 66, 72, 75, 78–80, 85–87, 89, 95, 97, 99–101, 103–104, 106, 111, 114–116, 119–123, 125–126, 128–132, 137, 139
Ferritin H subunit (*Fth*), 12
Ferrocene, 188
Ferroportin1, 5
Ferroxidase, 3
Fe-S cluster, 23
Flavonoids, 21
Flexed tail (*f*), mutation 11
Fluorescein-deferoxamine (FL-DFO, 69
Fluorophore, 57–58, 65
Frataxin, 10
Friedreich ataxia (FRDA), 10

GADD-45 (growth arrest and DNA damage gene), 218, 236, 238–240
Gallium(III), 141
Gamma amino butyric acid (GABA), 253
Gerbils, 32, 37, 42, 134, 140
G1/S phase, 218
Glucuronidation, 156–157
Glucuronidisation, 128, 131
Glutathione, 168
Glutathione and oxidized glutathione reductase, 21
Glutathione peroxidase, 21
Glutathione-S-transferase, 21
Glycans, 55
Growth hormone, 101, 102, 115, 123–126

Haber-Weiss reaction, 49, 80, 83, 93
Haemopexin, 45
Haptoglobin, 45–46
hbd mutants, 4
HBED, 127, 152, 163–164, 171, 189–191
Heat-damaged erythrocytes, 213–214
Hemochromatosis, 6–8, 13–17, 24, 29, 32, 37–38, 40, 42, 46, 48–49, 51–53, 55, 69, 72, 74, 76, 79–80, 88, 91, 107, 121
Hemosiderin, 29, 72

INDEX

Hepatic fibrosis, 28–29, 33, 37, 42–43, 92, 97, 134, 140
Hepatic iron stores, 97, 104
Hepatic stellate cells (HSC), 33
Hepatitis C virus (HCV), 134
Hepatocellular iron, 37, 79
Hepatocytes, 210–211, 213, 243–244, 247, 255
HepG2cells, 48
Hephaestin, 3
Hexadentate, 91, 93, 142–144, 146–148, 150, 152–154, 164, 189
HFE-associated hemochromatosis, 6
HL60 cells, 218, 245
Hydrogen peroxide (H_2O_2), 20
Hydroperoxidic products, 31
Hydrophilicity, 150
Hydroxamates, 93, 141, 143, 169, 170
Hydroxyl (OH·), 20
Hydroxypyridinones, 143, 149, 153, 159, 163, 165–166
Hydroxyurea, 231, 243
Hypertransfused rats, 215
Hypochlorous acid, 168
Hypogonadotrophic hypogonadism, 101
Hypothalamic-pituitary axis, 102, 120
Hypothalamic-pituitary-gonadal (HPG) axis, 101
Hypotransferrinemia, 8, 13, 47, 52, 53

ICL670A (4-[(3,5-Bis-(2-hydroxyphenyl)-1,2,4) triazol-1-yl]-benzoic acid), 155, 185–202
IFN-γ, 264
IL-4, 264
IMR32 cells, 218
Ineffective erythropoiesis, 77
Inositol(1,4,5) triphosphate, 22
Insulin receptor, 22
Insulin-like growth factor, 101
IRC11, 79
IREG1, 5
Iron absorption, 1–3, 7–8, 52, 77
Iron deficiency, 2
Iron disulfide, 167
Iron oxides, 167
Iron regulatory protein 2 gene (*Ireb2*), 11, 211
Iron responsive elements, IRE, 23
Iron(II), 141, 147, 149, 154, 156, 161
Iron(III), 141–145, 147–149, 151–156, 159–161, 164, 183
IRP, 211–212, 234
IRP1, 79
IRP-1, 24, 26–27, 39
IRP2, 79
IRP-2, 23
Ito cells, 33

Klebsiella, 112
Kupffer cells, 33–35, 37–38, 78, 107

Labile iron pool" (LIP), 19, 24–26, 30, 32, 55–56, 59, 61–63, 65, 67–68, 73, 78
Left ventricular ejection fraction, 101, 104, 108, 131
Ligand, 141–143, 147–148, 150, 152–157, 160, 172, 173–174, 181, 186, 189, 191, 199–200
Lipid peroxidation, 20, 29–32, 35–36, 41–42, 54, 78, 80–81, 83, 87–89
Lipocytes, 33
Lipophilicity, 145, 147, 149–151, 155, 160, 162, 178, 186, 209, 213, 245, 256
Lipoxygenase, 96, 111, 118, 149–150, 162, 233
Liver iron concentration, 40, 106–107, 119
LVEF, 104, 120
Lymphoblastoid, T-cell line, 217
Lysosomal, 29, 40, 49, 80, 88, 94, 96
Lysosomes, 29, 40, 47–48, 53, 88, 95

Macrophages, 210–211, 264–265
Magnetic resonance imaging (MRI), 95, 101–102, 107–109, 119, 122–123, 137
Magnetite, 167
Malaria, 250, 259, 260, 263, 268
Malondialdehyde, 47, 49, 83
MAP kinase, 22
Marmosets (*Callithrix jacchus*), 192, 193, 194, 196, 198, 199, 201, 202
mdm-2 (murine double minute gene), 236
MECAM, 153, 171
Metalloenzymes, 96, 111, 147, 167
Metaphyseal osseous defects, 112
Microcytic anemia (*mk*) mice, 2
Mitochondria, 10–11, 20–21, 30–32, 41–42, 49, 81, 84, 88
Mitochondrial respiratory control ratio, 31
Mitochondrial respiratory enzymes, 81, 137
Monoferric A, monoferric B, diferric transferrin, 45
MTP1, 5
Mucormycosis, 105, 113, 119
Multicopper proteins, 3
Multi-gated acquisition (MUGA), 108
Murine erythroleukemia (MEL) cells, 63
Myelodysplasia, 128–129, 133, 136
Myocardial cells, 52, 79–81, 87
Myocardiopathy, 84, 85

N,N'-bis(2-hydroxybenzyl)-ethylenediamine-N,N'-diacetic acid (HBED), 152
N,N'-bis(pyridoxyl)-ethylenediamine-N,N'-diacetic acid (PLED), 171
N,N-Dimethyl-2,3-dihydroxybenzamide (DMB) 144
Na,K,ATPase 81
NADH-cytochrome c oxidoreductase (complex I+III), 81
NADH-ferricyanide reductase, 81
NADPH supply transport system, 21

NADPH-quinone oxidoreductas, 21
Nannochelin, 170
Nannocystis exedens, 170, 182
Nephrotoxicity, 174–175, 184, 195, 202
Neurotoxicity, 151, 163, 183–184
Night blindness, 109
Nitric oxide - NO, 26
Nitric oxide (NO·), 20
Nitrilotriacetate (NTA), 46, 217
NO, 20, 26–27, 34, 39, 123–125, 264, 265
Nocardamine, 170
Non-transferrin-bound iron (NTBI), 30, 45–51, 56, 62, 68–70, 72–74, 78–80, 85–86, 95, 190
Nramp2, 2
Nuclear factor 1, 22
Nuclear factor kB (p50), 21–22
Nucleotidase, 81

Oligodeoxynucleotides (ODN), 66
Oligomenorrhoea, 102
Optic atrophy, 109
Ornithine, 170
Osteoporosis, 115
Ototoxicity, 89, 109–111, 125
Oxalate, 46, 71–73, 106
Oxyanions, 141

P. berghei, 252, 256
P. falciparum, 251,–260, 265, 267
P. vinckei, 256
p21/*WAF1*, 236, 238
p53 expression, 205, 246
Parabactin, 170, 183
Paracoccus denitrificans, 170
Parkinson's disease, 11
Partition coefficient, 146, 150, 163
PFBH, 211–213, 216
PFH, 212
Peroxyl (ROO·), 20
P-glycoprotein, 210
Phenanthroline, 58
Phosphatidylinositol 3-kinase, 22
Phospholipase L, D and A_2, 22
Placental syncytiotrophoblasts, 3
Platelet-derived growth factor receptor, 22
Pneumocystis carinii, 113, 123
Polycarboxylate ligands, 143
Polyunsaturated fatty acids (PUFAs), 31
Porphyria cutanea tarda (PCT), 8
Port-a-Cath venous access, 85, 89, 104–105, 108
Pregnancy, 103, 116, 118
Prochlorperazine, 111
Prodrugs, 152, 157–159, 165, 166
Protein poshatase 1 and 2A, 22
Protein serine/threonine kinase, 22

Protein serine/threonine phosphatase, 22
Protein tyrosine kinase, 22
Protein tyrosine phosphatase, 22
Proximal tubular epithelium, 195
PUFA, 35
Pulmonary embolism, 105
Pyridoxal 2-thiophenecarboxylic acid hydrazone (PTCH), 211
Pyridoxal benzoyl hydrazone (PBH), 207, 211–213, 216
Pyridoxal isonicotinoyl hydrazone (PIH), 127, 155, 165, 184, 206–218, 231–232, 235, 242–243, 246–247
Pyridoxal *m*-fluorobenzoyl hydrazone (PFBH), 211
Pyridoxal *p*-methoxybenzoyl hydrazone (PMBH), 211

Quinine, 257, 260–265, 268, 270

Radionuclide ventriculography, 100
Reactive oxidative intermediates (ROIs), 20
Reactive oxygen species (ROS), 20
Redox cycling, 147, 149
Respiratory distress syndrome, 46
Reticuloendothelial (RE), 78, 130, 213, 214
Retinal toxicity, 84, 92, 110, 115
Retinitis pigmentosa, 109
Retinopathy, 9, 104, 109, 110
Reversed siderophore, 64
Rhizoferrin, 144
Rhodotorula pilimanae, 170
Rhodotorulic acid, 170
Ribonucleotide reductase, 96, 120, 149, 143, 182, 205, 218, 231, 242–243, 247, 250–251, 267

Salicylaldehyde, 207–208, 211, 213, 230, 236, 242–243
Salicylaldehyde benzoyl hydrazone, 208, 219, 242
Salicylaldehyde isonicotinoyl hydrazone (SIH), 156, 207, 211, 217
Sarcolemma, 81
Serum ferritin, 86, 99, 100, 104, 106, 111, 114–115, 130–132, 137
Sex-linked anemia (*sla*) mice, 3
Shuttle effect, 136
Sideroblastic anemia, 10
Sideroflexin, 11
Siderophores, 146, 149, 151, 153, 161, 169, 170–171, 183
sIH, 253, 255–256
Singlet molecular oxygen ($^1O_2$), 20
Skeletal abnormalities, 112
SK-N-MC neuroblastoma cells, 210–211, 213, 218–219, 232, 237
Small G protein: Ras, 22

INDEX

Spermidine, 170, 183
Staphloferrin, 144
Staphylococcus aureus, 105
Streptomyces antibioticus, 154, 173
Streptomyces griseoflavus, 91
Streptomyces pilosus, 91, 151, 170
Succinate cytochrome c oxidoreductase (complex II+III), 81
Succinate dehydrogenase, 81
Succinate ubiquinone oxidoreductase (complex II), 81
Succinate-cytochrome c reductase (complex II+III), 81
Succinate-cytochrome-c-reductase activity (complex II to III), 31
Sulfonates, 55
Superconducting quantum interference device (SQUID), 107
Superoxide (O_2^-), 20
Superoxide dismutase, 21, 162, 168
Survival, 19, 22, 25, 84–85, 92, 96–97, 99–100, 107, 121, 130, 137

T2*, 95, 107–108, 118, 130–131, 137, 139
TfR1, 233–234
TfR2, 233–234
Th-1 cells, 260, 263–264
Th-2 cells, 264
Thalassemia, 17, 24, 29, 40, 49–50, 54–55, 69, 72, 74, 76–77, 84, 86, 89–90, 92, 96–97, 99–105, 107–108, 110–111, 113–115, 119–126, 138–140, 163, 165, 183, 185, 187, 194–195, 201, 207, 236, 242
Thiazoline methyl, 174
Thiosemicarbazones, 236, 246
Thyroid dysfunction, 103
Thyroid transcription factor 1, 238
Transcription factor E2F, 98
Transcription factor FNR, 23
Transferrin, 12, 15–16, 30, 38, 45–46, 68–69
Transferrin receptor (TfR), 23, 66
Transferrin receptor mutation (*Trfr*), 12
Transferrin receptor 1, 233

Transforming growth factor-β_1 (TGFβ_1), 36
Triazoles, 155
Tribasic cations, 141, 143
Tridentate, 142, 144, 146–148, 150, 154–156, 173, 186, 189
Trophozoites, 254, 267
Tyrosinase, 111
Tyrosyl free radical, 231

UDP-glucuronosyl-transferase, 21
Upstream stimulatory factor, 22
Urate, 21
Urinary iron excretion, 92, 94, 96, 104, 108, 124, 126, 129
Uroporphyrinogen decarboxylase, 8

Venous thromboses, 104
Vibrio cholera, 170
Vibriobactin, 170, 253
Vincristine, 210
Vinblastine, 210
Vitamin E (α-tocopherol), 21, 49

Weissherbst (*weh*) mutant fish, 5

xanthine oxidase ,168, 182
Xenobiotics, 24, 25, 27, 32, 34, 37
Xenopus oocytes, 2
X-linked sideroblastic anemia with ataxia, 10

Yersinia enterocolitica, 113, 120, 123, 125
Yfh, 10

Zebrafish (*Danio rerio*), 4
Zebrafish mutant, *chardonnay* (*cdy*), 6
Zinc deficiency, 133
Zymosan, 34

α-tocopherol (vitamin E) 21, 83, 87
β-hexosaminidase 80
β-carotene 21
β-thalassemia/Hb E disease, 49